Advanced Researches in Plant Pathology

Advanced Researches in Plant Pathology

Edited by **Chris Frost**

SYRAWOOD
PUBLISHING HOUSE

New York

Published by Syrawood Publishing House,
750 Third Avenue, 9ᵗʰ Floor,
New York, NY 10017, USA
www.syrawoodpublishinghouse.com

Advanced Researches in Plant Pathology
Edited by Chris Frost

International Standard Book Number: 978-1-68286-109-7 (Hardback)

Printed in the United States of America.

Contents

Preface

Plant pathology is an interdisciplinary subject area which incorporates theories and applications of botany, microbiology, genetics and ecology to study diseases of plants caused by pathogens and environmental factors. This book includes some of the vital pieces of work being conducted across the world, on various topics related to plant pathology. This book brings forth some of the most significant concepts and elucidates the unexplored molecular, physiological and ecological aspects of plant pathology. It covers disease epidemiology, clinical analysis of diseases, evaluation of pathogens and crop loss assessment. This book is a complete source of knowledge on the present status of this important field.

The researches compiled throughout the book are authentic and of high quality, combining several disciplines and from very diverse regions from around the world. Drawing on the contributions of many researchers from diverse countries, the book's objective is to provide the readers with the latest achievements in the area of research. This book will surely be a source of knowledge to all interested and researching the field.

In the end, I would like to express my deep sense of gratitude to all the authors for meeting the set deadlines in completing and submitting their research chapters. I would also like to thank the publisher for the support offered to us throughout the course of the book. Finally, I extend my sincere thanks to my family for being a constant source of inspiration and encouragement.

Editor

Super nitro plus influence on yield and yield components of two wheat cultivars under NPK fertilizer application

Rouhollah Rouzbeh[1], Jahanfar Daneshian[1,2] and Hossein Aliabadi Farahani[3*]

[1]Islamic Azad University of Takestan branch, 10 km Quzwin-Zanjan highway, Azad University of Takestan, Faculty of Agriculture, Iran.
[2]Seed and Plant Improvement Institute (SPII), P. O. Box: 4119, Mardabad Road, Karaj 31585, Iran.
[3]Islamic Azad University of Shahr-e-Qods Branch, Iran.

To evaluate the beneficial impact of biological and chemical fertilizers application on wheat (*Triticum aestivum*) some yield characters were investigated. Our objective in this study was the interactive effects of biofertilizer (BF) and chemical fertilizer (CF) applications on: quantity yield at Iran in 2006. In this respect, the experimental unit had designed by achieved treatments in factorial on the basis completely randomized block design with four replicates. Certain factors including five levels of fertilizer (100% CF application; 75% CF and 25% BF applications; 50% CF and 50% BF applications; 25% CF and 75% BF applications and 100% BF application respectively) and two wheat cultivars (Omid and Alvand) were studied. Our final statistical analysis was indicated that in the 50% CF and 50% BF applications, yield components were significantly higher. Those such as: biological yield, seed yield, stem dry weight, ear dry weight, awn dry weight, leaf dry weight, plant height, leaf number per main stem and spikelet number per spike were higher in unites by 50% CF and 50% BF applications together cultivated. Although the wheat cultivars treatment significantly increased yield components and harvest index. However highest yield components were produced by Alvand cultivar but highest plant height was obtained by Omid cultivar. Our finding may give applicable advice to farmers for management and concern on fertilizer strategy and carefully estimate chemical fertilizer supply by biofertilizer application.

Key words: Biofertilizer, chemical fertilizer, biological yield, seed yield, harvest index, wheat cultivars.

INTRODUCTION

Super nitro plus is a special compound of three bacterium with different effect on plant growth, soil boom diseases and nematode control. Bacterium active ingredients: 10^8 *Azospirillum* spp. g^{-1} or ml of the product. The combination different bacteria: 10^8 *Bacillus subtilis* per g or ml of the product. In super nitro plus produce different kinds of plant growth regulators, sidrophores, antibiotics and inhibitors for pathogenic agents and hydrogen cyanide for control of nematodes and finally increase the growth and yield of the crops. Many soil born disease agents like *Pythium, Fusarium, Rhizoctionia, Phytophthora, Sclerotinia* and *Verticillium* will be controlled by

application of super nitro plus (Sagi et al., 1988). Biofertilizers are products containing living cells of different types of microorganisms (Vessey, 2003; Chen, 2006) that have an ability to convert nutritionally important elements from unavailable to available form through biological processes (Vessey, 2003) and are known to help with expansion of the root system and better seed germination.

Biofertilizers differ from chemical and organic fertilizers in that they do not directly supply any nutrients to crops and are cultures of special bacteria and fungi. Some microorganisms have positive effects on plant growth promotion, including the Plant Growth Promoting Rhizobacteria (PGPR) such as *Azospirillum, Azotobacter, Pseudomonas fluorescens* and several gram positive *Bacillus* spp. (Chen, 2006). The diazotrophic rhizobiocoenosis is an important biological process that plays a

---
*Corresponding author. E-mail: farahani_1362@yahoo.com.

Table 1. The results of soil analysis.

Soil texture	Sand (%)	Silt (%)	Clay (%)	K (mg/kg)	P (mg/kg)	N (mg/kg)	Na (Ds/m)	EC (1: 2.5)	pH	Depth of sampling
Sa.L	45	30	25	142.2	5.2	38.7	0.05	0.18	7.9	0-30 cm

major role in satisfying the nutritional requirements of the commercial medicinal plants (Deka et al., 1992). The strong and rapidly stimulating effect of fungal elicitor on plant secondary metabolism in main crops has attracted considerable attention and research efforts (Zhao et al., 2005). *Azotobacter* and *Azospirillum* are free-living N_2-fixing bacteria that in the rhizospheric zone have the ability to synthesize and secret some biologically active substances that enhance root growth. They also increase germination and vigour in young plants, leading to improved crop stands (Chen, 2006).

Various *Pseudomonas* species have shown to be effective in controlling pathogenic fungi and stimulating plant growth by a variety of mechanisms, including production of siderophores, synthesis of antibiotics, production of phytohormones, enhancement of phosphate uptake by the plant, nitrogen fixation and synthesis of enzymes that regulate plant ethylene levels (Abdul-Jaleel et al., 2007). Nitrogen is the major nutrient that influences plants yield and protein concentration. When the amount of available soil N limits yield potential, additions of N fertilizers can substantially increase plants yield.

However, plants protein concentration can decrease if the amount of added N is not adequate for potential yield (Olson et al., 1976; Grant et al., 1985). Many researchers have found that late-season top-dress N additions as dry fertilizer materials were the most effective in attaining higher plants protein concentration (Fowler and Brydon, 1989; Vaughan et al., 1990; Stark and Tindall, 1992; Wuest and Cassman, 1992; Knowles et al., 1994). Good soil fertility management ensures adequate nutrient availability to plants and increases yields. High above-ground biomass yield is obviously accompanied by an active root system, which releases an array of organic compounds into the rhizosphere (Mandal et al., 2007).

It is well known that a considerable number of bacterial and fungal species possess a functional relationship and constitute a holistic system with plants. They are able to exert beneficial effects on plant growth (Vessey, 2003) and also enhance plant resistance to adverse environmental stresses, such as water and nutrient deficiency and heavy metal contamination (Wu et al., 2005). Therefore, the objective of this study was to evaluate the super nitro plus influence on yield and yield components of two wheat cultivars under NPK fertilizer application.

MATERIALS AND METHODS

This study was conducted on experimental field of Research Institute of Damavand (Absard station) at Iran (25°30' N, 52°15' W;

1860 m above sea level) during 2006, with sandy loam soil (Table 1), mean annual temperature (30°C) and rainfall in the study area is distributed with an annual mean of 321 mm. The experimental unit had designed by achieved treatment in factorial on the basis completely randomized block design with four replicates.

Certain factors including five levels of fertilizer (100% CF application; 75% CF and 25% BF applications; 50% CF and 50% BF applications; 25% CF and 75% BF applications and 100% BF application respectively) and two wheat cultivars (Omid and Alvand) were studied. Also we used 150 kg ha^{-1} chemical fertilizer of NPK and 1 lit ha^{-1} biofertilizer of super nitro plus. The soil consisted of 25% clay, 30% silt and 45% sand (Table 1) and further the field was prepared in a 15 m^2 area (5 × 3 m) totally 40 plots.

To determined biological yield and seed yield, stem dry weight, ear dry weight, awn dry weight, leaf dry weight, plant height, leaf number per main stem and spikelet number per spike 10 plants were selected randomly from each plot at maturity and then harvest index was determined by the following formula (Aliabadi-Farahani, 2006).

$$HI = \frac{\text{Seed yield (kg/ha)}}{\text{Biological yield (kg/ha)}} \times 100$$

The data were subjected to analysis of variance (ANOVA) using Statistical Analysis System (SAS) computer software at $P < 0.05$ (SAS institute Cary, USA 1988).

RESULTS

Final results of plants values showed that fertilizer levels significantly affected biological yield, seed yield, stem dry weight, ear dry weight, awn dry weight, leaf dry weight, plant height, leaf number per main stem and spikelet number per spike in $p \leq 0.01$ and harvest index and tiller number were not significantly affected due to by fertilizer levels (Table 2) which indicated the highest biological yield (14206 kg/ha), seed yield (4815 kg/ha), stem dry weight (5228 kg/ha), ear dry weight (605.2 kg/ha), awn dry weight (27.5 kg/ha), leaf dry weight (2476 kg/ha), plant height (104 cm), tiller number (4 tiller/plant), leaf number per main stem (7.2 leaf/main stem) and spikelet number per spike (19.1 spikelet/spike) under 50% CF and 50% BF applications and highest harvest index (34.72%) was obtained by 25% CF and 75% BF (Table 3).

Also, wheat cultivars significantly affected biological yield, seed yield, stem dry weight, ear dry weight, awn dry weight, leaf dry weight, plant height, tiller number, leaf number per main stem and spikelet number per spike in $p \leq 0.01$ and harvest index was not significantly affected due to by wheat cultivars (Table 2). However highest biological yield (15080 kg/ha), seed yield (5279 kg/ha),

Table 2. Analysis of variance.

Sources of variation	Df	Mean squares										
		Biological yield	Seed yield	Stem dry weight	Ear dry weight	Awn dry weight	Leaf dry weight	Plant height	Tiller number	Leaf number per main stem	Spikelet number per spike	Harvest index
Replication	3	0.007	15.931**	151.124**	189143.313	0.005	5.414**	0.084**	67.521	0.445	46.788	0.102
Fertilizer	4	1.423**	10.087**	112.006**	1520201.243**	0.071**	41.633**	0.041**	45.688	16.016**	211.387**	0.032
Cultivar	1	1.515**	11.797**	169.053**	1089514.354**	0.058**	67.903**	0.022**	14367.854**	19.007**	114.37**	0.021
Fertilizer × Cultivar	4	0.027*	0.008	0.961	211820.317*	0.056	0.267	0.004	78.354	2.836	14.323	0.036
Error	27	0.009	0.032	2.608	86199.201	0.003	1.541	0.002	310.654	1.299	16.201	0.031
CV (%)		6.92	4.54	6.99	8.22	2.77	9.64	8.08	7.57	7.13	7.28	8.48

* and ** : Significant at 5% and 1% levels respectively.

Table 3. Means comparison.

Treatments		Biological yield (kg/ha)	Seed yield (kg/ha)	Stem dry weight (kg/ha)	Ear dry weight (g/m2)	Awn dry weight (g/m2)	Leaf dry weight (kg/ha)	Plant height (cm)	Tiller number (tiller/plant)	Leaf number per main stem (leaf/main stem)	Spikelet number per spike (spikelet /spike)	Harvest Index (%)
Cultivar	Omid	12433 b	4352 b	4725 b	547 b	24.8 b	2238 b	112 a	3 b	5.3 b	16.1 b	35 a
	Alvand	15080 a	5279 a	5731 a	663.5 a	30.1 a	2714 a	78 b	5 a	7.1 a	18.6 a	35.1 a
Fertilizer	100% CF	12803 b	4433 b	4991 ab	560.5 b	26.5 b	2207 b	102 a	3 ab	6 b	17.8 b	34.6 a
	75% CF and 25% BF	12473 b	4303 b	4830 b	546.2 b	26.6 b	2181 b	98 a	3 ab	5.9 b	17.6 b	34.4 a
	50% CF and 50% BF	14206 a	4815 a	5228 a	605.2 a	27.5 a	2476 a	104 a	4 a	7.2 a	19.1 a	33.9 b
	25% CF and 75% BF	12217 b	4244 b	4861 b	537.9 b	25.7 c	1977 b	87 b	3 ab	5.7 b	17 b	34.72 a
	100% BF	10555 c	3673 c	4108 c	466 c	22.6 d	1787 c	83 b	2 b	4.3 c	15.2 c	34.7 a

Means within the same column and rows and factors, followed by the same letter are not significantly difference ($p < 0.05$).

stem dry weight (5731 kg/ha), ear dry weight (663.5 kg/ha), awn dry weight (30.1 kg/ha), leaf dry weight (2714 kg/ha), tiller number (5 tiller/plant), leaf number per main stem (7.1 leaf/main stem) and spikelet number per spike (18.6 spikelet/spike) were produced by Alvand cultivar but highest plant height (112 cm) was achieved by Omid cultivar (Table 3).

Interaction of the fertilizer levels and wheat cultivars had significant effect on ear dry weight in

$p < 0.05$ (Table 2) and highest biological yield (14643 kg/ha), seed yield (5047 kg/ha), stem dry weight (5479 kg/ha), ear dry weight (634.3 kg/ha), awn dry weight (28.8 kg/ha), leaf dry weight (2595 kg/ha), leaf number per main stem (7.2 leaf/main stem) and spikelet number per spike (18.8 spikelet/spike) were achieved under 50% CF and 50% BF applications and Alvand cultivar and high-est plant height (123 cm) was obtained under 50% CF and 50% BF applications and Omid cultivar

and also, the highest harvest index (39.71%) was produced by 75% CF and 25% BF applications and Alvand cultivar (Table 4).

DISCUSSION

In this study, increases in agronomic criteria were observed following inoculation with biofertilizer. This may be due to better utilization of nutrients in

Table 4. Means comparison of interaction.

Survey instance qualifications		Harvest Index (%)	Spikelet number per spike (spikelet /spike)	Leaf number per main stem (leaf/main stem)	Tiller number (tiller/plant)	Plant height (cm)	Leaf dry weight (kg/ha)	Awn dry weight (g/m2)	Stem dry Weight (kg/ha)	Ear dry weight (g/m2)	Seed yield (kg/ha)	Biological yield (kg/ha)
	100% CF	34.8 b	16.9 cd	4.9 c	3 cd	120 a	2222 c	25.6 c	553.7 c	4858 bcd	4392 cd	12618 d
	75% CF and 25% BF	39.7 a	16.8 d	5 c	3 cd	115 a	2209 c	25.7 c	564.6 c	4777 cd	4327 d	12453 d
Omid	50% CF and 50% BF	34.4 b	17.6 cd	6.5 b	3.4 bc	123 a	2357 b	26.1 c	576.1 c	4976 abcd	4583 ab	13319 c
	25% CF and 75% BF	34.8 b	16.5 d	4.8 c	3 cd	102 b	2107 c	25.7 d	542.4 c	4793 d	4298 d	12325 d
	100% BF	34.8 b	15.6 e	4 d	2 d	96 bc	2012 d	23.7 e	506.5 d	4416 e	4012 e	11494 e
	100% CF	34.8 b	18.2 b	6.4 b	4 a	83 de	2460 b	28.3 b	612 ab	5361 ab	4856 ab	13941 b
	75% CF and 25% BF	39.71 a	18.1 b	6.4 b	4 a	79 def	2447 b	28.2 b	604.8 ab	5280 abcd	4791 abc	13776 bc
Alvand	50% CF and 50% BF	34.4 b	18.8 a	7.2 a	4 a	86 cd	2595 a	28.8 a	634.3 a	5479 a	5047 a	14643 a
	25% CF and 75% BF	34.8 b	17.8 bc	4.7 c	4 a	72 ef	2345 c	27.9 b	600.7 b	5296 abc	4761 bc	13648 bc
	100% BF	34.8 a	16.9 cd	4.2 c	3.8 ab	69 f	2250 c	26.3 c	564.7 c	4919 d	4476 d	12718 d

Means within the same column and rows and factors, followed by the same letter are not significantly difference (p < 0.05).

the soil through inoculation of efficient micro-organisms. A positive effect of biofertilizer on yield and yield components has been reported in the literature (Migahed et al., 2004). In addition, higher dry matter production by the inoculated plant might be because of the augmented uptake of N, P and K which in turn was a consequence of the root proliferation. Also, the increased growth parameters in hyssop might be due to the production of growth hormones by the bacteria (Ratti et al., 2001).

The results showed that 50% CF and 50% BF applications and Alvand cultivar increased yield and yield components of wheat, because nitrogen of chemical fertilizer, which is a primary constituent of proteins, is extremely susceptible to loss when considering that average recovery rates fall in the range of 20 to 50% for dry matter production systems in plants. Nitrogenous fertilizers generally cause deficiency of potassium, increased

carbohydrate storage and reduced proteins, alteration in amino acid balance and consequently change in the quality of proteins and are a main element in chlorophyll production. Toxic concentrations of nitrogen fertilizers cause characteristic symptoms of nitrite or nitrate toxicity in plants, particularly in the leaves. Although pre plant fertilizer applications decrease the potential for nutrient deficiencies in early stages of growth, presence of residual soil NO_3-N (plant-available mineral N from the previous season) may pose a risk to the environment. The water of soil be salt by inordinate N application and increase its potential. Finally, the plant use high energy for absorb of salt water that be causes dry matter reduces in this condition. Therefore, dry matter reduced under application of high levels of chemical fertilizer application because injured roots and was reduced the absorption. Our results were similar to the findings of Stark and Tindall (1992);

Wuest and Cassman (1992) and Knowles et al. (1994).

Conclusion

In general it appears that, as expected, application of biofertilizer improved yield and other plant criteria. Therefore, it appears that application of these biofertilizers could be promising in production of wheat by reduction of chemical fertilizer application. Our finding may give applicable advice to farmers for management and concern on fertilizer strategy and carefully estimate chemical fertilizer supply by biofertilizer application.

ACKNOWLEDGEMENTS

The author thanks Dr. Valadabadi and Dr. Mahdavi Damghani for their field assistance and analytical support in processing the plant samples and collecting the data reported herein.

REFERENCES

Abdul-Jaleel C, Manivannan P, Sankar B, Kishorekumar A, Gopi R, Somasundaram R, Panneerselvam R (2007). *Pseudomonas fluorescens* enhances biomass yield and ajmalicine production in *Catharanthus roseus* under water deficit stress. Colloids Surf. B: Biointerf. 60: 7–11.

Aliabadi-Farahani H (2006). Investigation of arbuscular mycorrhizal fungi (AMF), different levels of phosphorus and drought stress effects on quantity and quality characteristics of coriander (Coriandrum sativum L.). M.Sc Thesis, Department of Agriculture, Islamic Azad University of Takestan branch, Iran p. 231.

Chen J (2006). The combined use of chemical and organic fertilizers and/or biofetilizer for crop growth and soil fertility. International Workshop on Sustained Management of the Soil-Rhizosphere System for Efficient Crop Production and Fertilizer Use. October, Thailand pp. 16-20.

Deka BC, Bora GC, Shadeque A (1992). Effect of Azospirillum on growth and yield of chilli (*Capsicum annuum* L.) cultivar Pusa Jawala, Haryana. J. Hortic. Sci. 38: 41- 46.

Fowler DB, Brydon J (1989). No-till winter wheat production on the Canadian prairies: Timing of nitrogen fertilizers. Agron. J. 81: 817-825.

Grant CA, Stobbe EH, Racz GJ (1985). The effect of fall-applied N and P fertilizer and timing of N application on yield and protein content of winter wheat grown on zero-tilled land in Manitoba. Can. J. Soil Sci. 65: 621- 628.

Knowles TC, Hipp BW, Graff PS, Marshall DS (1994). Timing and rate of top dress nitrogen for rainfed winter wheat. J. Prod. Agric. 7: 216-220.

Mandal A, Patra AK, Singh D, Swarup A, Ebhin Masto R (2007). Effect of long-term application of manure and fertilizer on biological and biochemical activities in soil during crop development stages. Biores. Technol. 98: 3585-3592.

Migahed HA, Ahmed AE, Abd El-Ghany BF (2004). Effect of different bacteial strains as biofertilizer agents on growth, production and oil of *Apium graveolense* under Calcareous soil. J. Agric. Sci. 12: 511-525.

Olson RV, Swallow CW (1984). Fate of labeled nitrogen fertilizer applied to winter wheat for five years. Soil Sci. Soc. Am. J. 48: 583-586.

Ratti N, Kumar S, Verma HN, Gautams SP (2001). Improvement in bioavailability of tricalcium phosphate to *Cymbopogon martini* var. motia by rhizobacteria, AMF and *Azospirillum* inoculation. Microbiol. Res. 156: 145-149.

Stark JC, Tindall TA (1992). Timing split applications of nitrogen for irrigated hard red spring wheat. J. Prod. Agric. 5: 221-226.

Vessey JK (2003). Plant growth promoting rhizobacteria as biofertilizers. Plant Soil. 255: 571- 586.

Vaughan B, Westfall DG, Barbarick KA (1990). Nitrogen rate and timing effects on winter wheat grain yield, grain protein, and economics. J. Prod. Agric. 3: 324-328.

Wuest SB, Cassman KG (1992). Fertilizer-nitrogen use efficiency of irrigated wheat: I. Uptake efficiency of pre-plant versus late-season application. Agron. J. 84:682- 688.

Wu SC, Caob ZH, Lib ZG, Cheung KC, Wong MH (2005). Effects of biofertilizer containing N-fixer, P and K solubilizers and AM fungi on maize growth: a greenhouse trial. Geoderma. 125: 155 -166.

Zhao J, Lawrence T, Davis C, Verpoorte R (2005). Elicitor signal transduction leading to production of plant secondary metabolites. Biotechnol. Adv. 23: 283 - 333.

2

The effect of fosetyl-Al application on stomata in tomato (*Lycopersicon esculentum* Mill.) plant

İlkay Öztürk Cali

Department of Biology, Faculty of Art and Science, Amasya University, Amasya-Turkey. E-mail:
ilkay.cali@amasya.edu.tr.

In this study, a fungicide namely Aliette WG 800 (80% Fosetyl-Al) was pulverized on tomato (*Lycopersicon esculentum* Mill.) grown under greenhouse conditions and the likely effects of this fungicide on stomata of tomato were examined. Applications of Fosetyl-Al were carried out at recommended dosage (200 g/100 L water) as given on the label and two fold higher (400 g/100 L water) dosages. The fungicide applications resulted in a decrease in stomatal index compared to untreated plants whereas the numbers of abnormal and closed stomata were increased. It is thought that this condition may indirectly cause a negative effect in physiological events of the plant.

Key words: Fosetyl-Al, stomata, fungicide, tomato, *Lycopersicon esculentum* mill.

INTRODUCTION

Plant diseases in tomato plants usually cause the loss of product. The diseases caused by fungus decrease yield in quantity and quality and bring about some problems in export. Pesticides are used frequently to prevent harmful organism in plants in Turkey.It was stated in The First National Ecology and Environment congress that pesticides were used ignorantly in Turkey and there were increase in the usage of these chemicals which damaged environment (Delen and Özbek, 1994).Pesticides used for eliminating various pests which are found in agricultural environments also cause harmful effects on agricultural plants. These substances result in a toxic effect on stomata which have the most important role in some crucial functions such as photosynthesis and transpiration.Salgare and Acharekar (1990) reported that stomata in leaves were indicator of industrial pollutants. In fact, it was found that stomata were affected negatively by some chemicals in various studies.It was found that the largeness of stomata pores were influenced by Triazole fungicide in china patoto (*Solenostemonrotundifolius*, Poir., J. K. Morton) (Bora et al., 2002).According to Gupta et al. (2004), Triazole fungicide affected the number of stoma in some plants.Besides, in another study when 40 and 80 g/100 L water dosages of Equation Pro 22.5% Famaxadone + 30% Cymoxanil) fungicide were applied to tomato plant, there was an increase in the

number of closed stoma in paralel to dosage increasing (Öztürk, 2006). In the present study, pesticide applications which were used frequently against diseases and harmful living organisms in agricultural areas were investigated. In the study, recommended dosage (200 g/100 L water) and double the recommended dosage (400 g/100 L water) of 80% Fosetyl-Al fungicide were applied to tomato plant which have economic importance for Turkey. The effects of the fungicide were studied on the stomata of tomato plant in the same study. Especially, the effect of high dosage of the fungicide on stoma structure was investigated.

MATERIALS AND METHODS

The study was carried out in a 970 m^2 greenhouse in the village of Karaçulha in Fethiye, Turkey. Two hundred and twenty eight seedlings were obtained from M-38 F$_1$ type domestic seeds. Seventy six seedlings were used per groups. The fungicide which was used in the study against *Phytophthora infestans* was Aliette WG 800 (80% Fosetyl-Al). A total of four applications were made at ten-day intervals. The applications were 200 g/100 L water as recommended by the manufacturing company on the label and 400 g/100 L water as double the recommended dosage. Plant materials for anatomical observation were obtained since the fourth application after seven days and fixed in 70% ethyl alcohol. Hand made superficial sections in leaves were obtained on upper and lower surfaces of epidermis that belonged to the control and fungicide groups.

Table 1. Effect of Fosetyl-Al on stomata number, epidermal cell number and stomatal index.

Stoma Parameters in the adaxial and abaxial surfaces of the leaf		Application groups		
		Control	% 80 Fosetyl-Al(200 g/ 100 L)	% 80 Fosetyl-Al(400 g/ 100 L)
Adaxial surface of the leaf	Stomata number (in 0.125 mm^2)	249 [bc]	120 [ac]	198 [ab]
	Epidermal cell number (in 0.125 mm^2)	2529 [bc]	2262 [ac]	3000 [ab]
	Stomatal index (SI)	8.963 $\pm$ 0.717 [bc]	5.037 $\pm$ 0.680 [a]	6.191 $\pm$ 0.692 [a]
Abaxial surface of the leaf	Stomata number (in 0.125 mm^2)	657 [bc]	540 [a]	516 [a]
	Epidermal cell number (in 0.125 mm^2)	2772	2736	2778
	Stomatal index (SI)	19.160 $\pm$ 0.790 [bc]	16.483 $\pm$ 0.583 [a]	15.664 $\pm$ 0.750 [a]

In the table "a" indicates the significant difference between "a" and control group. "b" indicates the significant difference between "b" and 200 ml/ 100 L group. "c" indicates the significant difference between "c" and 400 mL/ 100 L group.

Stomata were examined in the superficial sections of leaves. The number of epidermal cell and stomata in the 0.125 mm^2 area in 40 X 6.3' magnification were determined. Stomatal index was estimated according to according to Meidner and Mansfield's (1969) method:

$$\text{Stomatal index} = \frac{\text{Stomata number in unit area}}{\text{Stomata number} + \text{Epidermal cell in unit area} + \text{number in unit area}} \times 100$$

In the study, the number of opened-closed stomata and abnormal shape stomata were also determined. A total of 400 stomata was used for measurements. Statistical analyses of the values of stomatal index in the study were made on a SPSS 11.0 for Windows Statistical Program and Multiple Range Tukey Test (Tukey, 1954) was used for variance analyses. Statistical analyses for the number of stomata, epidermal cell, opened-closed stomata and abnormal shape stomata were made on a SPSS 11.0 for Windows Statistical Program and the variance analyses were made using the Chi-Square Test, a nonparametric test widely utilized in such procedure.

RESULTS

The values of stomatal index, the number of stomata and epidermal cell in the control and the fungicide groups are given in Table 1. According to these results, stomata index for adaxial and abaxial surfaces of leaf were found to be lower than those of the control group in treated plants. This decrease was found to be statistically signifi-cant in all fungicide groups as compared to the control group. When the numbers of stoma and epidermis cell on upper and lower surfaces of leaf are examined, it is seen that the stomatal values are lower again according to control As for the values of epidermis cell for upper surface of leaf in 200 g/100 L dasage, they are lower as compared to the control whereas higher in 400 g/100 L water.Opened-closed stomata value number and results

in percentage are shown in Table 2. An examination of the values of opened-closed stoma percentage in the control and application groups showed that the opened stomatal percentage decreased in both layer surfaces of leaf, but closed stoma percentage increased.When the results of abnormal stomata percentage are examined, it is be seen that the values are higher parallel to the dosage (Table 3).

DISCUSSION

In the study, it was found that 80 % Fosetyl-Al reduced stomatal indices which belong to upper and lower surfaces of leaf in all application groups according to the control. Besides, the number of stomata in the fungicide groups was decreased as compared to the control group. The decrease in the number of stomata reduced the values of stomatal index in the application groups as compared to the control. The reduction in the number of stomata in the fungicide groups resulted from the fungicide that prevented the division of main stomata cell. Tort et al. (2004) stated that Acrobat (9% Dimethomorf + 60% Mancozeb) and Sandofan (10% Oxadixyl + 56% Mancozeb) which were pulverized at label dosages on tomato plant reduced stomatal index as compared to the control. According to Prakash et al. (1978), the applications of Alachlor and Flurochloridone decreased stomatal index in fungicide groups as compared to the control. On the other hand, Tort and Dereboylu (2003) found that 2.5, 5 and 7.5 g/L dosages of Captan fu-ngicide caused a reduction in the values of stoma index in pepper (Capsi cum annuum L.) plant as the dosage increased. According to Öztürk (2006), the reduction which was determined in the values of stoma index in the fungicide group as compared to the control will affect some important

Table 2. Effect of Fosetyl-Al opened-closed stomata number and percentages.

Stoma Parameters in the adaxial and abaxial surfaces of the leaf		Application groups		
		Control	% 80 Fosetyl-Al (200 g/ 100 L)	% 80 Fosetyl-Al (400 g/ 100 L)
Adaxial surface of the leaf	Total stomata number	249 [bc]	120 [ac]	198 [ab]
	Opened stomata number	234 [bc]	42 [ac]	24 [ab]
	Opened stomata %	93.98	35	12.12
	Closed stomata number	15 [bc]	78 [ac]	174 [ab]
	Closed stomata %	6.02	65	87.88
Abaxial surface of the leaf	Total stomata number	657 [bc]	540 [a]	516 [a]
	Opened stomata number	597 [bc]	60 [a]	63 [a]
	Opened stomata %	90.87	11.11	12.21
	Closed stomata number	60 [bc]	480 [a]	453 [a]
	Closed stoma %	9.13	88.89	87.79

In the table "[a]" indicates the significant difference between "[a]" and control group. "[b]" indicates the significant difference between "[b]" and 200 mL/ 100 L group. "[c]" indicates the significant difference between "[c]" and 400 mL/ 100 L group.

Table 3. Effect of Fosetyl-Al on abnormal stomata number and percentage.

Stoma Parameters in the adaxial and abaxial surfaces of the leaf		Application groups		
		Control	% 80 Fosetyl-Al (200 g/ 100 L)	% 80 Fosetyl-Al (400 g/ 100 L)
Adaxial surface of the leaf	Total stomata number	249 [bc]	120 [ac]	198 [ab]
	Abnormal stomata number	9 [bc]	90 [ac]	192 [ab]
	Abnormal stomata %	3.61	75	96.96
Abaxial surface of the leaf	Total stomata number	657 b[c]	540 [a]	516 [a]
	Abnormal stomata number	15 [bc]	240 [ac]	426 [ab]
	Total stomata number	2.28	44.44	82.55

In the table "[a]" indicates the significant difference between "[a]" and control group. "[b]" indicates the significant difference between "[b]" and 200 mL/ 100 L group. "[c]" indicates the significant difference between "[c]" and 400 mL/ 100 L group.

physiological events such as ph-otosynthesis and respiration negatively.

When results of opened-closed stoma percentage are examined, it was determined that there was a reduction in the percentage of opened stoma in all application group as compared to the control, but an increase in the values of the percentage of closed one. An increase in the values of closed stomata is going to affect photosynthesis of the plant negatively. It was reported that Triazole fungicide resulted in a closing in stomata which belong to Phaseolus vulgaris (Fletcher and Hofstra, 1988). According to Asere-Boamah et al. (1986), there was a decrease in stoma pore in Phaseolus vulgaris applied with Thiapenthenol. The results of the studies above are in aggrement with the results of the present study.

In the study, when percentage of abnormal stoma results were evaluated, it was seen that values increased in parallel with the dosage in all application groups accor-

ding to the control. It was found that various chemicals affected stomata negatively in other studies. Turunen and Huttunen (1991) observed abnormal stomata in some plants which were affected from acid rain. Furthermore, the toxic effect of ozone brought about the loss of resistance in stomata (Moldau et al., 1990). Cireli and Önür (1983), stated that Stomp 330 E caused the development of abnormal stomata. It was also found that abnormal stomata were found in the present study.

It was established that Fosetyl-Al fungicide which was used for preventing fungous diseases brought about some changes in the structure of stoma in tomatoes in the study. The value of stomatal index and the percentage of opened stoma were lower than those in the control group while the percentage of abnormal stoma was higher. It is believed that such a negative effects in stomata which have an important role in plant life influenced physiological events in plant negatively. In consideration of the fact that ignorant use of fungicides is by no means at minimal levels, studies that are dealing with a number of problems caused by application of fungicides at excessive dosages have gained far greater importance.

REFERENCES

Asare-Boamah NK, Hofstra G, Fletcher RA, Dumbroff EB (1986). Triadimefon protects bean plants from water stress through its effects on ABA. Plant Cell Physiol. 27: 383-390.

Bora KK, Mathur SR, Ganesh R, Bohra SP (2002). Effect of paclobutrazol on water loss of excised groundnut seeds. Bioregulants Applied Plant Biotechnol. Pointerpublishers, India. pp. 58-64.

Cireli B,Önür MA (1983). The Effect of Stomp 330 E (herbisit) application on anatomical structure of Vicia faba leaf. J. Nat. Sci. 7: 297-307.

Delen N, Özbek T (1994). The role of pesticides in natural pollution. Ege. Univ. J. 16:67-75.

Fletcher RA, Hofstra G (1988). Triazoles as potential plant protectants. In Berg D, Plemple M, eds., Sterol biosynthesis inhibitors. Ellis Horwood Ltd., Cambridge, England pp.321-331.

Gupta SK, Raghava RP, Raghava N (2004). Stomatal studies of cowpea (Vigna unguiculata) L. walp. Cultivars in relation to bromiconazole. J. Ind. Bot. Soc. 83: 116-119.

Moldau H, Sober J, Sober A (1990). Differential sensitivity of stomata and mesophyll to sudden exposure of bean shoots to ozone. Photosynhetica (Progue) 24: 446-458.

Meidner H, Mansfield TA (1969). Physiology of stomata. Mc Graw-Hill, Newyork, USA.

Öztürk İ (2006). The effect of Equation Pro (Fungicide) Application on stomata in tomato (Lycopersicon esculentum Mill.) plants. Ankara Univ. J. of Agric. Sci. 12 (2): 195-202.

Prakash J, Barber S, Pahwa SK (1978). Effect of some herbicides on the epidermis of Vicia sativa L. Weed Res. 18: 379-380.

Salgare SA, Acharekar C (1990). Effect of industrial air pollution (from chembur, India) on the micromorphology of some wild plants. II. Adv. Plant. Sci. 3: 1-7.

Tort N, Dereboylu AE (2003). The effect of Captan on stomata and photosyntetic pigment matters in pepper plant (Capsicum annuum L.) Anadolu, J. of AARI 13(1):142-157.

Tort N, Öztürk İ, Tosun N (2004). The effect of fungicide applications on anatomical structure and physiology of tomato (Lycopersicon esculentum Mill.) J. Agric. 41(2): 111-122.

Tukey JW (1954). Some selected quick and easy methods of statistical analysis.. Trans of New York Acad. Sci. pp. 88-97

Turunen M, Huttunen S (1991). Effect of simulated acid rain on the epicuticular wax of Scots pine needles under northerly conditions. Can. J. Bot. 69:412-419.

3

Allelopathic and antifungal potentialities of *Padina pavonica* (L.) extract

Faten Omezzine[1], Rabiaa Haouala[1]*, Asma El Ayeb[1] and Neziha Boughanmi[2]

[1]Department of Biology, Higher Institute of Biotechnology of Monastir, University of Monastir, 5000, Tunisia.
Unit of search for Mycotoxins, Phycotoxins and associated Pathologies 03/UR/08-14.
[2]Department of Biology, Faculty of Sciences of Bizerte, Tunisia.

In this study, potential allelopathic of a brown alga *Padina pavonica* (L) was evaluated. Aqueous extracts of the alga obtained at room temperature / 24 h (E1), 50 °C / 4 h (E2) and 100 °C / 2 h (E3) were tested on the germination and early growth of crop plants and growth of three fungal strains: *Fusarium graminearum*, *Penicillium expansum* and *Alternaria alternata*. Also, a fractional, of the alga in three organic solvents with increasing polarity: hexane, chloroform and acetone, was carried out and estimated as well as that the seedling growth soil composition was estimated. Results revealed a perceptible allelopathic capacity of *P. pavonica*. Although the percentage of germination of seeds was not influenced or slightly stimulated compared to the control in the presence of the three kind extracts and the dry powder, root and shoot growth was clearly improved. Results varied according to extracts and the vegetable material dose. Extract prepared at room temperature (E1) was most favorable. Stimulation percentage varied between 15 and 78% for roots and between 1 and 67% for shoots. Fungal growth was strongly inhibited in the presence of extract E3 compared with E1. Thus, growth inhibition percentages were 95.83, 80.76 and 63.33% for *F. graminearum*, *P. expansum* and *A. alternata* respectively, in presence of E1, against 41.66, 72.30 and 51.66% in presence of E3. For organic extracts, the most spectacular seedling growth stimulations were recorded in the presence of the chloroform extract. This indicated that the active molecules had an average polarity. The incorporation of algal powder in the soil was shown very beneficial for the target species in particular with low dose (50 g/Kg). Moreover, the effect of the algal powder was similar to that obtained by the addition of chemical fertilizers. Finally, algal powder allowed a clear improvement of the chemical composition of the soil concerning its richness on calcium, magnesium and organic matter and did not affect its pH. Results might be considered very interesting, since molecules of this species could have a double effect in the crops: a fertilizing effect and an antifungal effect without deteriorating the physico-chemical properties of the soil.

Key words: Anti-fungal activity, aqueous extracts, chemical fertilizers, fertilizing capacity, growth, *Padina pavonica*, seeds.

INTRODUCTION

Molisch (1937) was the first to define the term allelopathy, in a broad sense to describe either positive or negative biochemical interactions among all plant kinds. Rice (1984) included microorganisms and restricted the conceptual content of allelopathy exclusively to negative effects arising by the production and excretion of chemical compounds from plants and microorganisms. Many plants proved to have the allelopathic potentials; generally they had negative effects, and could be used in biological pesticide production. Majority of works present the harmful effects of plant extracts and rare were works which indicate beneficial effects of plants. Marine plants and macro-algae constitute a richness to explore and exploit in several regions of the world (Ben Said et al., 2002). Recently, Inderjit and Dakshini (1995) gave an overview of allelopathic activities in aquatic habitats with particular emphasis on algae. Allelopathy is a prevalent natural phenomenon in aquatic ecosystem; however, it is difficult to study its effects among aquatic organisms under natural conditions because factors such as nutrient

*Corresponding author. E-mail: rabiahaouala@yahoo.fr.

and light competition, temperature and pH change could totally mask allelopathic effects (Keating, 1977).

Padina pavonica (L.), a brown alga, was widespread in Tunisian littoral. It belonged to the family Dictyotaceae, order Dictyotales, subclass Isogeneratae, class Phaeophyceae (Kamenarska et al., 2002). Studies conducted on this specie were particularly interested in the nature and rate of carbohydrate, lipids, vitamins, mineral salts and other active ingredients (Al Easa et al., 1995; Wahheb, 1997; Ktari and Guyot, 1999; Kamenarska et al., 2002). The richness of this specie let think of its exploiting for a positive purpose on the crops. Indeed, the over use of synthetic agrochemicals often caused environmental hazards, imbalance of soil microorganisms, nutrient deficiency, and change physicochemical properties of soil, resulting in a decrease of crop productivity. So, incorporation of allelopathic substances into agricultural management might reduce the use of industrial pesticides and fertilizers and lessen environmental deterioration (Chou, 1999).

In this chapter, assessment of fertilizing and antifungal potentialities of brown algae *P. pavonica* (L.) was investigated.

MATERIALS AND METHODS

Algal sample collection

The tested macro-algae *(P. pavonica)* was collected from Tunisian littoral (Mahdia, Tunisia) (35°30'N, 11°04' East) in July 2007, identified and carefully washed with distilled water immediately to remove attached remains and organisms.

Aqueous extracts

Fresh tissue of macro-alga was air dried for 6 days at room temperature and then ground into a fine powder using a mortar. Thirty grams of dry powder was mixed with 1000 ml distilled water (Einhellig et al., 1993) and shaken for i) 24 h at room temperature (E1) (Khanh et al., 2005), ii) 4 h at 50°C (E2) (Delabays et al., 1998) and iii) 2 h at 100°C (E3) (Mao et al., 2006). The mixture was filtered through a filter paper several times and kept at 4°C in the dark for further use.

Organic extracts

Sequential extraction was carried out in organic solvents with rising polarity: hexane, chloroform and acetone. Forty grams of algal powder were immersed in the organic solvent for 7 days at room temperature. Organic extracts were evaporated to dryness under reduced pressure in a rotary evaporator at 45 - 50°C. The residue was weighed and yield was determined (0.153, 0.467 and 0.277% for respectively hexane, chloroform and acetone). The extracts were tested at concentration 3000 ppm in the biological assays.

Petri Plate assay

Effect of aqueous extracts: Aqueous extracts were tested on four crop species: Dicot species: *Lens culinaris* L. (lens), *Lactuca sativa* L. (lettuce), and Monocot species: *Triticum aestivum* L. (wheat) and *Hordeum vulgare* L. (barley). Lettuce was known to be very sensi-

tive to allelochemicals (Leather and Einhellig, 1987; Ervin and Wetzel, 2003), whereas other species were selected because they represent the first food source in Tunisia. All plant seeds were surface sterilized with 0.525 g/L sodium hypochlorite for 15 min. The seeds were rinsed four times with deionized water, imbibed in it at 22°C for 12 h and carefully blotted using a folded paper towel (Chon et al., 2005). Without any dilution, ten milliliters of each extracts, contained 0.3 g of algal dry powder, were pippetted onto filter paper. The seedlings watered with distilled water were used as control. Thirty imbibed seeds of target species were separately placed on the filter paper in Petri dishes. They were covered and placed in a growth chamber at 24°C during the 14 h light period and 22°C during the 10 h dark period. The plates were illuminated with 400 µmol photons.m^{-2}s^{-1} photosynthetically active radiation (PAR), provided by a mixture of incandescent and fluorescent lamps. Treatments were arranged in a completely randomized design with three replications.

Cumulative germination was determined by counting the number of germinated seeds at 24 h intervals over a 144 h period and transformed into germination percentage. Shoot and root length and dry weight of recipient species were measured on all seedlings in each Petri dish on day 7 after placing seeds on the medium. Data were transformed to percent of control for analysis. The inhibitory or stimulatory percent was calculated using the following equation given by Chung et al. (2001): Inhibition (-)/stimulation (+) % = [(extract – control)/control] x100 with: Extract: parameter measured in presence of *P. pavonica* extract Control: parameter measured in presence of distilled water.

Effect of organic extracts

The three extracts concentrated from hexane, chloroform and acetone were dissolved in acetone (15 mg in 2 ml) to estimate their effect on germination and early growth of crop species. Two controls were considered, one in the presence of distilled water and another one was in presence of acetone in order to eliminate the effect of organic solvent. Filter paper placed in Petri plates, were soaked in distilled water, acetone or various organic extracts. The solvents were evaporated for 24 h at 24°C. After that, 5 ml of distilled water were added (final concentration: 3000 ppm) and 30 soaked seeds were put to germinate for 7 days. Germination, shoot and root length and dry weight of target species were determined and expressed as percentage of the control.

Antifungal activity assay

The antifungal activity was tested on *Fusarium gramineaum, Penicillium expansum* and *Alternaria alternata*. These fungi were provided by the phytopathology laboratory of Higher Institute of Agriculture - Chott-Mariem in Tunisia.

100 µl of extracts at 100 µg/ml were put in dug wells (5 mm broad and 20 mm length) of sterilized Petri dished containing Potato Dextrose Agar (PDA). The fungal plugs (0.4 mm in diameter) were placed opposite the well with 1 cm of the edge limps. Limps control consisted distilled water in the well. After an incubation for 72 h at 24 ± 2°C, the mycelium development of pathogenic fungi in each Petri dish and the phytotoxic effect of *P. pavonica* extracts by measuring the distance covered by the mycelium. The percentage of growth inhibition was calculated as follows (Khanh et al., 2005): I (%) = (1- Cn/Co) * 100 Where Cn was the distance covered by fungi in the presence of extracts and Co was the distance covered in the presence of distilled water. The antifungal effect was measured under a totally random design with three replications.

Glasshouse assay

For incorporation treatment, biomass of *P. pavonica* was mixed with

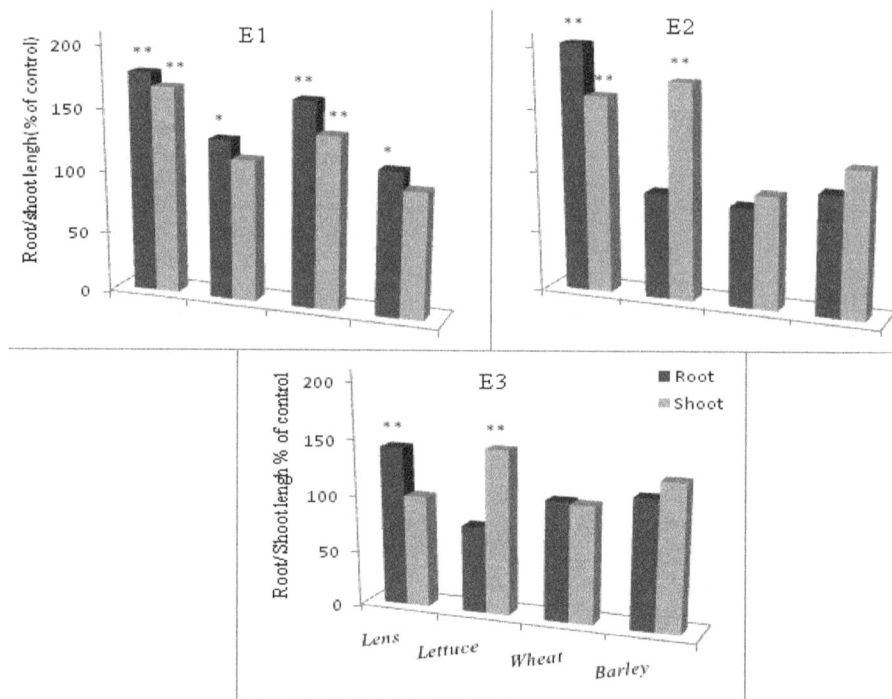

Figure 1. Root and shoot length of seedlings grown in the presence of aqueous extract prepared at room temperature (E1), at 50°C (E2) and at 100°C (E3). Differing statistically (Duncan test) from control are marked with one (P ≤ 0.05) or two (P ≤ 0.01) asterisks.

soil by vigorous shake in 20 cm diameter plastic bags. Four trials were carried out: soil without fertilization, fertilization with algal powder at two doses (50 and 100 g.kg^{-1} of biomass (Khanh et al., 2005)) and fertilization with chemical fertilizer 0.5 g/Kg (Ammonitre). The pots were filled with the soil and residue mixture and twenty seeds per pot were planted. The growing medium was maintained near field capacity by sub-irrigation without nutrition solution (Fischer et al., 2000). The experiments were conducted in greenhouse for 15 days at 28 / 22°C day/night temperature. Germination percentage and seedling growth were measured as described previously. Data were transformed to percent of control for analysis.

Effect of *P. pavonica* residue incorporation on soil composition

After 15 days of germination, the mixture of soils and algal residue were analyzed for calcium, magnesium, carbon, organic matter and pH. Soil samples were air dried and then crushed. The sum of ions (Ca^{2+} + Mg^{2+}) was titrated with EDTA. The proportioning of calcium was determined by flame spectrophotometry and magnesium was calculated by the formula Mg (Méq/g) = ((Ca + Mg) − Ca). Carbon was proportioned by spectrocolorimetry at 590 - 600 nm and organic matter was calculated according to the formula: m.o. =1.72 x C (Aubert, 1978). The pH was measured by a pH-meter.

Statistical analysis

The laboratory bioassays and pot culture were conducted in a complete randomized design with three replications. Duncan – tests and ANOVA were performed on SPSS 13.0, for Windows program, to analyze treatment differences. The means were separated on the basis of least significant differences at the 0.05 probability level.

RESULTS

Effect of aqueous extracts on the germination and growth of test crops

The variance analysis showed that *P. pavonica* extracts had no significant effect on germination (P < 0.001). Indeed, germination percentage did not vary in presence of all extracts and kinetic was not influenced or slightly accelerated. However root and/or shoot growth was improved. This improvement was more significant when the seedlings were exposed to the extract obtained at room temperature (E1) and the stimulation percentage varied between 15 and 78% and between 1 and 67% for roots and shoots, respectively (Figure 1). Root growth was more or less inhibited in presence of the extract prepared at 50°C (E2) with a non significant respective reduction of 18.39, 13.23 and 1.75% for wheat, lettuce and barley, however, this extract improved growth of lens roots by inducing stimulation of 101.34%. The effect of the extract prepared at 100°C (E3), was marked on root growth of lens which was stimulated by 42.5%. For other species, a non significant difference compared to the control was recorded, except for the lettuce whose growth was inhibited by 23%.

Since, root growth of various target species was near to the control or was strongly stimulated in the presence of various kinds of *P. pavonica* aqueous extracts. The sensitivity of shoots varied according to target species and extracts (Figure 1). Globally, the behavior of shoots was

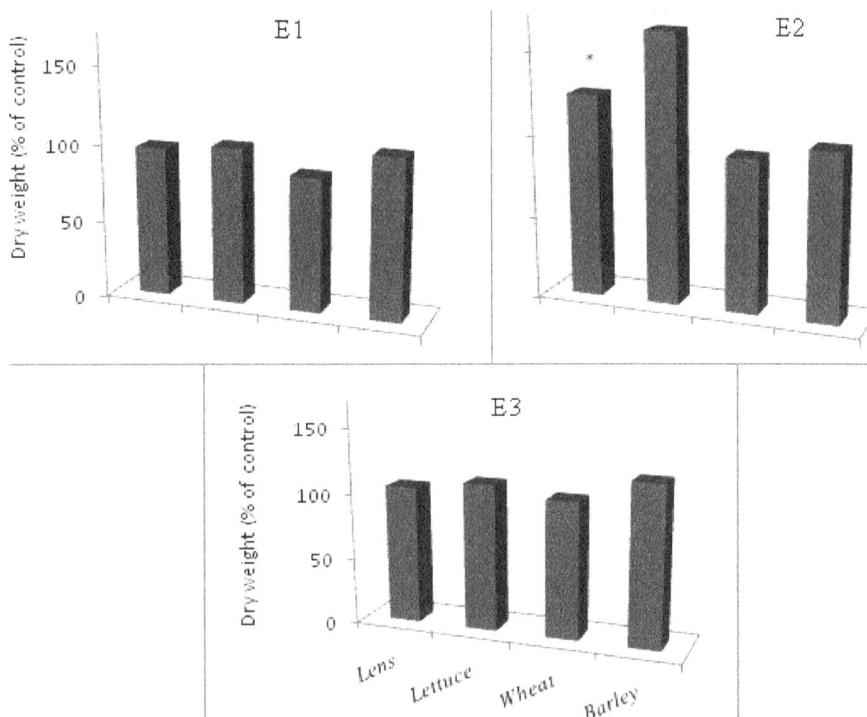

Figure 2. Biomass production by target species grown in presence of aqueous extract prepared at room temperature (E1), at 50°C (E2) and at 100°C (E3). Differing statistically (Duncan test) from control are marked with one (P ≤ 0.05) or two (P ≤ 0.01) asterisks.

similar to that of roots except for the lettuce which showed a better growth of shoots in the presence of E2 and E3. Shoot growth of the various target species was near to the control or was stimulated by an average factor of 1.2 in the presence of several of *P. pavonica* (Figure 1).Biomass production was significantly ameliorated for lens and lettuce in presence of E2 and E3. For other species, it was similar to the control (Figure 2).

Effect of organic extracts on the germination and growth of test crops

To determine the chemical group to which bioactive molecules of *P. pavonica* extracts could be owned three organic extracts: hexane, chloroform and acetone. Organic residues were dissolved in acetone, which required an acetone control. Results showed that this solvent did not affect germination and growth and the recorded results would be allotted to the allelopathic compounds. Germination of the four species was similar to control in all cases (data no shown). Nevertheless, results indicate a stimulation of root/shoot growth for all species and in presence of the three organic extracts (Figure 3). The most important stimulations were recorded in the presence of chloroform extract which showed a rise of root growth of 5.17, 80.32, 108.67 and 121.67% for lens, lettuce, wheat and barley, respectively. Similarly, this extract improved shoot growth of seedlings and the percentage of stimulation varied between 13 and 164%

(Figure 3). Acetone extract ameliorated significantly the growth of barley.

The effect of organic extracts on biomass production was not significant (Figure 4). Indeed, the quantity of dry matter was near to the control or slightly higher in all cases. However, an improvement of the biomass production in lettuce was recorded in the presence of the three organic extracts with a better stimulation in the presence of the chloroform extract (23.23%). For water content, the seedling hydration was near to the control or was improved in the presence of various organic extracts (Figure 4). In the presence of hexane extract a respective stimulation of 105 and 243% for lettuce and lens. A stimulation of an average of 51% was recorded for the monocotyledons in the presence of the same extract.

Effect of residue incorporation of *P. pavonica* in soil on germination and growth

The initial bioassay was necessary and often used to evaluate the allelopathic potentialities of plant species (Chon et al., 2005; Kato-Noguchi and Tanaka, 2003); however, pot test was desired in order to indicate the effects that could be reproduced under natural conditions (Corrêa et al., 2008). Thus, the incorporation of algal powder to the soil was carried out to evaluate its effect on growth of test crops under natural conditions and to compare with those of the extracts. Moreover algal powder was compared with the chemical fertilizer used usually in

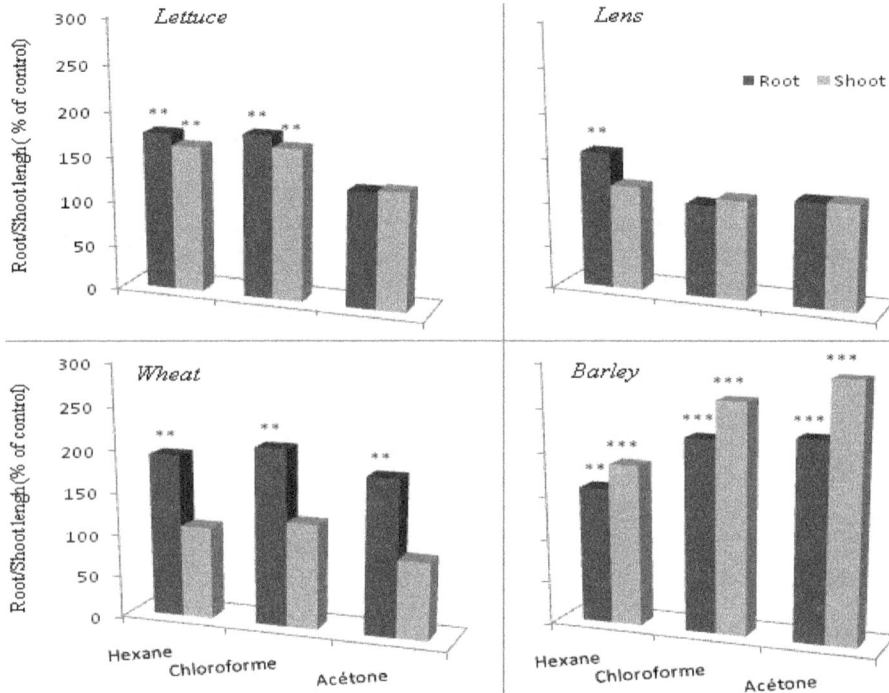

Figure 3. Root and shoot length (% control) of target species in the presence of three organic extracts. Differing statistically (Duncan test) from control are marked with one (P ≤ 0.05), two (P ≤ 0.01) or three asterisks (P ≤ 0.001).

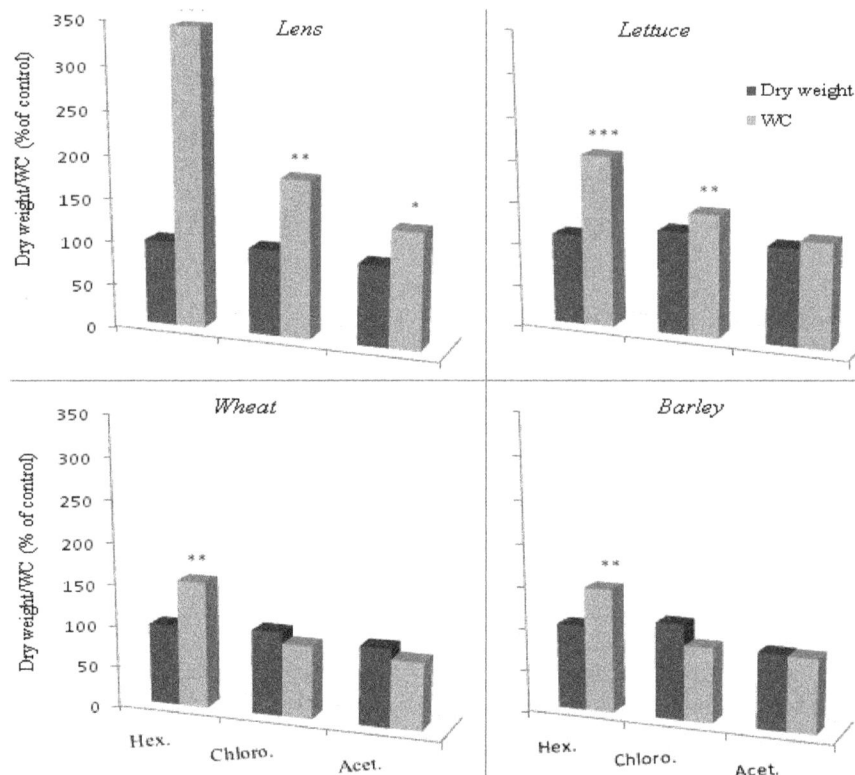

Figure 4. Biomass production and water content (% of control) of target species in the presence of organics extracts: hexane (Hex.), chloroform (Chloro.) and acetone (Acet.). Differing statistically (Duncan test) from control are marked with one (P ≤ 0.05), two (P ≤ 0.01) or three asterisks (P ≤ 0.001).

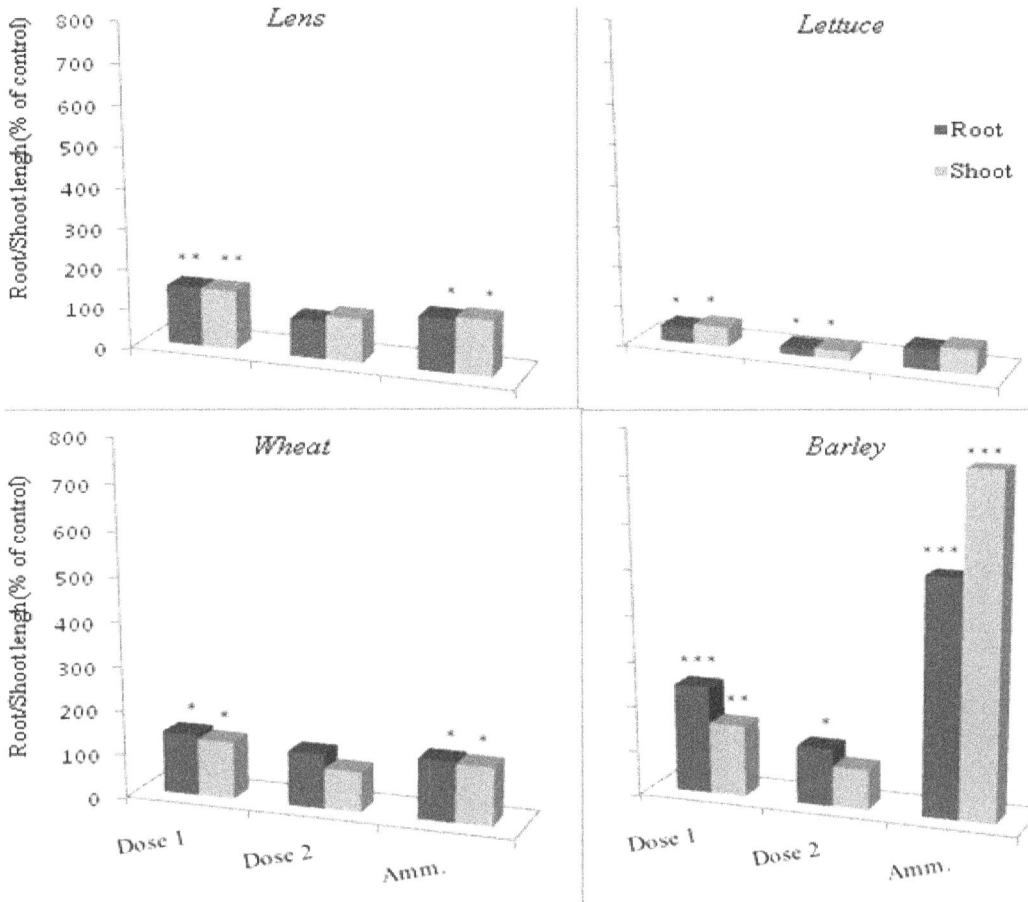

Figure 5. Effects of residue incorporation with peat, from thalli of *P. pavonica* at 50 g/Kg (dose 1) and 100 g/Kg (dose 2) and incorporation of industrial fertilizer (Amm.) on root and shoot length of test crops, 15 days after germination, expressed in % of control. Differing statistically (Duncan test) from control are marked with one (P ≤ 0.05), two (P ≤ 0.01) or three asterisks (P ≤ 0.001).

agriculture. Incorporation of dry algal powder in the soil gave similar results to those obtained by the incorporation of a synthetic fertilizer (Ammonitre). Indeed, we have registered an average stimulation of 1.5 times of seedling growth in the presence of low doses of algal biomass or fertilizer (50 g/Kg) and a slightly inhibition (an average of 15%) at 100 g/Kg. An inhibiting effect of algal biomass was registered on lettuce growth especially at 100 g/Kg (Figure 5). Biomass production was significantly ameliorated in presence of *P. Pavonica* residue at dose 1 for wheat, whereas similar results were obtained from other test crops and control (Figure 6). In the presence of dose 1, lens seedlings produced the same quantity of dry matter as the control, whereas in the presence of dose 2 and synthetic fertilizer, a reduction of 26.97 and 21.58% respectively. However, lettuce had slightly benefited from the algal matter at dose 2.The water content was near to the control for all target species except wheat which presented a hydration improvement in presence of algal matter at dose 1 and the lettuce whose water content significantly decreased in presence of dose 2 of the biological material (Figure 6).

Effect of residue incorporation of *P. pavonica* on soil composition

After 15 days of experiment, the chemical properties of soil with algal powder and with fertilizer were analyzed and compared. Algal incorporation did not influence the soil pH, which was comparable with the control and with that in the presence of chemical fertilizer. An average value of 8.3 in all cases (Table 1).

Concerning the nutritive elements, the quantity of calcium and magnesium which were very significant elements for the improvement of the soil texture were dosed. Results showed an increase in the quantity of calcium compared with the control in the presence of the algal biomass: 0.064 meq/g with dose 1 and 0.052 meq/g with dose 2. In the presence of industrial fertilizer the quantity of calcium as 0.09 meq/g. The important quantities of magnesium were also recorded in the presence of the algal powder compared with the control and chemical fertilizer. An increase of 0.114 meq/g and 0.176 meq/g in the presence of dose1 and dose 2 respectively was recorded compared with the control (0.0013 meq/g). The

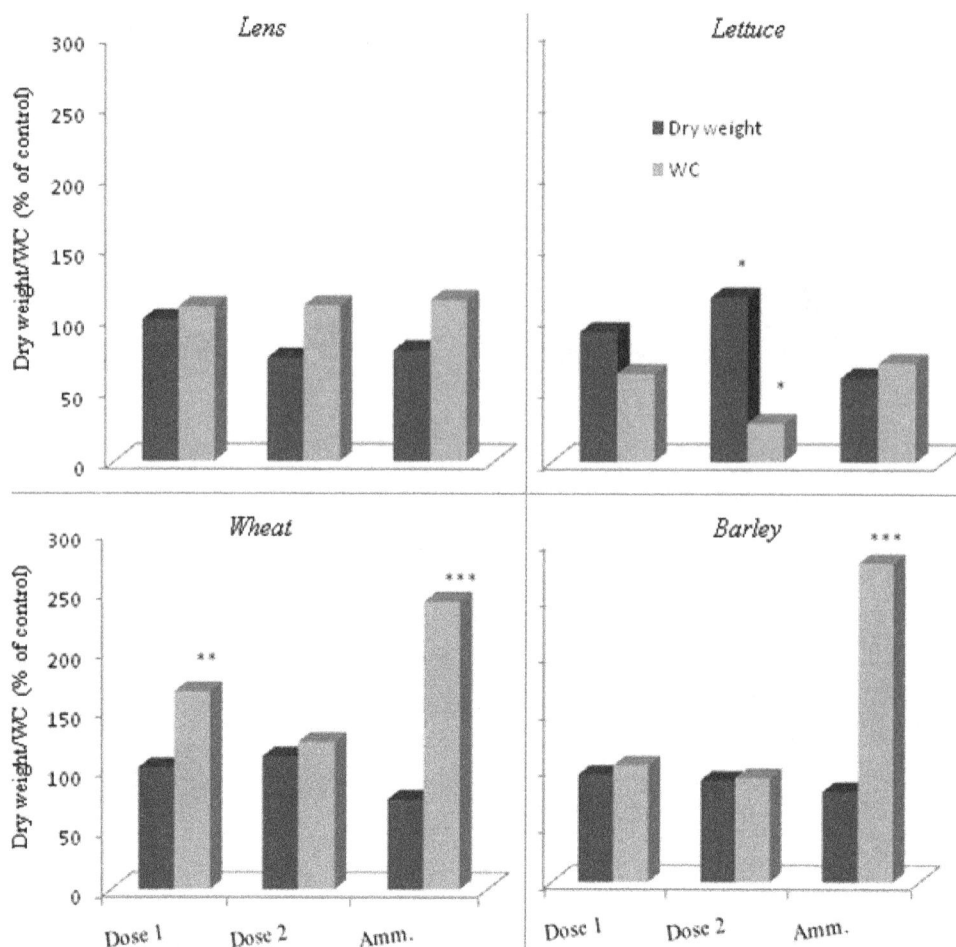

Figure 6. Effects of residue incorporation with soil, from thalli of *P. pavonica* at 50 g/Kg (dose 1) and 100 g/Kg (dose 2) and incorporation of industrial fertilizer (Amm.) on dry weight and water content (WC) of test crops, 15 days after germination, expressed in % of control. Differing statistically (Duncan test) from control are marked with one (P ≤ 0.05), two (P ≤ 0.01) or three asterisks (P ≤ 0.001).

Table 1. Contents in calcium, magnesium, carbon and organic matter and pH of the soil without (control) and in the presence of *P. pavonica* powder with two doses (50 g/Kg and 100 g/Kg) and of a chemical fertilizer (0.5 g/Kg) (Ammonitre), 15 days after germination.

	Ca^{2+} (Meq/g)	Mg^{2+} (Meq/g)	carbon (%)	organic matter (%)	pH
Control	0.041	0.0013	0.37	0.6	8.3
Dose 1	0.064	0.116	0.57	1	8.4
Dose 2	0.052	0.178	0.65	1.1	8.3
Ammonitre	0.09	0.066	0.5	0.9	8.5

Fertilizer was of a less significant contribution (0.066 meq/g) compared with algal powder. Also, powder of *P. pavonica* improved the in organic matter (Table 1), which was essential to retain the nutritive elements and moisture in the soil as well as to nourish and shelter the organizations of the ground.

Antifungal activity of extracts

Fungicides are chemical compounds having toxicological properties, used by the farmers to fight against the phytopathogenic fungi. These fungicides neutralize and reduce the activity of fungi but they remain a worrying and

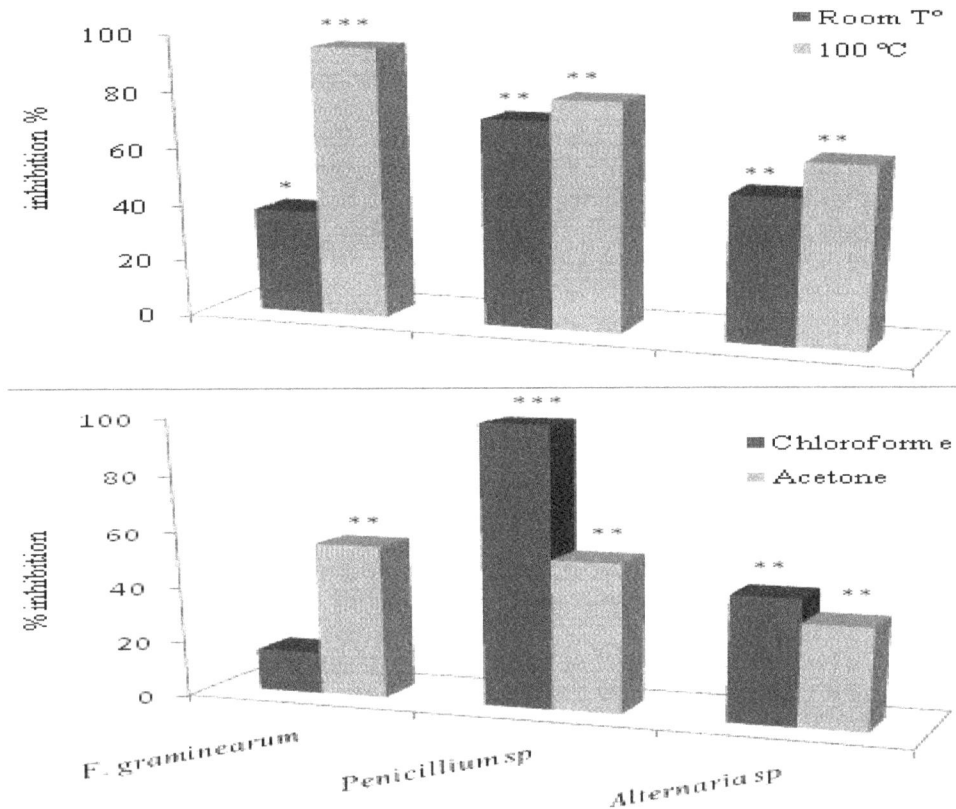

Figure 7. Effects of aqueous extracts (a) prepared at room temperature (Room T °C) and at 100 °C, and organic extracts (b) of *P. pavonica* on fungal growth. Differing statistically (Duncan test) from control are marked with one (P ≤ 0.05), two (P ≤ 0.01) or three asterisks (P ≤ 0.001).

frightening source of pollution and toxicity. To search for biological molecules with antifungal potential, *P. pavonica* extracts were evaluated on growth of three fungi strains: *F. graminearum, P. expansum* and *A. alternata*. Results revealed a strong toxicity of aqueous extracts on fungal growth (Figure 7). This toxicity was more significant when strains were exposed to the extract prepared at 100 °C, thus the percentages of inhibition were 95.83, 80.76 and 63.33% for *F. graminearum, P. expansum* and *A. alternata* respectively (Figure 7).

The screening indicated that chloroform extract was most toxic for *P. expansum*. As shown by its complete inhibition. This was in contrast to *F. graminearum* whose growth inhibition was not significant in presence of this extract. Acetone extract presented an average inhibition percentage of 48% for the three receptive strains. The difference in behavior of the various fungal species indicated that various organic extracts did not contain the same molecules and those bioactive molecules with antifungal potentialities against *P. expansum* were extracted by chloroform and had a medium polarity.

DISCUSSION

Germination and growth of target species organs were not varied or more or less improved by *P. pavonica*

aqueous extracts. Differences in behavior of the organs and the target species could be allotted to the nature of allelochemicals present in extracts. Sanna et al. (2004) recorded a difference in sensitivity of micro-algae exposed to various types of filtrates of algae. Growth improvement of seedlings would be ascribable to the presence of bio-active substances of the growth like amino acids, nutriments, phytohormones, enzymes and vitamins released by *P. pavonica* thalli. Similar results were reported by Klarzynski et al. (2005, 2006) who recorded a growth stimulation of tomato and wheat plants by filtrates of *Ascophyllum nodosum* (brown alga). Morot-Gaudry (1997) reported that algal filtrates stimulated the enzymes responsible for minerals absorption, such enzymes responsible for minerals absorption, such reductase nitrate which was a significant element of plant nitrogen nutrition.

In presence of organic extracts, the most important stimulations were recorded in the presence of chloroform extract which showed a percentage stimulation reaching an average of 142% for the two target organs. Kamenarska et al. (2002) recorded that *P. pavonica* was rich in sterol and lipid compounds which were significant components of cellular membranes and responsible for a great number of cellular functions. The principal sterol of *P. pavonica* in Mediterranean area was fucosterol (24-hydroperoxy-24-vinyl-cholesterol) (Ktari and Guyot, 1999). Moreover, *P. pavonica* contained a weak concentration of toxic allelochemical compounds like terpenoids and phenolic acids (Kamenarska et al., 2002). Indeed, these compounds were known to have inhibiting effects on germination and growth of various plants (Vyvyan, 2001). This could explain root growth inhibition recorded in lettuce exposed to the extract prepared at 100°C. This extraction probably allowed the release of a sufficient quantity of allelochemicals which induced the inhibition known by its sensitivity to allelochemicals (Chon et al., 2005; Abdelgaleil and Fumio, 2007). However, stimulation of germination and growth of target species under various *P. pavonica* extracts would be due to the presence of the stimulating allelochemicals. In fact, number of chemical compounds of terpenoid group were identified like stimulating germination such as strigol isolated from cotton (Cook et al.,1972), and from mays (Yoneyama et al.,2004), the sorgolactone isolated from sorghum (Hauck et al., 1992), and alectrol and orobanchol from red clover (Yokota et al., 1998).

Biomass production and growth were significantly ameliorated in presence of *P. pavonica* residue at low doses and a slightly inhibition at 100 g/Kg. Growth inhibition could be attributed to a release of toxic compounds following a degradation of the algal matter by soil microorganisms (Schmidt and Ley, 1999) or the oxidation by the soil (Ohno, 2001). On the other hand, Andre (1994) reported that plants which pushed in a mixed ground of marine algae had a faster growth than those which pushed in a ground with a comparable quantity of industrial fertilizer. The richness of the algae in mineral matter, basic elements for plants nutrition was well documented. In addition, Klarzynski et al. (2006) reported that algae powder was rich in carbohydrates, mainly brown algae were shown rich in mannitol, which was able to activate reductase nitrate enzyme and to increase the chlorophyll content.

Analysis of the soil mixed with the algal powder showed that this biological material did not have any effect on the pH, whereas it strongly enriched calcium and magnesium. It was well known that these two elements played a key role in the improvement of the soil texture. They returned the structure of the more movable and more stable ground, regularized the pH and supported the exchanges of ions necessary to the nutrition of the plants and created a medium favorable to the useful microbes of the ground (Soliner, 1983). Moreover, the powder of *P. pavonica* enriched the soil organic matter, compared with the control and with that enriched by chemical fertilizer. The richness of the organic matter soil was a great important character since it represented a source of nutriaments. In addition, they enhanced the soil permeability to water and air, ensuring its stability and being used of support for fauna and flora (Soliner, 1983).

Results revealed a strong toxicity of aqueous extracts on fungal growth and the screening indicated that chloroform extract was most toxic. Similar results were recor-

ded by Bennamara et al. (1999) who showed that the methoxybifurcarenone (molecule extracted from a brown alga *Cystoseira tamariscifolia)* had an antifungal activity against three pathogenic tomato fungal strains: *Botrytis cinerea, Fusarium oxysporum. F.* sp. *lycopersici* and *Verticillium alboatrum.* Also, Kamenarska et al. (2002) reported an antifungal activity of ethanolic extract of *P. pavonica* against *Candida albicans*

Conclusion

Our research showed that *P. pavonica* was rich in stimulating natural substances for test crops growth without causing damage to the soil. *P. pavonica* could be used as a natural fertilizer and would be used in the development of biological manures in order to reduce the dependence of the chemical fertilizers in the agricultural production. Furthermore, the aqueous extracts of this alga proved endowed with a strong antifungal capacity against the fungi tested in this work. A result could be considered very interesting, since *P. pavonica* molecules had a double effect in the crops: a fertilizing and an antifungal effect without deteriorating the physico-chemical properties of the soil. The differences in behavior of seeds and fungi in the presence of various organic extracts used enabled to have a preliminary idea on the nature of the bioactive molecules. Further studies are necessary to separate, purify and identify responsible molecules for the antifungal effects. The industrialization possibility of these natural bioactive products is a challenge that must be faced.

REFERENCES

Abdelgaleil AMS, Fumio H (2007). Allelopathic potential of wo sesquiterpene lactones from Magnolia grandiflora L. Biochem. Systematics Ecol. 35: 737-742.
Al Easa HS, Komprobst J, Rizk AM (1995). Major sterol composition of some algae from Qatar. Phytochemistry 39: 73-374.
André PM (1994). Algue marine : Acadie. « www.distrival.qc.ca ».
Aubert G (1978). Méthodes d'analyses des sols. Centre régional de documentation pédagogique de Marseille. Crop Marseille.
Ben Said R, El Abed A, Romdhane MS (2002). Etude d'une population de l'algue brune *Padina pavonica* (L.) Lamouroux à Cap Zebib (Nord de la Tunisie). Bull. Inst. Natn. Scien. Tech. Mer de Salammbô 29: 95-103.
Bennamara A, Abourriche A, Berrada M, Charrouf M, Chaib N, Boudouma M, Xavier Garneau F (1999). Methoxybifurcarenone: An antifungal and antibacterial meroditerpenoid from the brown alga Cystoseira tamariscifolia. Phytochemistry 52: 37-40.
Chon SU, Jang HG, Kim DK, Kim YM, Boo HO, Kim YJ (2005). Allelopathic potential in lettuce (Lactuca sativa L.) plants. Scientia Horticulturae 106: 309-317.
Chou CH (1999). Roles of Allelopathy in Plant Biodiversity and Sustainable Agriculture. Crit. Rev. Sci. Plant 18(5): 609-636.
Chung IM, Ahn JK, Yun SJ (2001). Assessment of allelopathic potential of barnyard grass (Echinochloa crus-galli) on rice (Oryza sativa L.) cultivars. Crop Prot. 20: 921-928.
Cook CE, Whichard LP, Wall ME (1972). Germination stimulants 2. The structure of strigol- a potent seed germination stimulant for wichweed (Striga lutea Lour.). J. Am. Chem. Crop Prot. 94: 6198-6199.
Corrêa LR, Soares GLG, Fett-Neto AG (2008). Allelopathic potential of Psychotria leiocarpa, a dominant understorey species of subtropical

forests. South Afr. J. Bot. Article in press.

Delabays N, Ançay A, Mermillod G (1998). Recherche d'espèces végétales à propriétés allélopathiques. *Revue suisse Vitic. Arboric. Hortic.* 30 (6): 383-387.

Einhellig FA, Rasmussen JA, Hejl AM, Souza IF (1993). Effects of root exudate sorgoleone on photosynthesis. J. Chem. Ecol. 19: 369–375.

Ervin GN, Wetzel RG (2003). An ecological perspective of allelochemical interference in land–water interface communities. Plant Soil 256: 13–28.

Fischer AJ, Beyer DE , Carriere MD, Ateh CM, Yim KO (2000). Mechanisms of resistance to bispyribac-sodium in an Echinochloa phyllopogon accession. Pestic. Biochem. Physiol. 68: 156-165.

Hauck C, Muller S, Schildtknecht H (1992). A germination stimulant for parasitic flowering plants from *Sorghum bicolor,*a genuine host plant. J. Plant Physiol. 139: 474-478.

Inderjit, Dakshini KMM (1995). On laboratory bioassays in allelopathy. Bot. Rev. 61: 28-44.

Kamenarska Z, Gasic MJ, Zlatovic M, Rasovic A, Sladic D, Kljajic Z, Stefanov K, Seizova K, Najdenski H, Kujumgiev A, Tsvetkova I, Popov S (2002). Chemical composition of the brown algae *Padina pavonica* (L.) Gaill. from the adriatic sea. Bot. Mar. 45: 339-345.

Kato-Noguchi H, Tanaka Y (2003). Allelopathic potential of citrus fruit peel and abscisic acid-glucose ester. Plant Growth Regul. 40: 117-120.

Keating KI (1977). Allelopathic influence on blue-green bloom sequence in a eutophic lake. Sci. 196: 885-887.

Khanh TD, Hong NH, Xuan TD, Chung IM (2005). Paddy weeds control by medicinal and leguminous plants from Southeast Asia. Crop Prot. 24(5): 421-431.

Klarzynski O, Esnault D, Euzen M, Joubert JM (2005). Mécanismes d'action de l'extrait d'algue GA7. Phytoma, la défense des végétaux 585.

Klarzynski O, Fablet E, Euzen M, Joubert JM (2006). État des connaissances sur leurs effets sur la : physiologie des plantes = The primary physio-activators of a marine Alga extract . Phytoma, la défense des végétaux 597: 10-12.

Ktari L, Guyot M (1999). A cytoxic oxysterol from the marine red sea alga *Padina pavonica (L.)* Thivy. J. Appl. Pycol. 11: 511-513.

Leather GR, Einhellig FA (1987). Bioassays of naturally occuring allelochemicals for phytotoxicity. J. Chem. Ecol. 14: 1821-1828.

Mao J, Yang L, Shi Y, Hu J, Piao Z, Mei L, Yin S (2006). Crude extract of *Astragalus mongholicus* root inhibits crop seed germination and soil nitrifying activity. Soil Biol. Biochem. 38(2): 201-208.

Molisch H (1937). Der Einfluss einer Pflanze auf die andere-Allelopathie. Fischer Verlag, Jena, Germany p. 106.

Morot-Gaudry (1997). Assimilation de l'azote chez les plantes. INRA Editions.

Ohno T (2001). Oxidation of phenolic acid derivatives by soil and its relevance to allelopathic activity. J. Environ. Qual. 30: 163-1635.

Rice EL (1984). Allelopathy. Second Edition. Academic Press, Inc., Orlando 1-7.

Sanna S, Giovana S, Edna O (2004). Allelopathic effects of the Baltic cyanobacteria *Nodularia spumigena*, *Aphanizomenon flos-aquae* and *Anabaena lemmermannii* on algal monocultures. J. Exp. Mar. Biol. Ecol. 308: 85-101.

Schmidt SK, Ley RF (1999). Microbial competition and soil structure limit the expression of allelochemicals in nature. In: Principles and Practices in Plant Ecology Allelochemicals Interactions, CRC Boca Raton, FL pp. 339-351.

Soliner D (1983). Les bases de la production végétale. Collection Sciences et Techniques Agricoles. Tome 1: Le sol. 12ème édition.

Vyvyan JR (2001). Allelochemicals as leads for new herbicides and agrochemicals. Tetrahedron 58: 1631-1646.

Wahheb MI (1997). Amino and fatty acid profiles of four species of macroalgae from Aquaba and their suitability for use in fish diets. Aquaculture 159: 101-109.

Yokota T, Sakai H, Okuno K, Yoneyama K, Takeuchi Y (1998). Alectrol and Orobanchol. Germination stimulants for Orobanche minor, from its host red clover. Phytochemistry 49: 1967-1973.

Yoneyama K, Takeuchi Y, Sato D, Sekimoto H, Yokota T (2004). Determination and quantification of strigolactones. In: Proceedings of the 8th International Prasitic Weed Symposium (ed DM Joel), 9. Inter. Parasitic Plant Soc. Amsterdam.

4

Interactive effects of P supply and drought on root growth of the mycorrhizal coriander (*Coriandrum sativum* L.)

Hossein Aliabadi Farahani[1], Sayed Alireza Valadabadi[1] and Mohammad Ali Khalvati[2]

[1]Islamic Azad University of Shahr-e-Qods branch, Iran.
[2]Department of Plant-Microbe Interactions, German Research Centre for Environment and Health, Munich (HelmholtzZentrum München), Ingolstaedter 1, Neuherberg Germany.

Arbuscular mycorrhizal fungi (AMF) are among the most important plant root symbiosis fungi enhancing plant P uptake. This research was conducted in Iran during 2006. The experimental design was a split factorial on the basis of completely randomized block design with four replicates. The control and AM species, *Glomus hoi* were assigned to the main plots and the combination of P fertilizer including 0, 35 and 70 kg ha^{-1} of triple super phosphate and the drought treatment including control (30 mm evaporation) and 60 mm evaporation from the evaporation pan were factorially assigned to the subplots. AM and P fertilizer significantly increased the root yield, root length, primary root dry weight and primary root length of coriander. Although the non-drought stress treatment significantly increased the root of coriander, the longest root length was achieved under the drought stress. It can be stated that AM is able to enhance the growth of coriander under drought stress through enhancing P uptake. This can have very important environmental impact through decreasing the amount of P fertilizer under control and drought stress conditions and also through enhancing the coriander resistance when subjected to the drought stress. Findings may interest farmers and agricultural researches to consider carefully on huge amount of soil phosphorus with interaction to water restriction as a challenge in environmental issues.

Key words: AM fungi, drought stress, phosphorus, root growth, coriander.

INTRODUCTION

Mycorrhizal fungi live in a 'symbiotic' relationship with plants. They grow in close association with the roots and play an important role in the concentration and transfer of soil nutrients to the plant. In exchange, the plant supplies the fungus with sugars. Although specific fungus-plant associations with respect to drought tolerance are of great interest (Ruiz-Lozano et al., 1995), the exact role of arbuscular mycorrhizal fungi (AFM) in drought resistance remains unclear (Auge et al., 1992a). More studies are therefore needed to determine the direct or indirect mechanisms which control plant-water relations in AMF-plant symbiosis. Although the effects of AM fungi on plant water status have been ascribed to the improved host nutrition (Graham and Syverten, 1987; Nelsen and Safir,

1982; Fitter, 1985), there are reports that drought resistance of AMF plants is somewhat independent of plant P nutrition status of plants (Sweatt and Davies, 1984; Auge et al., 1986a; Bethenfalvay et al., 1988; Khalvati et al., 2005). Although improved host nutrition has been ascribed to AM fungi effects on plant water status, there are reports that the drought resistance of AMF plants is somewhat independent of phosphorous levels. For example, Vivas (2003) reported that the increased metabolically active fungal biomass in inocu-lated plants was independent of phosphorous levels and was not related to phosphorous uptake from the poor nutrients soil (Khalvati, 2005). Baon et al. (1993) reported that extent of colonization of different barley cultivars was not consistently affected by *Glomus intraradices* and was only variably sensitive to the addition of phosphorous. Therefore, we examined the effects of *Glomus hoi* inocu-

*Corresponding author. E-mail: farahani_1362@yahoo.com.

Table 1. The results of soil analysis.

Soil texture	Sand (%)	Silt (%)	Clay (%)	K mg/kg	P mg/kg	N mg/kg	Na Ds/m	EC 1: 2.5	pH	Depth of sampling cm
Sa	49	30	21	147.2	6.2	34.7	0.04	0.19	8.1	0-15cm
Sa.c.L	56	25	19	124.3	3.7	28.2	0.03	0.16	7.9	15-30cm

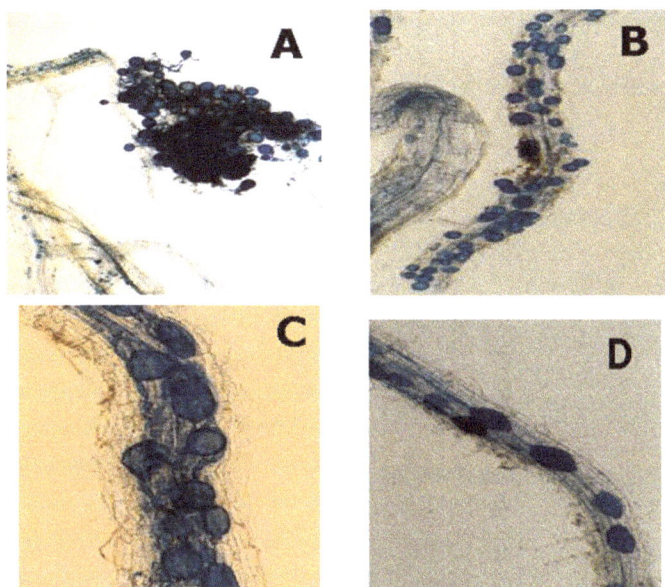

Figure 1. Symbiosis of AM fungi and coriander root. A) Spores of *Glomus hoi* B) Penetration of spores in coriander root C) Vesicles of *Glomus hoi* in main root D) Vesicles of *Glomus hoi* in lateral root.

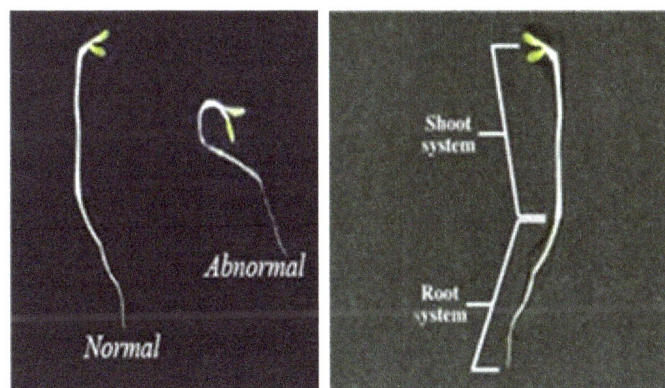

Figure 2. Normal and abnormal seedling (A) and seedling system (B).

lation of coriander on the underground organ components produced under different phosphorus availability. The objective of this investigation was to examine the degree of improvement in root growth of mycorrhizal plants in a soil with different P-content under simulated drought conditions.

MATERIALS AND METHODS

The experimental design was a split factorial on the basis of completely randomized block design with four replicates. The control and AM species, *G. hoi* were assigned to the main plots and the combination of P fertilizer including 0, 35 and 70 kg ha^{-1} of triple super phosphate and the drought treatment including control (30 mm evaporation) and 60 mm evaporation from the evaporation pan were factorially assigned to the subplots. The irrigation system was a piping system and determined water used in each irrigation period and amount of water for each irrigation treatments was 17.28 m^{-3} in each period and soil consisted of 48% clay, 30% silt and 21% sand (Table 1).

G. hoi was consisting of root fragments and adhering spores was mixed with soil (90 - 110 propgules per 10 g soil). A total number of 48 plots, each measuring 15 m^2 area (5 m × 3 m), were prepared in the field and a native variety named Shahabdolazimi was used in this experiment. Coriander seeds inoculated by inoculums in the planting time and the seeds were allowed to germinate, after which the seedlings were thinned to achieve 11 cm spacing within rows. Nitrogen fertilizer was added in two periods; 75 kg ha^{-1} urea in the beginning of steming stage and 75 kg ha^{-1} urea in beginning of flowering stage and also 150 kg ha^{-1} potash and phosphorus fertilizers were added in planting time. The degree of internal colonization produced by each entophyte in the roots of plants was rated by observation under a microscope. Root segments were clarified and stained with trypan blue. The extent of root colonization was estimated by comparison of root length between inoculated and non-inoculated plants as well as root length having hyphae. The results illustrated that coriander inoculated with *G. hoi* displayed 60 - 71% root colonization with *G. hoi* (Figure 1. A - D). At the end of growth, 100 plants were randomly collected from each plot for the study of plant characteristics and 100 seeds were collected from plants. Seeds of coriander were first sterilized in a 0.5% NaOCl solution for 15 min and then washed three times in sterile distilled water. Samples were placed in a germinator between two sheets of paper in Petri dishes. Seeds germinated after 21 days and we collected 10 normal seedlings (Figure 2). A and B) from each Petri dishes and determined radial dry matter and radial length (Abdul-Baki and Anderson, 1973).

Data were subjected to analysis of variance (ANOVA) using Statistical Analysis System and followed by Duncan's multiple range tests. Terms were considered significant at $P < 0.05$ (SAS institute Cary, USA, 1988).

RESULTS

The results showed that drought stress significantly affects root yield, primary root dry weight, root diameter, highest plant, number of branch and stem diameter (α = 1%), root length and primary root length (α = 5%) and the highest upon charactristics appeared under without stress but longest root length was achieved under the drought stress (Tables 2 and 3). The results also showed

Table 2. Variance analysis.

Value Sources	df	PRL	PRDW	Root diameter	Root length	Root yield	Number of branch	Stem diameter	Highest plant
Replication	3	0.018	0.0001	0.036*	4.802	162755.694	4.347	0.023*	346.65
Mycorrhiza	1	0.505**	0.001*	0.001	136.603**	360317.291*	0.047	0.004	209.167
Error a	3	0.014	0.0001	0.004	3.792	17703.73	2.701	0.001	37.723
Phosphorus	2	0.136**	0.001**	0.003	36.91**	291690.374**	1.97	0.015	31.72
Mycorrhiza × phosphorus	2	0.033	0.0001	0.003	8.964	102190.046	2.183	0.001	17.671
Drought stress	1	0.158*	0.035**	0.694**	42.779*	14969087.337**	92.13**	0.715**	7874.563**
Mycorrhiza × drought stress	1	0.066	0.0001	0.0001	17.977	1736.263	3.05	0.002	5.88
Phosphorus × drought stress	2	0.008	0.0001	0.002	2.247	142939.147*	1.258	0.008	16.459
Mycorrhiza × phosphorus × drought stress	2	0.16	0.0001	0.0001	4.204	3072.814	4.421	0.004	4.107
Error bc	30	0.025	0.0001	0.009	6.745	39184.251	2.73	0.006	33.233
CV (%)		14.05	9.2	21.93	14.05	9.2	17.37	17.72	11.89

PRDW= Primary root dry weight
PRL= Primary root length

* and ** : Significant at 5% and 1% levels respectively.

Table 3. Variance analysis.

Treatments		Root yield (kg ha^{-1})	Root diameter (cm)	Root length (cm)	PRDW (g)	PRL (cm)	Highest plant (cm)	Stem diameter (cm)	Number of branch
mycorrhiza									
	Application	2693.7 a	0.75 a	20.02 a	0.1101 a	1.25 a	50.6 a	0.44 a	9.5 a
	Non application	2008 b	0.73 a	16.13 b	0.1 b	0.94 b	46.4 a	0.42 a	9.5 a
	Non application	1915 c	0.69 b	15.93 c	0.1 c	0.95 c	46.9 a	0.39 a	9.1 a
Phosphorus									
	35 kg ha^{-1}	2153 b	0.71 a	18.27 b	0.106 b	1.2 b	49.2 a	0.44 a	9.8 a
	70 kg ha^{-1}	2203 a	0.72 a	20.56 a	0.1109 a	1.32 a	49.4 a	0.45 a	9.6 a
	Normal irrigation	2659 a	0.91 a	15.25 b	0.1362 a	1.12 a	61.3 a	0.55 a	11 a
Irrigation according									
	Drought stress	1510 b	0.47 b	18.89 a	0.947 b	0.73 b	35.7 b	0.31 b	8.1 b

PRDW= Primary root dry weight
PRL= Primary root length
Means within the same column and rows and factors, followed by the same letter are not significantly difference (P < 0.05) using Duncan´s multiple range test.

that root yield, primary root dry weight (α = 5%), root length and primary root length (α=1%) among non-inoculated and *G. hoi*-inoculated coriander seedlings had significantly higher values for these characteristics among mycorrhizal plants (Tables 2 and 3). In addition, phosphorous application had a significant effect on root yield, primary root dry weight, root length and primary root length (α =1%) and highest values were appeared under 70 kg h^{-1} P fertilizer (Tables 2 and 3). Data of drought and interactive effect between P available and mycorrhizal plants on underground organs products demonstrated (Table 4) no significant difference between mycorrhizal plants treated with 35 and 70 kg h^{-1} P under sever water restriction (60 mm eva.). Therefore highest upon characteristics appeared under application of *G. hoi*, 70 kg ha^{-1} P and without drought stress conditions but longest root length was achieved under application of *G. hoi*, 35 kg ha^{-1} P and drought stress conditions (Table 4).

DISCUSSION

Plants colonized by mycorrhizal fungi have been shown to deplete soil water more thoroughly than non-mycorrhizal plants (Auge, 2001). One reason for this is the fact that the shoots of plants with AMF usually have a larger biomass (more evaporative leaf surface area) than non-AMF plants (Fitter, 1985; Nelsen, 1987). Also the root systems of plants with AMF are often more finely divided and thus have more absorptive surface area (Allen et al., 1981; Busse and Ellis, 1985; Ellis et al., 1985; Huang et al., 1985; Sharma and Srivastava, 1991; Osonubi et al., 1992; Osonubi et al, 1994; Okon et al., 1996). Furthermore, the roots of plants with AMF dry the soil more quickly than non-AMF plants of similar size (For example, Bryla and Duniway, 1997). In our experiment, mycorrhizal coriander significantly root dry matter throughout the improvement of plant water relations under drought conditions, corresponding of mycorrhiza's contribution in P uptake to AMF-plants and act to synthesis certain phytohormones such as ABA and cytokinin. In the present study, mycorrhizal (*G. hoi*) treatment of coriander significantly improved root dry matter through improvement of plant water relations under drought conditions. These improvements were likely achieved via the mycorrhizal contribution to phosphorous uptake and the ability of AM fungi to stimulate plant synthesis of certain phytohormones such as ABA and cytokinins. Consequently, plants with AMF had higher phosphorous content in shoots than non-AMF plants. In agreement with the observation of Labour et al. (2003) and Dhanda et al. (2004) our data also revealed that roots of AMF-inoculated coriander were longer with increasing fungal hyphae growth, similar to the findings of Ruiz-Lozano et al. (1995). In addition, we found longer root length among AMF-inoculated plants under drought conditions compared to well-watered plants. Drought stress causes increased root length due to restriction of water content in plant organs, which leads plants to make an effort to uptake water from further destinations (Mayaki et al., 1976). In fact, shoot growth is more restrained by drought than root growth and because drought causes a decrease in root dry matter and length, corresponding to a reduction of nutrient transfer to seeds. Therefore, in drought conditions, plants product lower strong seeds when compared to well-watered plants seeds. Importantly, seeds from plants grown under water stress in turn grow plants with low root dry matter. We have also observed that the highest root production under drought conditions occurred when plants were treated with 35 kg h^{-1} phosphorous, in agreement with Shubhra et al. (2004). However, increasing the phosphorous content to 70 kg h^{-1} did not significantly improve root production or phosphorous uptake. Therefore, mycorrhizal plants showed a low necessity of phosphorus fertilizer in terms of contribution of P uptake to plants by hyphae and produced a same amount of dry matter with 35 kg h^{-1} available phosphorus as compared to 70 kg h^{-1}. This observation has illustrated in the produced of root dry matter of mycorrhizal plants in the treated level by 35 kg h^{-1} phosphorus. It was no significant different with the plants which has treated by 70 kg h^{-1} phosphorus under drought conditions. However, similar produce of root dry matter in both 35 and 70 kg h^{-1} phosphorus correspond to the contribution of mycorrhizal fungi to uptake water and P for inoculated plants. In fact, under drought conditions, AMF-inoculated plants grown with 35 kg h^{-1} phosphorous produced root dry matter similar to that found among inoculated plants grown with 70 kg h^{-1} phosphorous, indicating the contribution of hyphea to phosphorous uptake regardless to soil phosphorus levels. Our findings indicate that AMF-inoculation improves root dry matter and root length decreases the phosphorous requirement for coriander plants subjected to water stress.

Conclusion

The investigation showed that AM is able to enhance the growth of coriander under water stress through enhanceing P uptake. This can have very important environmental implications through decreasing the amount of P fertilizer under control and water stress conditions and also through enhancing the coriander resistance when subjected to the water stress.

ACKNOWLEDGEMENTS

The authors are indebted to Dr. Sharifi-Ashoorabadi and Dr. Sefidkon for providing financial assistance and continuous encouragement to carry out in the investigation.

REFERENCES

Allen MF, Smith WK, Moore TS, Chiristensen M (1981). Comparative

Table 4. Means comparison.

Number of branch	Highest Plant (cm)	Stem Diameter (cm)	PRL (cm)	PRDW (g)	Root diameter (cm)	Root Length (cm)	Root yield (kg ha-1)		survey instance qualifications
9.2 a	44.7 b	0.4 a	0.9d	0.1 c	0.63 a	14.53d	1868 c	Non application of phosphorus	Non application
10.2 a	46.2 ab	0.44 a	1 cd	0.102b	0.59 a	15.79cd	2058 d	35 (kg ha-1) phosphorus	
9.3 a	48.4 ab	0.47 a	1.2b	0.11a	0.57 a	18.46 b	2201 a	70 (kg ha-1) phosphorus	
Mycorrhiza									
9.1 a	49.1 ab	0.39 a	1.1bc	0.103b	0.63 a	18.07bc	2062 b	Non application of phosphorus	Application
9.4 a	52.1 a	0.44 a	1.3a	0.11a	0.63 a	21.06a	2279 a	35 (kg ha-1) phosphorus	
10 a	50.4 ab	0.43 a	1.3a	0.11a	0.61 a	21.4 a	2239 a	70 (kg ha-1) phosphorus	
11.2 a	58.9 b	0.56 a	1.1b	0.11a	0.53 a	17.47b	2106 b	Non stress	
Non application									
8 b	34 c	0.32 b	0.9c	0.071c	0.27 b	15.25c	1416 d	Stress	
Mycorrhiza									
10.6 a	63.8 a	0.55 a	1.3 a	0.11a	0.56 a	20.51a	2190 a	Non stress	
Application									
8.4 b	37.4 c	0.29 b	1.2 b	0.08b	0.29 b	19.83a	1603 c	Stress	
10.8 a	58.6 a	0.51 a	1.15b	0.12 c	0.51 a	16.83bc	2319 c	Non application of phosphorus	Non stress
11.1 a	62.2 a	0.59 a	1.2 a	0.14 b	0.56 a	18.85ab	2760 b	35 (kg ha-1) phosphorus	
10.8 a	63.1 a	0.56 a	1.3 a	0.144a	0.56 a	20.99a	2908 a	70 (kg ha-1) phosphorus	
Drought stress									
7.4 b	35.2 b	0.28 b	1d	0.07 e	0.27 b	15.76c	1412 e	Non application of phosphorus	Stress
8.5 b	36.3 b	0.3 b	1.1c	0.08 d	0.29 b	18bc	1577 d	35 (kg ha-1) phosphorus	
8.5 b	35.7 b	0.35 b	1.1c	0.08 d	0.28 b	18.87ab	1542 d	70 (kg ha-1) phosphorus	
10.9 a	56.5 a	0.5 a	1.2bc	0.11 c	0.5 a	18.6abc	2216 c	**Non stress**	
11.3 a	58.9 a	0.58 a	1.1 c	0.13 b	0.56 a	17.2bcd	2661 b		
11.3 a	61.3 a	0.59 a	1.2 b	0.15 a	0.62 a	19.38ab	2890 a		
Non application									
7.4 b	32.9 b	0.3 b	0.9f	0.07 f	0.28 b	13.77d	1319 f	**Stress**	
7.1 b	33.5 b	0.31 b	0.95f	0.072ef	0.33 b	14.43cd	1454 ef		
7.3 b	35.6 b	0.36 b	1 ef	0.073ef	0.31 b	17.6bcd	1474 ef		
7.4 b	32.9 b	0.3 b	0.9f	0.07 f	0.28 b	13.77d	1319 f		

Table 4. Contd.

Mycorrhiza								
10.7 a	60.7 a	0.52 a	1.2 b	0.13 b	0.53 a	18.4abc	2621 b	Non stress
10.8 a	65.6 a	0.53 a	1.3 a	0.13 b	0.54 a	20.55ab	2558 a	
10.9 a	65.9 a	0.62 a	1.32 a	0.14 a	0.6 a	20.83 a	2927 a	
Application								
7.4 b	35.5 b	0.26 b	1.05de	0.075e	0.32 b	17.7bcd	1505 e	Stress
8 b	39.1 b	0.28 b	1.07d	0.08 d	0.25 b	20.8a	1699 d	
7.8 b	35.8 b	0.34 b	1.07d	0.08 d	0.33 b	20.5a	1608 d	

PRDW= Primary root dry weight.
PRL= Primary root length.
Means within the same column and factors, followed by the same letter are not significantly difference (P < 0.05) using Duncan´s multiple range test.

water relation and photosyntesis of mycorrhizal and non-mycorrhizal Boutelua Gracilis H.B.K. New Phytol. 88: 683-693.

Anonymus (1999). International rules for seed testing. Internationl Seed Testing Association (ISTA). Seed Science and Technology. 27, Supplement.

Auge RM, Stodola AJ, Brown MS, Bethlenfatvay G J (1992a). Stomatal response of mycorrhizal cowpea and soybean to short-term osmotic stress. New phytol. 120: 117-125.

Auge RM, Schekel KA, Wample, RL (1986). Osmotic adjustmen in leaves VA mycorrhizal nonmycorrhizal rose plant in response to drought stress. Plant Physiol. 82: 765-770.

Auge RM (2001). Water relation, drought and VA mycorrhizal simbiosis. Mycorrhiza. 11: 3-42.

Baon JB, Smith SE, Alston AM (1993). Mycorrhizal responses of barley cultivars differing in P efficiency. Plant Soil. 157(1): 97-105.

Bethenfalway GJ, Brown MS, Ames RN,Thomas RS (1988). Effects of drought on host and endophyte development in mycorrhizal soybeans in relation to water use and phosphate uptake. Plant Physiol. 72: 565-571.

Bryla DR, Duniway JM (1997). Effects of mycorrhizal infection on drought tolerance and recovery in safflower and wheat. Plant and Soil. 197: 95-103.

Busse MD, Ellis JR (1985). Vesicular-Arbuscular Mycorrhizal (Glomus fasciculatum) Influence on Soybean Drought Tolerance in High Phosphorus Soil. Can. J. Bot. CJBOAW. 63(12): 2290-2294.

Dhanda SS, Sethi GS, Behl RK (2004). Indices of Drought Tolerance in Wheat Genotypes at Early Stages of Plant Growth. J. Agron. Crop. Sci., 190(1): 6-12.

Ellis JR, Larsen HJ, Boosalis MG (1985). Drought resistance of wheat plants inoculated with vesicular-arbuscular mycorrhi-zae. Plant and Soil. 86: 369-378.

Fitter AH (1985). Functioning of vesicular-arbuscular mycor-rhizas under field conditions. New Phytol. 99: 257-265.

Graham JH, Syvertsen JP, Smith ML (1987). Water relations of mycorrhizal and phosphorus-fertilized non-mycorrhizal Citrus under drought stress. New phytol. 105: 411-419.

Huang RS, Smith WK, Yost RS (1985). Influence of vesicular-arbuscular mycorrhiza on growth, water relations and leaf orientation in Leucaena leucocephala (Lam.) de Wit. New Phytol. 99: 229-243.

Khalvati MA (2005). Quantification of Water Uptake of hyphae contributing to barely subjected to drought conditions. Technical University of Munich. pp 8-11.

Khalvati MA, Mozafar A and Schmidhalter U (2005). Quantification of Water Uptake by Arbuscular Mycorrhizal Hyphae and its Significance for Leaf Growth, Water Relations, and Gas Exchange of Barley Subjected to Drought Stress. Plant Biology –Stuttgart. 7(6): 706-712.

Labour K, Jolicoeur M, St-Arnaud M (2003). Arbuscular mycorrhizal responsiveness of in vitro tomato root lines is not related to growth and nutrient uptake rates. Can. J. Bot. 81(7): 645-656.

Mayaki WC, Teare ID, Stone LR (1976) Top and root growth of irrigated and non-irrigated soybeans. Crop Sci., 16: 92-94

Nelsen CE (1987). The water relations of vesicular-arbuscular mycorrhizal systems. In Ecophysiology of VA Mycorrhizal Plants. Ed. G RSafir. CRC Press, Boca Raton, FL. pp. 71-91.

Okon IE, Osonubi O and Sanginga N (1996). Vesicular-arbuscular mycorrhiza effects on Gliricidia sepium and Senna siamea in a fallowed alley cropping system. Agro. Sys. 33(2): 165-175.

Osonubi O, Bakare ON, Mulongoy K (1992). Interactions between drought stress and vesicular-arbuscular mycorrhiza on the growth of Faidherbia albida (syn. Acacia albida) and Acacia nilotica in sterile and non-sterile soils. Bio. Fer. Soils. 14(3): 159-165.

Osonubi O (1994). Coperactive effects of visicular arbuscular mycorrhizal inoculation and phosphrus fertilization on growth and phosphorus uptake of maize and sorgum plant under drought stressed conditions. Bio. Fer. Soils. 14: 159-165.

Ruiz-Lozano JM, Azcon R, Gomez M (1995). Effects of Arbuscular-Mycorrhiza Glomus Species on Drought Tolerance: Physiological and Nutritional Plant Responses. Appl. Environ. Microbiol. 61(2): 456–460.

Sharma AK, Srivastava PC, Johri BN, Rathore VS (1991). Kinetics of zinc uptake by mycorrhizal (VAM) and non-mycorrhizal corn (Zea mays L.) roots. Bio. Fer. Soils. 13(4): 206-210.

Shubhra K, Dayal J, Goswami CL, Munjal R (2004). Influence of Phosphorus Application on Water Relations, Biochemical Parameters and Gum Content in Cluster Bean Under Water Deficit. Bio. Planta. 48(3): 445-448.

Vivas A (2003). Physiological characteristics (SDH and ALP activities) of arbuscular mycorrhizal colonization as affected by Bacillus thuringiensis inoculation under two phosphorus levels. Soil. Bio. Biochem. 35(10): 987-996.

5

Biplot analysis of grain yield in barley grown under differing management levels in years of contrasting season-end drought

Woldeyesus Sinebo*, Berhane Lakew and Abraham Feyissa

Ethiopian Institute of Agricultural Research, P. O. Box 2003, Addis Ababa, Ethiopia.

Variance analysis and graphical biplots were used to understand the nature of genotype × environment interaction (G × E) in a grain yield data set obtained from 39 barley (*Hordeum vulgare L.*) genotypes grown in 18 environments (a combination of three sowing dates, two crop protection treatments and three years) at Holetta, central highlands of Ethiopia. Genotype × year interaction was much more important than genotype × management interaction. Season-end drought was the environmental variable and time to maturity was the genotypic variable responsible for the high G × year interaction variance. An elite breeding line gave the highest mean yield and was the best under low but not under high season-end drought stress. Sasa, an early maturing landrace, was the best in a year of high season-end drought. Biplots enabled visual identification of compromise genotypes such as 3304 - 11 and 3381 - 04 that yielded reasonably well under both low and high season-end drought conditions. Selection for post-anthesis drought tolerance may result in high and stable yields across years and wider geographical adaptation in Ethiopian barley. The importance of unique landraces for stress situations is ascertained.

Key words: Barley landraces, genotype × environment interaction, GGE biplot, season-end drought, sowing date, insecticidal seed treatment, fungicidal disease control.

INTRODUCTION

Raising crop yield in subsistence rain fed farming systems in Ethiopian highlands is constrained by a host of problems including unpredictable weather, low use of chemical inputs and unavailability or poor adoption of improved varieties. Integration of genetic and management approaches that optimize yield under variable environmental conditions and resource endowments can enhance incomes and minimize risk.

In Ethiopia, climate trend analysis reveals increase in temperatures and reduction in rainfall and season length (IGAD, 2007). With climate change looming so large, crops face an array of biophysical challenges such as increased temperature, unpredictable moisture and increased disease and insect pest pressures (Tubiello et al., 2007). Crop production can be adapted to climate

change through, *inter alia*, selection of appropriate genotypes in tandem with modulation of management practices to fit into the changing circumstances (Howden et al., 2007). However, little information is available on how crop varieties that differ in phenology and morphology interact with management practices and season under the increasingly unpredictable environment in Ethiopia.

Crop genotypes respond differently to environments giving rise to complex genotype-by-environment (GE or G × E) interaction. The environmental factors inducing the G × E interaction can arise from predictable or unpredictable variations in the form of location, management levels or years. Biplots have been used to visualize differential response of genotypes to environments and identify winning cultivars for target production environments (Kempton, 1984). Two contemporary GE interaction analysis tools that make use of biplots are additive main effect and multiplicative interaction (AMMI) and genotype main effect plus GE interaction (GGE)

---

*Corresponding author. E-mail: wsinebo@hotmail.com.

(Zobel et al., 1988; Yan et al., 2000).

Broadly, barley exhibits specific adaptation, therefore, GE interaction particularly in stress-prone environments (van Oosterom et al., 1993; Ceccarelli, 1994; Sinebo, 2005). Barley in Ethiopia is grown mainly as a low input staple food crop in the higher altitudes, on steep slopes, eroded lands or in moisture stress areas (Gebre and van Leur, 1996). Ethiopian barleys display specific adaptation to variable stresses such as low nitrogen and drought (Gróny, 2001; Sinebo, 2002) owing to the large diversity of agroecology including marginal environments apparent in the country.

In many barley growing parts of Ethiopia, rainfall is perhaps the single most important factor determining crop growing season length, cultivar choice and grain yield. Reportedly, farmers have increasingly opted to growing low yielding but early maturing varieties instead of long cycle high yielding varieties in response to a perceived shortening trend of crop growing season length (Yirga et al., 1998).

Change of sowing date can be used as an adaptive strategy to climate change-induced shortening of the season length. When the season is favorable, early sowing may allow longer crop growth duration leading to larger biomass accumulation perhaps resulting in greater economic yield. In some barley growing areas such as Holetta, the main season rain may arrive at about early June but barley is sown from about late June to late July leading to the loss of potential crop growth duration. Despite this, a few sowing date trials conducted on barley at Holetta, Ethiopia, failed to establish the superiority for grain yield of early sowing (Mola et al., 1996). However, these studies were conducted with released varieties that had not been selected for early sowing. In addition, high incidence and damage by scald disease [Rhynchosporium secalis (Oud.) Davis.] was observed with early sowing (Mola et al., 1996), but whether this disease was the major cause of low yields was not established. Furthermore, early sowing is known to dispose young seedlings to attack by insects particularly of barley shoot fly (Delia arambourgi (Seguy)). It was hypothesized that subjecting a large number of genotypes to varying sowing dates in the presence or absence of insecticidal seed treatment plus fungicidal disease control would lead to differential genotype grain yield response enabling the exploitation of specific adaptation to the management levels particularly sowing date. This, in the end, is hoped to increase crop growth duration, biomass accumulation and grain yield and minimize yield penalty arising from season-end moisture stress in intermediate to late maturing varieties.

The objectives of this study are (i) to assess patterns and causes of GE interaction for grain yield, and (ii) to examine the relative contribution of genotype × management and genotype × season interaction to GE in a data set generated from 39 barley genotypes grown under three sowing dates and two crop protection treatments over three years in the central highlands of Ethiopia.

MATERIALS AND METHODS

Site, design and treatments

Thirty-nine barley cultivars and experimental lines (referred hereafter as genotypes) representing different phenological and morphological groups sampled from field books of the barley breeding program at Holetta were tested in a factorial combination of three sowing dates and two crop protection treatments for three years (2002 - 2004) on a red brown clay (a Eutric Nitosol) at Holetta Agricultural Research Center (9°03'N, 38°31'E, elevation 2400 m), 28 km west of Addis Ababa, Ethiopia. The environment is seasonally humid with long term (1976 - 2005) average annual rainfall of 1055 mm, 85% of which is received between the months of June and Sept., and mean max. and min. temperatures of 22.2 and 6.1°C, respectively. The three sowing dates were early (at about the on-set of main season rain), normal (15 days after the on-set of main season rain), and late (15 days after the normal sowing date). The two crop protection treatments were insecticidal seed treatment plus fungicide application vs. no insecticide seed treatment plus no fungicide application. The genotypes included 13 improved released varieties, nine landrace cultivars grown in different parts of the country, four experimental lines developed from local crosses, three introduced experimental lines, and 10 experimental lines developed from landrace populations (Table 1).

The experiment was planted after Ethiopian mustard (Brassica carinata A. Braun) rotation in 2002, after faba bean (Vicia fabaea L.) rotation in 2003 and after potato (Solanium tuberosum L.) rotation in 2004. A split-plot design with a factorial combination of sowing dates and crop protection treatments in the main plots and the genotypes in the sub-plots in three replications was used. The sub-plots consisted two rows of 2.5 m length separated by a blank row. The details of sowing and fungicide application dates are given in Table 2. Crop protection treatment included absence or presence of seed treatment with Gaucho (Imidacloprid) 70% WS at a rate of 1 g product per kg seed followed by the foliar application of Propiconazole (1–[2–(2,4–dichlorophenyl)–4–propyl–1,3–dioxolan–2–yl–methyl]–1H–1,2,4–triazole; Ciba-geigy, Whittlesford, Cambridge) at a rate of 125 g a.i. ha^{-1} depending on the incidence of fungal diseases (Table 2). Fertilizer was drilled in rows and slightly incorporated with sticks after which seeds were drilled at a rate of 80 kg ha^{-1} as uniformly as possible and lightly covered by hand. The total plot area was harvested for yield determination.

Data were collected on grain yield, yield components, vegetative growth and growth durations. Vegetative shoot height, from the ground level to the tip of the shoot, was measured as a proxy for early vegetative vigor (Sinebo, 2002) on the dates and mean growing degree days (GDD) given in Table 2. Mature plant height was measured from the ground level to the tip of the spike excluding the awns after physiological maturity had been reached. Heading date was recorded as when the spikes of 50% of the culms in a plot had fully extruded out. Physiological maturity was recorded when the plants had almost lost their green color from both vegetative and reproductive tissues. Grain filling duration was calculated as a difference of time to heading and time to maturity. Scald, net blotch (Helminthosporium teres Sacc.) and spot blotch (H. sativum Pam., King and Bakke) diseases were scored on a 0 to 9 scale (Loegering, 1959). In this scale 0 indicates free from infection, 1 indicates resistant with few isolated lesions on lower most leaves, 5 indicates moderate susceptibility with severe infection of lower leaves and 9 high susceptibility with severe infection on all leaves. Likewise, leaf rust (Puccinia hordei Otth.) infection was scored on a similar scale but on percentage basis.

Variance analysis

Analysis of variance was carried out with PROC MIXED of the SAS

Table 1. The list of genotypes used in the study.

Variety	Code	Description	Origin†
208038-90	2038	Landrace line	N. Shewa /Bita Belew
1829-76	1829	Landrace line	W. Shewa /Ambo
3381-04	3381	Landrace line	Arsi /Digelu & Tijo
3304-11	3304	Landrace line	Arsi /Kofele
3371-18	3371	Landrace line	Arsi /Sude
Tolese S8.2H.2	th2	Landrace line	N. Shewa
Tolese S8.SP.1	ts1	Landrace line	N. Shewa
Tolese S8.B.7	tb7	Landrace line	N. Shewa
Baleme S1.2H.2	bh2	Landrace line	W. Shewa
Baleme S1.3H.1	bh1	Landrace line	W. Shewa
EH 1682/F7.1H	eh82	Akalase × IBON 93/91	HARC
EH 1642/F7.3H	eh42	Baleme × IBON 93/91	HARC
EH 1665/F7.1H.28.40.16	eh65	White barley (W. Shewa) × Composite 29	HARC
EH 1507	eh07	White Sasa × EH538/F2-12B ×× White Sasa × 3336-03	HARC
ARDU 12/60B	ard	Landrace line selection, released	Arsi
Beka	bka	Introduction, released	HARC
HB-42	hb42	Released, IAR/H/81 × Composite 29 ×× Composite 14/20 × Coast	HARC
Shege	shg	Landrace line selection, released	Arsi /Guna
HB-120	hb20	Released, EH11/F3.A.1.A.L × Beka	HARC
HB-1533	hb33	Released, B.F2 (S×W) × 3284-11	HARC
HB-52	hb52	Released, Compound 29 × Beka	HARC
Holkr	hkr	Released, Holetta Mixed × Kenya Research	HARC
Ahor 880/61	ahr	Introduction, released	HARC
IAR/H/485	iar	Landrace line selection, released	Arsi
Abay	aby	Landrace line selection, released	Arsi /Sude
Dimtu	dim	Landrace line selection, released	Arsi /Sude
Misratch	mis	Landrace line selection, released	Arsi
EMBSN 13/98	em13	Introduced breeding line	ICARDA
EMBSN 44/98	em44	Introduced breeding line	ICARDA
EMBSN 42/98	em42	Introduced breeding line	ICARDA
Semereta	sem	Landrace cultivar	Gojam, Shewa
White Sasa	sas	Landrace cultivar	Tigray
Ehilzer	ehil	Landrace cultivar	Wollo
Shasho	sho	Landrace cultivar	Bale
Black barley T.Inchini	bbti	Landrace cultivar	W. Shewa
Chare - Degem	cha	Landrace cultivar	N. Shewa
Ginbote	gin	Landrace cultivar	W. Shewa
Feresgama	fer	Landrace cultivar	N. Shewa
Baleme	bal	Landrace cultivar	W. Shewa

† N. = north, W. = west, HARC = Holetta Agricultural Research Center, ICARDA = International Center for Agricultural Research in the Dry Areas, HARC = Holetta Agricultural Research Center.

statistical package version 8.12 (SAS Institute INC, Cary, NC) using the following model:

$$T_{ijklm} = \mu + Y_l + S_m + C_k + G_i + R(Y)_{jl} + YS_{lm} + YC_{lk} + CS_{lm} + YSC_{lkm} + SCR(Y)_{jklm} + GY_{il} + GS_{im} + GC_{ik} + GYS_{ilm} + GYC_{ikl} + GSC_{ikm} + GYSC_{iklm} + e_{ijklm},$$

where T is the observation of the ith variety G in the lth year Y of the mth sowing date S and kth crop protection treatment C in the jth replication R within year l; μ is the general mean, e is the variation

Table 2. List of sowing and fungicide application dates by sowing date treatments and year.

Sowing date treatments	Year 2002	Year 2003	Year 2004	
		Sowing dates		
Early	17 June	16 June	14 June	
Normal	1 July	30 June	28 June	
Late	15 July	14 July	12 July	
		Fungicide application dates		
Early	2 Sept	18 Aug and 8 Sept	24 Aug	
Normal	23 Sept	28 Aug and 17 Sept	8 Sept	
Late	23 Sept	28 Aug and 17 Sept	21 Sept	
	Vegetative height measurement dates and growing degree days in brackets (base T° = 5°C)			**Mean (GDD)**
Early	22 July (370)	25 July (349)	23 July (372)	363
Normal	5 Aug (355)	11 Aug (370)	9 Aug (387)	371
Late	21 Aug (361)	22 Aug (360)	20 Aug (360)	360
Mean (GDD)	362	360	373	

due to random error or the residual, and *YS, YC, CS, YSC, GY, GS, GC, GYS, GYC, GSC, GYSC,* and SCR(Y) are the interactions. In the analysis, Y, S, C, and all possible interactions among these three factors were considered fixed, and all the remaining effects were considered random. Genotypes were considered as a random sample of germplasm handled by the breeding program at Holetta in order to be able to draw broad inferences on the patterns of response of the barley materials in the breeding program with respect to the management levels and the years tested. Incidentally, the test years were contrasting manifesting features apparent in short and long season barley growing ecologies of the country. As a result, years were considered fixed representing short cycle and long cycle barley growing locations of the country.

Genotype plus genotype × environment interaction (GGE) biplot analysis

For the GGE analysis, grain yields of the 39 genotypes in each of the 18 environments (3 years × 3 sowing dates × 2 crop protection treatments) were expressed as best linear unbiased predictions (BLUPs). Environment centered residuals were obtained as:

$$y_{ij} - \hat{y}_{\cdot j},$$

for the genotype *i* and environment *j* cell.
The residuals were subjected to singular value decomposition using the PROC IML in SAS. The resulting singular values for the first and second principal components were partitioned to the respective genotype and environment eigenvectors using a factor of 0.5 (symmetric scaling; Yan *et al.*, 2000) as:

$$g_{il} = \lambda_l^{0.5}\xi_{il} \text{ and } e_{jl} = \lambda_l^{0.5}\eta_{jl}$$

where g_{il} and e_{jl} are PC *l* scores (*l* = 1 or 2) for genotype *i* and environment *j*, respectively. The resulting genotype and environment PC scores were plotted using Microsoft® Excel 2000 Software (Microsoft Corporation).

RESULTS

Weather

The main crop-growing season started between 6 and 8 June in the three years. There was a 7 to 9 days delay for the first sowing from the presumed sowing with the onset of rainfall. Rainfall for the crop growing months of June to September were nearly similar amounting 668, 656 and 672 mm for the years 2002, 2003 and 2004, respectively (Table 3). Nonetheless, year 2002 was the most stressful because of early cessation of rainfall. The last shower of rain in 2002 was a 5.3 mm rain received on the 22nd of September. Total rainfall for September was 77.4 mm in 2002, 107.4 mm in 2003 and 119.7 mm in 2004. Total rainfall for the month of October was 0 mm in 2002, 10 mm in 2003 (received on the 12th of October) and 3.6 mm in 2004 (received on three dates within the first week of October) (Table 3). There were three rainy days with a total fall of 22.5 mm during the last week of September in 2003. There was a single rainy day of 6.1 mm during the same period in 2004. Maximum temperature and sunshine hours were greater and relative humidity lesser in 2002 than in 2003 or 2004 for the grain filling months of September, October and November (Table 3). Pan evaporation measurements for the same months in 2002 were comparable with those in 2003 but were greater than those in 2004 (Table 3).

Analysis of variance

Significance of grain yield variances for the fixed effects

Table 3. Mean monthly rainfall, minimum and maximum temperatures, Mean pan evaporation, sunshine hours and relative humidity during the cropping months of June - November for the test years 2002 - 2004 at Holetta, Ethiopia.

Year	June	July	Aug	Sept	Oct	Nov	Total
			Rainfall (mm)				
2002	123.2	273.1	194	77.4	0	0	667.7
2003	117.1	194	237.2	107.4	10	0	665.7
2004	121.4	204	226.6	119.7	3.6	0.7	676
			Minimum (°C)				Mean
2002	8.0	9.1	8.3	6.8	4.2	2.4	6.5
2003	7.9	9.3	9.1	7.8	3.8	2.2	6.7
2004	8.1	8.7	8.7	7.7	4.3	2.5	6.7
			Maximum (°C)				Mean
2002	22.9	21.0	20.3	21.2	23.3	23.9	22.1
2003	21.6	18.1	18.7	19.8	22	22.4	20.4
2004	21.2	19.4	19.1	19.8	20.9	22.5	20.5
			Pan evaporation (mm)				
2002	3.79	2.95	2.86	3.56	5.19	5.71	4.0
2003	5.00	3.52	3.01	3.18	5.60	5.73	4.3
2004	4.17	3.56	3.22	2.84	3.87	4.23	3.6
			Sunshine hours				
2002	5.8	3.4	2.8	5.8	7.8	10.6	6.1
2003	4.3	2.0	1.9	3.3	8.2	8.9	4.8
2004	3.6	2.5	2.7	3.8	6.3	8.7	4.6
			Relative humidity (%)				
2002	54	72	80	68	45	39	59.5
2003	63	83	85	82	57	52	70.3
2004	67	75	76	74	59	51	67.0

is given in Table 4. Year (Y), sowing date (S), crop protection treatment (C), Y × S and Y × C effects were highly significant (Table 4). Grain yield was significantly lower in 2002 (244 g m^{-2}) than either in 2003 or 2004. Mean grain yield difference between the years 2003 and 2004 was not significant, averaging 457 g m^{-2}. Grain yield was significantly lower for early sowing (362 g m^{-2}) than for normal or late sowing dates (each averaged 399 g m^{-2}). Grain yield averaged 332 g m^{-2} without crop protection treatment and 441 g m^{-2} with crop protection treatment. Mean grain yields in individual environments ranged from 178 g m^{-2} to 555 g m^{-2} (data not shown) and mean genotype grain yield ranged from 323 g m^{-2} to 517 g m^{-2} (Table 5). Mean genotype days to heading ranged from 56 to 89 days and time to maturity from 112 to 135 days(Table 5). Mature plant height ranged from 75 to 123 cm (Table 5).

Variance component estimates for genotype and interaction of genotype with year, sowing date and crop protection treatment are given in Table 6. Only genotype × year and genotype × crop protection treatment interaction variances were significant ($P < 0.05$). Genotype × year interaction was by far the largest variance making up for 80% the total G × E variance. (Table 6) The sum of genotype × management interaction variance was only 20% of the total G × E variance estimate.

GGE biplots

In GGE biplot analysis, the first PC accounted for 68.3% and the second PC for 17.1% of the GGE sum of squares making up for 85.4% of the total variation contained in the GE grain yield matrix. Environments aggregated much more based on year than based on either sowing dates

Table 4. Significance of variances for year, sowing date, crop protection treatment and interaction among these factors for grain yield of 39 barley genotypes tested under three sowing dates and two crop protection treatments for three years at Holetta, Ethiopia.

Source	df†	F-value	Probability
Year (Y)	2 (8.75)	22.53	0.0004
Sowing date (S)	2 (37)	7.84	0.0015
Crop protection (C)	1 (44.6)	119.19	< 0.0001
Y × S	4 (31)	14.67	< 0.0001
Y × C	2 (32.2)	16.76	< 0.0001
S × C	2 (31.9)	0.03	0.9661
Y × S × C	4 (29.7)	2.65	0.0529

† Numerator df with the denominator df (in brackets) estimated using the DDFM=SATTERTH option in the model statement of the SAS mixed procedure.

Table 5. Genotype mean grain yield expressed as best linear unbiased predictions for 39 barley genotypes tested under three sowing dates and two crop protection treatments in 2002, 2003 and 2004 at Holetta, Ethiopia.

Genotype	YLD	STR	SPK	KPS	KWT	HI	VHT	HT	HED	DMT	GFD	FLY	SCD	NET	SPT	RUS
	---g m^{-2}---		----no.---		mg		---cm---		--------days--------			no.	------(0-9)------			%
gin	323	780	242	40	41	0.28	37	107	77	125	48	13.6	3.8	4.2	2.3	36
fer	333	679	275	31	42	0.33	43	105	62	114	51	9.0	4.7	3.3	2.3	9
ehil	334	632	299	35	37	0.35	40	99	63	112	49	11.2	5.3	3.8	2.3	13
3371	335	709	301	29	43	0.32	44	107	62	113	51	7.4	4.2	2.5	1.4	8
em13	343	441	267	41	38	0.44	38	75	56	114	58	12.8	4.1	2.0	1.3	1
bal	344	1124	297	24	53	0.24	39	116	81	126	45	11.9	3.2	4.2	3.1	11
sem	345	975	306	28	46	0.27	38	111	78	124	46	10.6	2.6	3.9	2.8	19
bh2	346	1001	310	23	53	0.26	39	118	80	125	45	12.1	3.1	4.5	3.3	10
bh1	355	1111	307	25	52	0.25	39	119	82	125	43	10.3	3.1	4.4	3.0	15
1829	357	1222	315	22	57	0.23	42	123	81	126	45	9.0	3.1	3.7	2.7	11
em44	358	526	238	48	41	0.40	33	96	64	119	55	14.4	5.2	2.4	1.3	1
2038	361	792	281	32	45	0.32	40	108	73	118	46	7.6	2.8	3.5	2.2	25
tb7	364	788	255	39	48	0.32	39	108	72	125	53	10.3	2.4	3.2	2.5	16
sho	368	712	266	32	46	0.34	42	111	66	116	50	10.5	4.1	2.6	1.4	8
eh65	369	686	259	41	47	0.36	35	102	67	120	53	10.2	3.0	3.0	2.3	20
bbti	372	990	278	39	46	0.27	40	119	77	125	48	10.5	4.3	4.3	3.1	17
aby	372	856	287	48	37	0.31	37	111	72	120	47	9.3	4.3	3.0	1.9	4
cha	373	1017	309	24	51	0.27	39	114	77	124	47	9.8	3.4	4.0	3.0	16
eh07	381	1012	291	29	51	0.27	35	118	80	127	47	12.4	2.0	2.2	3.1	3
ahr	384	812	209	54	46	0.30	28	107	89	135	46	12.5	1.4	2.1	3.1	1
hb52	390	858	302	28	45	0.32	33	115	76	125	49	16.4	1.3	1.1	1.0	8
eh82	390	426	227	46	40	0.47	31	80	62	118	56	14.9	2.8	2.4	2.2	2
ard	390	959	266	45	43	0.29	39	115	82	126	45	10.5	4.0	4.2	2.7	8
hkr	392	749	299	26	49	0.34	34	100	76	124	49	13.7	2.3	2.3	1.6	4
hb42	393	861	240	44	49	0.31	35	115	82	128	47	11.7	2.4	3.2	2.3	5
th2	394	823	278	36	46	0.32	41	112	69	122	52	10.8	3.6	4.1	2.9	23
ts1	399	993	272	47	43	0.29	41	112	82	124	42	9.0	1.7	2.4	2.3	35
sas	401	627	316	24	49	0.39	44	94	60	112	52	9.8	5.1	2.6	1.5	6
bka	409	1004	315	28	43	0.29	36	115	77	126	49	14.3	1.6	1.3	1.5	10
em42	411	538	248	49	41	0.43	33	89	65	121	57	15.5	1.7	1.5	1.5	1
mis	416	755	287	44	41	0.36	41	104	68	117	49	9.6	4.0	3.3	1.9	19
hb33	418	1233	313	27	50	0.25	38	123	82	126	45	10.5	1.2	1.6	2.1	5

Table 5. Cont'd

shg	418	811	258	47	45	0.34	33	107	78	126	47	10.6	2.5	2.8	2.2	6
3381	421	725	272	46	42	0.38	40	103	68	117	49	11.5	3.4	2.9	1.9	21
hb20	423	961	324	28	45	0.31	34	116	76	126	50	13.8	1.0	1.2	1.4	11
iar	436	980	288	46	43	0.31	34	116	82	129	47	13.2	1.7	2.5	2.5	16
3304	440	762	286	43	43	0.38	40	104	68	116	48	12.1	3.7	3.2	1.9	15
dim	490	1124	307	40	47	0.31	41	120	80	126	46	8.0	3.2	3.2	2.8	9
eh42	517	991	336	27	56	0.34	38	107	78	128	50	11.8	2.0	1.8	2.1	23
SE (±)	18	33	11	1	0.9	0.01	0.4	2	0.4	0.6	0.6	0.8	0.2	0.2	0.3	2.1

YLD = grain yield; STR = straw yield; SPK = spike per square meter; KPS = kernels per spike; KWT = kernel weight; HI = harvest index; VHT = vegetative shoot height; HT = mature plant height; HED days to heading; FLY = number of seedlings attacked by shoot fly per unit area; DMT = days to maturity; GFD = grain filling duration; PHI = phase index; SCD = scald disease score; NET = net blotch disease score; SPT = spot blotch disease score; RUS = rust disease score.

Table 6. Variance components for genotype and interaction of genotype with year, sowing date and crop protection treatment for 39 barley genotypes tested at three sowing dates and two crop protection treatments for three years at Holetta, Ethiopia.

Variance Component	Estimate	Z value	Probability	% G × E variance
Genotype (G)	36.09	0.07	0.4737	
G × Year (Y)	4436.95	5.35	< 0.0001	80.4
G × Sowing date (S)	141.65	0.83	0.2038	2.6
G × Crop protection (C)	399.43	1.86	0.0314	7.2
G × Y × S	133.5	0.56	0.2887	2.4
G × Y × C	208.37	0.97	0.1663	3.8
G × S × C	180.25	0.85	0.1966	3.3
G × Y × S × C	20.73	0.06	0.4757	0.4
Residual	8388.39			

or crop protection treatments. In the biplot, the vertex genotypes were *eh42, dim, sas, 3371, fer, gin* and *ahr* (Figure 1). According to Yan et al. (2000), these vertex genotypes are either universal winners or universal losers in environments towards which they project the most. For instance, *eh42* is the vertex genotype in the sector in which the 2003 and 2004 environments are located. Hence this genotype is the winner in all the management-year combination of environments that involve the highest yielding years of 2003 and 2004. The genotype *sas* projected the most and is the vertex genotype in year-management combinations that involve the most stressful year of 2002. Hence this genotype was the highest yielding in this set of environments. *Fer, gin* and *ahr*, although are vertex genotypes, did not project to sectors were any of the environments are located and are, therefore, universal losers. What is also interesting in this biplot is the possibility of selecting genotypes that give one of the highest yields under high yielding conditions such as those in the years 2003 and 2004 but at the same time giving better yields under stress conditions than any other high yielding genotype. For instance, if high yielding years are more common in a given location

while there are some occurrences of low yielding years and farmers are risk-averse, recommending cultivars such as *dim* may be more appropriate than opting for cultivars such as *eh42*. On the other hand, in areas where the probability of short seasons are high, but the occurrence of extended season are also somewhat common, farmers may opt for growing cultivars such as *mis, 3304* or *3381* for high yields under stressful conditions while also harvesting substantial yields under high yielding non-stressful conditions. But in areas where seasons are short and season-end moisture stress is most likely, the ultimate choice is a cultivar such as *sas*. The two rays at the top of Figure 1 indicate the direction for choosing varieties with increasing (right to left) and decreasing (left to right) intensity of stress. Genotypes such as *ahr* have negative response to stress environments.

Figure 2 depicts the mean vs. stability view of the GGE biplot (Yan, 2001). *Eh42* and *dim* are the most yielding with *eh42* more yielding than *dim* but *dim* being more stable than *eh42*. Genotypes such as *3304, mis, 3381* and *sas* were the best under stress conditions but also responsive to high yielding conditions. *Dim* was the

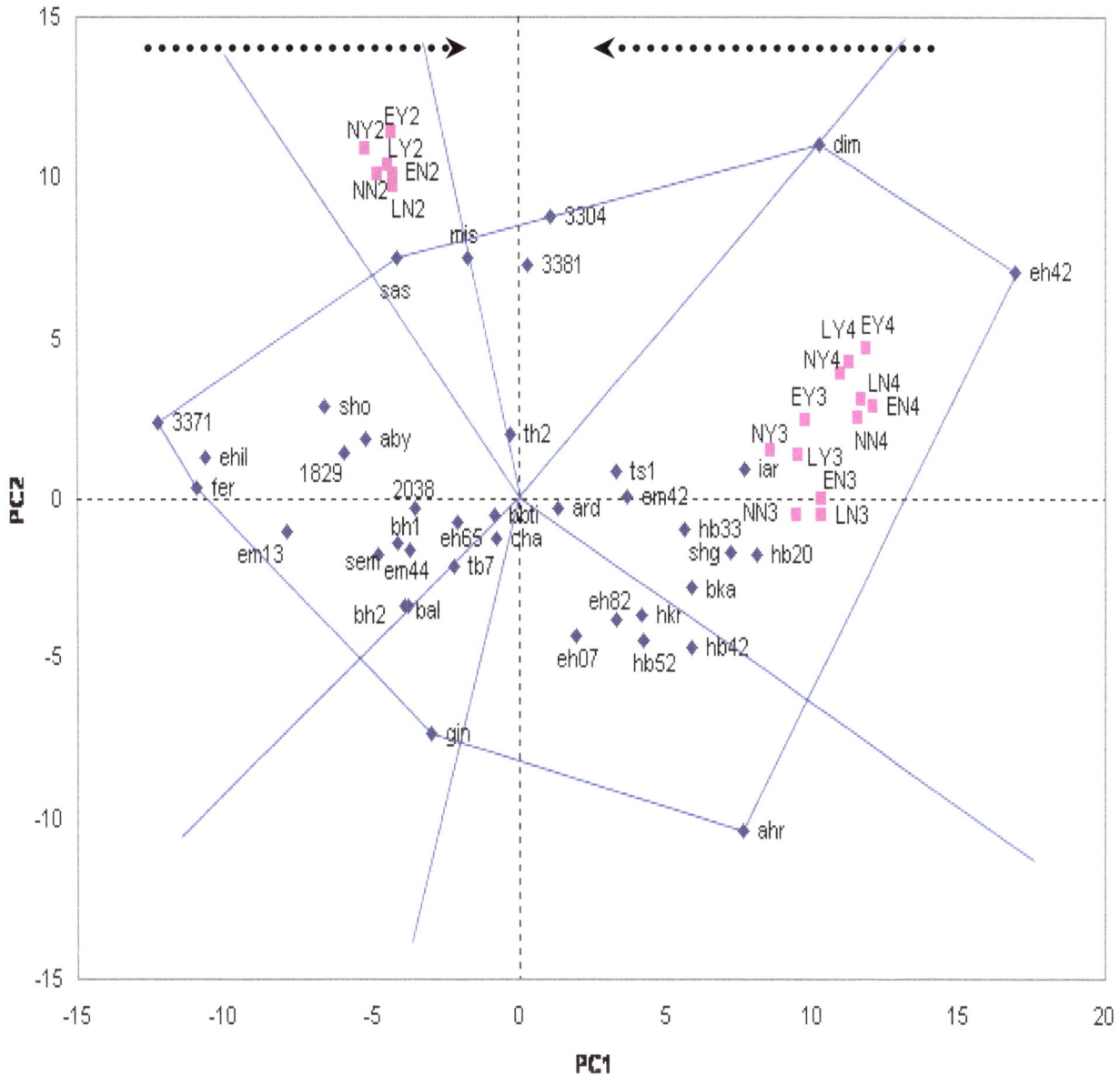

Figure 1. GGE biplot of the first and the second principal components resulting from singular value decomposition of environment centered grain yield expressed as best linear unbiased predictions for 39 genotypes grown in 18 environments at Holetta, Ethiopia.

second best under high yielding conditions but better than other high yielding genotypes in stress environments. In general, each of the four sectors partitioned by the average environment coordinate (AEC)-c and AEC-y axis has unique distinguishing features (Figure 2). The first sector encompasses stress environments and stress tolerant genotypes such as *3304*, *mis*, *3381* and *sas* which also responded to high yielding conditions. Note the positive score for these genotypes on both AEC-c and AEC-y axis.

The second sector includes high yielding environments and genotypes with high yields in these environments as typified by *eh42*, *iar*, *hb20*, etc. The third sector includes

genotypes such as *3371*, *ehil*, *fer*, *em13*, *sho*, etc. that did well under high stress condition but that did not respond to high yielding condition. The fourth sector encompasses genotypes such as *ahr* and *gin* which were relatively inferior under both high and low yielding conditions (Figure 2).

Genotypes such as *ahr* and *3371*, which lie furthest away from the center when projected perpendicular to the AEC-y axis, changed ranks the most and, therefore, were the most unstable (Yan et al., 2007). However, if stability is defined as risk reduction through ensuring a minimum yield level as is common in subsistence agriculture, stability of the genotypes decrease as one goes from left

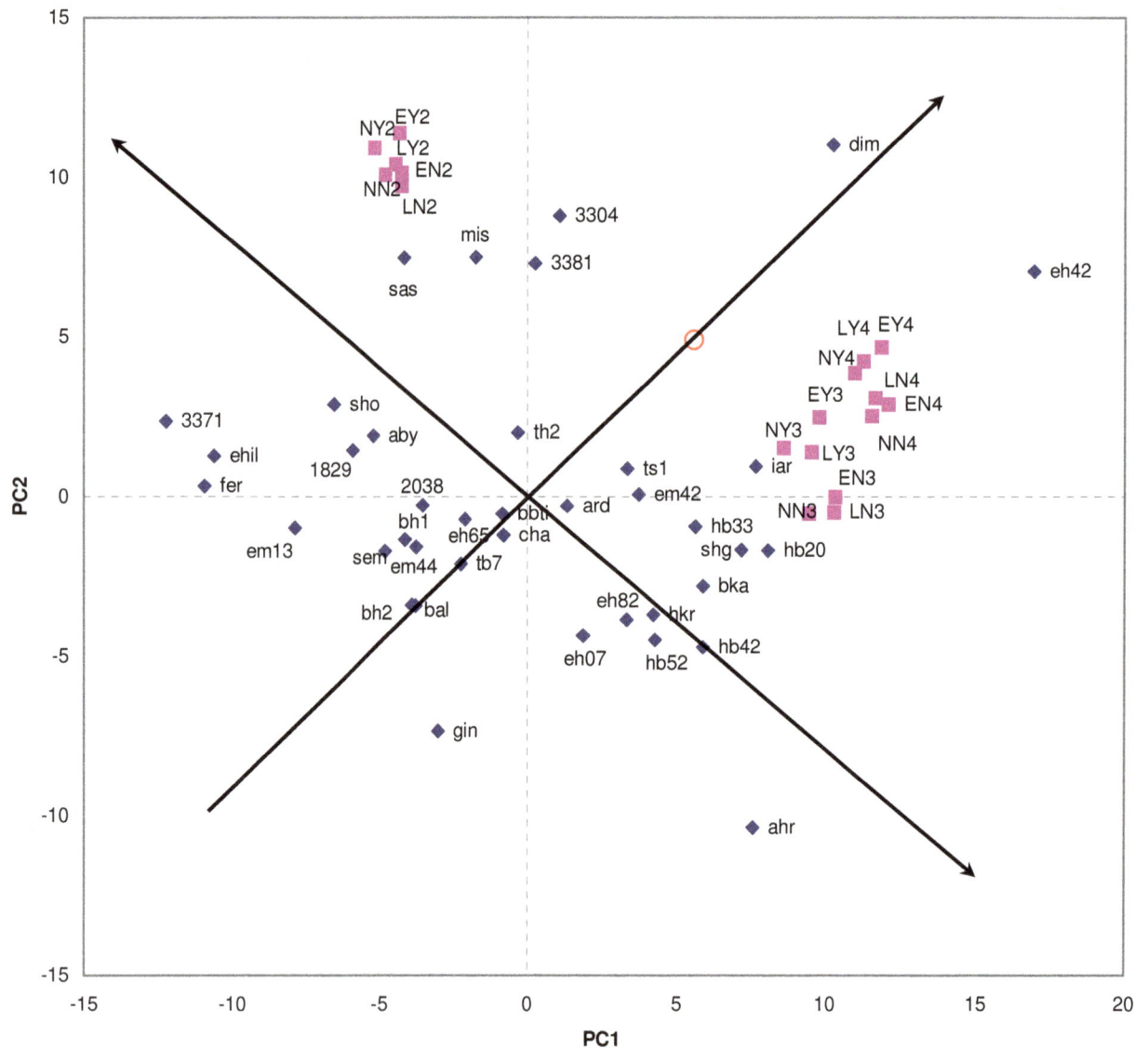

Figure 2. Mean yield vs. stability view of a GGE biplot resulting from singular value decomposition of environment centered grain yield expressed as best linear unbiased predictions for 39 genotypes grown in 18 environments at Holetta, Ethiopia.

to right along the AEC y-axis (Figure 2). This can be confirmed by the generally low regression coefficients for genotypes such as *3371*, *ehil*, *fer, sas, mis*, etc. and high regression coefficients for genotypes such as *eh42, ahr, iar*, etc. (Data not shown).

Correlation of PC scores with genotypic and environmental variables

The relationship among environmental PC scores, grain yield and some environmental variables are given in Table 7. Table 8 presents correlation coefficients among genotypic PC scores and genotypic variables. Environmental

mean grain yield was highly positively correlated with PC1 score. Environmental PC2 score was highly negatively correlated with environmental mean grain yield and environmental PC1 score (Table 7). Environmental mean grain yield was highly negatively correlated with mean maximum temperature and sunshine hours measured during the vegetative stage and with maximum temperature measured during the grain filling stage. Environmental mean grain yield was positively correlated with relative humidity measured during grain filling period (Table 7).

The correlations of PC1 with environmental variables follow those of the correlations of environmental mean grain yield with environmental variables. Environmental

Table 7. Correlation coefficients between principal components for environments, environmental mean grain yield (YLD), and environmental variables.

Parameters	YLD	PC1	PC2
PC1	0.79***		
PC2	-0.68**	-0.91***	
	During vegetative period		
Heat units	-0.77***	-0.62**	0.66**
Rainfall	-0.42	-0.30	0.28
Minimum temperature	0.21	0.39	-0.52*
Maximum temperature	-0.77***	-0.87***	0.96***
Sunshine hours	-0.74***	-0.89***	0.95***
Relative humidity	0.42	0.42	-0.72***
Pan evaporation	0.54*	0.73***	-0.56*
	During grain filling period		
Heat units	0.34	0.37	-0.45
Rainfall	0.25	0.42	-0.28
Minimum temperature	0.01	0.20	0.06
Maximum temperature	-0.64**	-0.86***	0.62**
Sunshine hours	-0.33	-0.55*	0.25
Relative humidity	0.56*	0.77***	-0.64**
Pan evaporation	-0.26	-0.44	0.06

*, ** and *** indicate significance at $P \leq 0.05$, $P \leq 0.01$ and $P \leq 0.001$, respectively.

PC2 score, a measure of performance under high stress condition, was highly positively correlated with maximum temperature of both vegetative and grain filling stages and with sunshine hours of vegetative period. Environmental PC2 score was also negatively correlated with relative humidity of both periods and with minimum temperature and pan evaporation of vegetative period (Table 7).

Mean genotype grain yield was highly positively correlated with each of genotypic PC1 and PC2 scores (Table 8). Mean genotype grain yields under high season-end stress (year 2002) was negatively correlated with PC1 score and positively correlated with PC2 score (Table 8). Mean genotype yields under low season-end stress (years 2003 and 2004) was positively correlated with PC1 score.

Overall mean genotype grain yield was highly positively correlated with mean genotype grain yield under low stress but not with mean genotype grain yield under high stress (Table 8). Genotypic PC1 score was positively correlated with time to heading and time to maturity and with the number of seedlings damaged by shoot fly but negatively correlated with vegetative vigor and scald and net blotch diseases score (Table 8). Genotypic PC2 score was positively correlated with the number of spikes per square meter, vegetative vigor and scald disease

score.

Correlation among environmental mean grain yields

Correlation coefficients among environmental mean grain yields (data not shown) revealed relationships that were also apparent from GGE biplot. Mean grain yields in individual environments within the same year were highly positively correlated. Grain yields in each of the 2002 environments were negatively correlated with environmental mean grain yields in the 2003 and 2004 environments. Mean grain yields in individual environments in the year 2003 and 2004 were highly positively correlated. This and observations in the biplots (Figures 1 and 2) imply a crossover interaction between the high yielding low stress years (2003 and 2004) and low yielding high stress year (2002).

Mean genotype grain yield and other trait relationships

Mean genotype grain yield was correlated with scald and net blotch diseases score only (Table 8). However, splitting overall mean genotype grain yield into mean

Table 8. Correlation among genotype principal component scores, grain and straw yields, grain yield under high stress, grain yield under low stress, yield components, growth durations and disease and insect pest data for 39 barley genotypes grown in 18 environments at Holetta, Ethiopia.

Trait	PC1†	PC2	YLD	YST	YNST	STR	SPK	KPS	KWT	HI	VHT	HT	HED	FLY	DMT	GFD	SCD	NET	SPT
PC2	0.00																		
YLD	0.84***z	0.53***																	
YST	-0.52***	0.84***	0.02																
YNST	0.99***	0.14	0.91***	-0.39*															
STR	0.31	-0.01	0.22	-0.21	0.29														
SPK	0.04	0.41**	0.24	0.33*	0.09	0.56***													
KPS	0.20	0.00	0.18	-0.1	0.2	-0.45**	-0.78***												
KWT	0.26	-0.06	0.16	-0.19	0.23	0.66***	0.44**	-0.63***											
HI	-0.01	0.27	0.18	0.27	0.05	-0.89***	-0.42**	0.47**	-0.57***										
VHT	-0.55***	0.63***	-0.16	0.78***	-0.45**	0.16	0.46**	-0.39*	0.10	-0.19									
HT	0.20	-0.05	0.11	-0.17	0.17	0.91***	0.44**	-0.38*	0.56***	-0.91***	0.19								
HED	0.54***	-0.36*	0.22	-0.62***	0.46**	0.82***	0.14	-0.10	0.55***	-0.78***	-0.29	0.77***							
FLY	0.36*	-0.45*	0.10	-0.54***	0.31	-0.28	-0.23	0.14	-0.16	0.31	-0.72***	-0.31	0.00						
DMT	0.68***	-0.48***	0.30	-0.78***	0.59***	0.65***	-0.03	0.03	0.51***	-0.58***	-0.54***	0.58***	0.92***	0.25					
GFD	-0.16	0.08	-0.05	0.17	-0.11	-0.82***	-0.34*	0.25	-0.44**	0.83***	-0.16	-0.80***	-0.81***	0.36*	-0.52***				
SCD	-0.72***	0.37*	-0.42**	0.67***	-0.66***	-0.41**	-0.08	0.05	-0.32*	0.25	0.55***	-0.31	-0.60***	-0.37*	-0.71***	0.26			
NET	-0.47**	0.05	-0.43**	0.22	-0.48**	0.22	-0.01	-0.11	0.19	-0.38*	0.46**	0.26	0.13	-0.47**	-0.06	-0.37*	0.53***		
SPT	0.01	-0.20	-0.14	-0.23	-0.02	0.54***	-0.01	-0.08	0.46**	-0.59***	0.09	0.48**	0.58***	-0.32*	0.48**	-0.55***	-0.03	0.70***	
RUS	0.00	0.19	0.03	0.08	-0.01	0.22	0.13	-0.02	0.04	-0.26	0.38*	0.24	0.16	-0.30	0.03	-0.01	-0.01	0.38*	0.24

† PC = principal component; YLD = grain yield; YST = grain yield with high stress; YNST = grain yield with low stress; STR = straw yield; SPK = spike per square meter; KPS = kernels per spike; KWT = kernel weight; HI = harvest index; VHT = vegetative shoot height; HT = mature plant height; HED = days to heading; FLY = number of seedlings attacked by shoot fly per unit area; DMT = days to maturity; GFD = grain filling duration; PHI = phase index; SCD = scald disease score; NET = net blotch disease score; SPT = spot blotch disease score; RUS = rust disease score.

*, ** and *** indicate significance at P ≤ 0.05, P ≤ 0.01 and P ≤ 0.001, respectively.

genotype yield under low stress and mean genotype yield under high stress revealed a different pattern (Table 8). Mean genotype yield under high stress was positively correlated with the number of spikes per square meter, early shoot vigor and scald score but negatively correlated with both days to heading and maturity, and with the number of shoots damaged by shoot fly. Mean genotype grain yield under low stress was negatively correlated with early shoot vigor and scald and net blotch diseases score but positively correlated with both days to heading

and maturity (Table 8).

DISCUSSION

Nature and causes of G × E interaction

The negative correlations among some groups of environments, the different signs for environmental PC1 scores and the large angle between the environments in the biplot (Figures 1 and 2) indicate that the GE interaction was because of

rank changes rather than scale effects (Yan et al., 2000; Dehghani et al., 2006). In this study, year effect was the largest source of environmental variance and the GY interaction effect the largest contributor to GE interaction variance. However, the total rainfall amounts during the crop growing period in the three years were similar. Therefore, the major cause of grain yield differences among the years and the underlying cause for the significant GY interaction for grain yield was the variability among the years in the degree of season-end moisture stress (Table 3). This indicates that a slight deviation of rainfall towards season-end would cause a large sway in yield and a large crossover type of GY interaction. In this study, the G × management interaction component as a whole was small relative to GY interaction component. The large GY interaction indicates the greater importance of working to stabilize performance of genotypes across years than across management levels. Of the G × management interaction components, the GS was important only in stress years such as the year 2002. This may indicate lack of advantage from early sowing of late maturing genotypes in years of low season-end moisture stress.

The environmental and genotypic causes of the GE interaction can be visualized by relating information on environmental variables and PC scores in Table 7 with the corresponding information on genotypic variables and PC scores in Table 8.

Environmental (Table 7) and also genotypic (Table 8) mean yield were positively associated with PC1 scores indicating PC1 to be a measure of overall mean yield to some extent. As the association of both yield and PC1 scores with environmental variables indicates (Table 7), environments with high yield had lower daytime temperature, lesser sunshine hours and higher humidity. By inference, these environments tend to be cooler, cloudier and wetter (Table 3). High yielding genotypes in these environments had high PC1 score and more likely high overall yield.

On average, these genotypes had slower early vegetative shoot height growth, longer growth duration, higher shoot fly damage and lower incidence of scald and net blotch diseases (Table 8). These genotypes, as depicted by strong positive association of PC1 score with grain yield under low stress and strong negative association between PC1 score with grain yield under high stress (Table 8), are suitable for low but not high season-end moisture stress environments.

In barley grown in moisture stress environments, early vigor is associated with early heading. These early maturing genotypes when grown under low stress environments exhibit higher incidence of scald and net blotch diseases but lower shoot fly damage. PC2 was negatively associated with night temperature and relative humidity and positively associated with daytime temperature and sunshine hours.

In the highlands of Ethiopia, nights are increasingly cooler, days hotter and relative humidity lower as the rain recedes and the weather gets drier towards season-end. Environments with such features and genotypes that do well in these environments had large PC2 score (Figure 1). PC2 score was positively associated with vegetative shoot height but negatively associated with time to heading and time-to-maturity (Table 8). In effect, time to maturity and moisture status during the critical period of grain filling was the major factors contributing to the observed GE interaction.

Beyond a unique winning genotype

From the present study, it is evident that improved varieties selected for high mean yield were beaten by early maturing landraces in years of season end-moisture stress. This confirms rational choice by risk-averse subsistence farmers in preferring relatively low yielding early maturing varieties to high yielding late maturing varieties in order to minimize risk in bad years rather than maximize average production over the years. Because of this, in subsistence agriculture with a fairly unpredictable weather, a crop breeding program is tasked with providing options in the form of more than one "winning" genotypes rather than a unique winning genotype (for each mega-environment) – for instance, one genotype for risk taking farmers whose goal is high yield and another for risk averse farmers whose goal is to ensure harvest stability year after year.

A feature of merit with the present GGE analysis is the possibility for spotting in the biplot of genotypes that may not be winners in any of the mega-environments but that are appealing to farmers that want to trade certain level of risk-taking for some degree of yield reward. In our case, such genotypes that could easily be spotted from the GGE biplot are *3304*, *3381* and *mis* (Figure 1). These same genotypes could as well be grown in a barley production system where the season length allows late maturing barley variety but food shortages towards season-end necessitates leveraging the need for high yield with the requirement for earliness.

In Ethiopia, despite the diversity of barley growing agro-ecologies and availability of some multi-location variety trial data, no comprehensive assessment of G × location interaction has been reported. In this study, our familiarity with barley growing agroecologies in the country and our insights into cultivars with unique adaptations in contrasting agroecologies offered us an opportunity to infer the likely pattern of G × location interaction from G × Y interaction. The longer growing season with a relatively better season-end moisture regime observed in 2003 and 2004 is typical of the central, southeastern and north-western barley growing highlands whereas the shorter growing season with a high season-end moisture stress of the year 2002 is usually representative of the North-eastern and Northern highlands, namely Wollo and Tigray. In this study, the year with greater season-end moisture stress amplified the relative merits of early maturing varieties including prominent landrace varieties such as *sasa* that are adapted to short season drought prone highlands of northern Ethiopia, notably the Tigray region. Barely variety selections made for high yield in relatively high rainfall areas of central Ethiopia have yielded less than local landraces in Tigray (Abay and Bjørnstad, 2008).

Bearing in mind the unpredictable weather and the changing climate, be it the low rainfall areas of northern Ethiopia or the relatively better rainfall regimes of Central

Ethiopia, barley varieties that make best use of environmental resources and give high yield in favorable years while tolerating yield reductions in years of season-end drought are required. To this end, a still pending question is how to combine high yielding ability in favorable years with relative yield stability in years of early cessation of rainfall. One approach is breeding for post-anthesis drought tolerance while maintaining or even improving yielding abilities under favorable environments.

In the GGE biplot, genotypes that did well under high stress conditions were located further along the AEC y-axis (Figure 2) indicating their greater instability relative to many genotypes in the study. It is important to note that the AEC y-axis measures rank stability (Yan et al., 2007) rather than yield stability. In subsistence agriculture, perhaps stability of yields is more important than stability of ranks. In the future, the analysis of multi-environment yield data for risk-averse subsistence farming needs to closely examine the relevance of rank stability vis-à-vis absolute yield stability.

ACKNOWLEDGMENTS

Misa Demisse, Dhabata Mideksa and Tiruwork Amogne assisted with the field work. Seid Ahmed and Asnakech Dubale facilitated access to seeds of the barley genotypes. Dr. Bayeh Mulatu facilitated insecticidal treatment of the seed lots. This work results from experiment number 01/01/06/BA/Ag-HO(02-1) of Holetta Agric. Res. Center.

REFERENCES

Abay F, Bjørnstad A (2008). Specific adaptation of barley varieties in different locations in Ethiopia. Euphytica, doi: 10.1007/s10681-008-9858-3.

Ceccarelli S (1994). Specific adaptation and breeding for marginal conditions. Euphytica, 77: 205-219.

Dehghani H, Ebadi A, Yousefi A (2006). Biplot analysis of genotype by environment interaction for barley yield trial in Iran. Agron. J., 98: 388-393.

Gebre H, Van Leur J (1996). Barley research in Ethiopia: Past work and future prospects. Proceedings of the First Barley Research Review Workshop, Addis Ababa. 16-19 Oct. 1993. IAR/ICARDA. Addis Ababa, Ethiopia.

Gróny AG (2001). Variation in utilization efficiency and tolerance to reduced water and nitrogen supply among wild and cultivated barleys. Euphytica, 117: 59-66.

IGAD (Intergovernmental Authority on Development) Climate Prediction and Applications Centre (ICPAC) (2007). Climate change and human development in Africa: Assessing the risks and vulnerability of climate change in Kenya, Malawi and Ethiopia. Occasional paper, UNDP Human Development Report Office.

Howden SM, Soussana J-F, Tubiello FN, Chhetri N, Dunlop M, Meinke H (2007). Adapting agriculture to climate change. Proc. Natl. Acad. Sci. USA, 104: 19691-19696.

Kempton RA (1984). The use of biplot in interpreting variety by environment interactions. J. Agric. Sci. (Cambridge), 103: 123-135.

Loegering WQ (1959). Method for recording cereal rust data. A note on the "USDA International Spring Wheat Rust Nursery, 1959".

Mola A, Gorfu A, Ashagrie Y, Yilma Z (1996). Food barley agronomy research. In H. Gebre and J. van Leur. (eds.) Barley research in Ethiopia: Past work and future prospects. Proceedings of the First Barley Research Review Workshop, Addis Ababa. 16–19 Oct. 1993. IAR/ICARDA. Addis Ababa, Ethiopia. p. 68-84.

Sinebo W (2002). Yield relationships of barleys grown in a tropical highland environment. Crop Sci., 42: 428-437.

Sinebo W (2005). Trade off between yield increase and yield stability in three decades of barley breeding in a tropical highland environment. Field Crops Res., 92: 35-52.

Tubiello FN, Soussana JF, Howden SM (2007). Crop and pasture response to climate change. Proc. Natl. Acad. Sci. USA, 104: 19686-19690.

Van Oosterom EJ, Kleijn D, Ceccarelli S, Nachit MM (1993). Genotype-by-environment interactions of barley in the Mediterranean region. Crop Sci., 33: 669-674.

Yan W (2001). GGEbiplot—A windows application for graphical analysis of multienvironment trial data and other types of two-way data. Agron. J., 93:1111-1118.

Yan W, Hunt LA, Sheng Q, Szlavnics Z (2000). Cultivar evaluation and mega-environment investigation based on the GGE biplot. Crop Sci., 40: 597-605.

Yan W, Kang MS, Ma B, Woods S, Cornelius PL (2007). GGE biplot vs. AMMI analysis of genotype-by-environment data. Crop Sci., 47: 643-653.

Yirga C, Alemayehu F, Sinebo W (1998). Barley-based farming systems in the highlands of Ethiopia. Ethiopian Agricultural Research Organization. Addis Ababa, Ethiopia.

Zobel WR, Wright MJ, Gauch HG (1988). Statistical analysis of a yield trial. Agron. J., 80: 388-393.

6

Tn5 tagging biomonitoring of *Rhizobium* inoculants

T. Neeraj[1]*, Sachin[2], S. S. Gaurav[1] and S. C. Chatterjee[3]

[1]Department of Biotechnology, C. C. S. University, Meerut -250004, India.
[2]Department of Genetics and Plant Breeding, S. V. Patel University of Agriculture and Technology, Meerut, India.
[3]Department of Plant Pathology, Indian Agricultural Research Institute (IARI), Pusa Campus, New Delhi-110001, India.

Sinorhizobium freddi R0132, which forms nitrogen fixing nodules in association with *Vigna radatia* L. and fix nitrogen symbiotically, was the bioinoculant used in this study. Drug resistant *S. freddi* R0132::*Tn*5 remained constant at 40°C higher levels temperature for about 35 days (5 weeks) from the date of inoculation in solarzied and autoclaved biomanure VC-MA. The *Tn*5 mutagenesis was carried by the random transposon mutagenesis. The population of drug resistant *S. freddi* R0132::*Tn*5 in these treatments was significantly higher as compared to control. The present study attempted to evaluate the role of the said bio-manure VC-MA as carrier for survival of improved drug resistant *S. freddi* R0132::*Tn*5 in all the carrier material tested as compared to sample at higher (37 - 40°C) temperature, and higher population of drug resistant *S. freddi* R0132::*Tn*5 were observed when stored at relatively lower temperature (28°C). It is therefore not recommended to store the mass culture of biofertilizers particularly at higher temperature.

Key words: *Rhizobium*, transposon Tn*5* mutagenesis, biomanure.

INTRODUCTION

Indiscriminate and excessive use of chemical fertilizers have badly affected the soil properties with growing concern over food security and sustainable agriculture, the use of ecofriendly technologies for agriculture production has become imperative. Biofertilizers may play important role in reducing environmental hazards by supplementing organic / biological nitrogen means and thereby reducing the inputs of chemical fertilizers. Over use of chemical fertilizers have seriously deteriorated the soil properties. The procedure use for random transposon mutagenesis was that given by Selvaraj and Iyeri (1983) and was modified by Khanuja (1987) because it is one of the most powerful tools for initial localization of gene or gene cluster and for preliminary analysis of the organization of genome, efficiently in many gram negative bacteria including Pseudomonas. *Sinorhizobium freddi* R0132::*Tn*5, which forms nitrogen fixing nodules in association with *Vigna radatia*. L. and fix nitrogen symbiotically, was the bioinoculant used in this study. The over all objective of the present study was to develop a low cost as well as growth supporter carrier for *S. freddi* R0132::*Tn*5. NTG mutagenesis was carried out for development of marker drug resistant of *S. freddi* R0132::*Tn*5 and was used to tag *Rhizobium* cells by insertion of transposon in to genome which was used as a marker gene for monitoring the population of *S. freddi* R0132::*Tn*5. This biomanure could not efficiently support the bioinoculant *S. freddi* R0132::*Tn*5 as such, probably because of high microbial load existing in the biomanure. Biomanure VC-MA has been proved to be the most efficient carrier material compared with other carrier material tested viz; charcoal, vermiculite and FYM (Tomer et al., 1998). Further studies are needed to standardize the inoculums load, moisture content and temperature 40°C for improving the shelf life of biofertilizers. Transposable (Tn) elements conferring drug resistance, e.g. Kanamycin, nornycin etc can be simple transposons and also more complex ones. Simple transposons, or insertion sequences (IS), which carry only the genes necessary for their own transposition that is the genes encoding for the transposes protein and the inhibitor protein. They can be detected in two ways: first, the insertion sequences interrupt and inactivate gene into which they insert, and the second, they may contain

---

*Corresponding author. E-mail: neerajtandan@gmail.com.

promoters that allows RNA polymerase to transcribe and thus turn on adjacent and even distal genes. Complex transposons, on the other hand, contain one to several genes in addition to those encoded by the insertion sequences which are essential to the process of transposition; typically they carry genetic markers for antibiotics resistance (Watsons et al., 1992).

Transposition

The process of transposition is the movement of a transposable element from one locus (donor site) to another locus on a DNA molecule (target site) which typically has little or no homology with the transposable element. Clearly, transposition gives rise to rearrangement of the host genome, including deletions and inversions. Transposition is a rare recombination event since excessive rearrangement of the genomic DNA is lethal to the host. Transposition occurs at frequencies which are comparable to spontaneous mutation rates. For example, the frequency of transposition is about 1/100,000 per generation for Tn5 in *E. coli* compared to spontaneous mutation rates of about 1/1 million per generation. The low frequency can be attributed to tight regulation, inefficient transcription signals, inefficient translation signals, and or inefficiencies which are intrinsic to the transposition process itself. Generally the frequency of transposition depends on the nature of the transposon and especially the relative amounts of transposase protein to inhibit protein which are present in the host cell. The DNA encoding for these two proteins constitutes the simplest transposon structure. The mechanism of transposition may involve cut-and-paste mechanism in which the transposon is transferred to a different locus, leaving a gap, which is potentially lethal, at the donor site on a DNA molecule or by a conservative two-step replicative mechanism in which a co integrate (fused replicon) structure is formed as an intermediate.

Transposon Tn*5*

The transposon Tn5 is a complex transposon of about 5818 base pair. There are two nearly identical insertion sequences (IS50L and IS50R, both 1533 bp in length) which are inverted with respect to a central region of DNA encoding three antibiotics resistance genes: kanamycin (kan), streptomycin (str) and bleomycin (bleo) resistance. On the other hand, the IS50R is a fully functional transposable element, encoding two proteins essential for transposition. The transposase protein (Tnp) and an inhibitor of transposase (Inh). The transposase consists of 476 ammo acids while the inhibitor is lacking the first 55-amino acids on the N-terminal end of the transposase such that the inhibitor has only 421-amino acids in length; both polypeptides are read from the same reading frame

and share the same carboxyl-terminal. The IS50L differs from the IS50R at a single base pair such that it contains an ochre codon that results in synthesis of inactive proteins which are counterparts to the transposase and the inhibitor proteins (Wiegand et al., 1992). These are thought to have some inhibitory effect on the transposition effect, perhaps by binding to the Tnp-Inh heterooligomers.

In *Rhizobium*, Tn5 confer resistance to kanamycin. Two 1533 bp inverted repeats at the end of Tn5 (IS50) contain the genes necessary for transposition of Tn5 and its regulation.

Transposon mediated mutagenesis or tagging

Tn5 mutagenesis can greatly be divided in two major categories: "random Tn5 mutagenesis" and "site directed mutagenesis". The first category involves the introduction of Tn5 in to *Rhizobium* of interest via transformation, transduction, or conjugation with plasmid or phage vectors carrying Tn5. These Tn5 containing vectors are called "suicidal vectors" because of their instability (or inability to be suitably maintained) in the recipient bacteria.

The second category involves the Tn5 mutagenesis of DNA segments cloned in to (multicopy) plasmids in *Escherichia coli*. This is followed by the physical mapping of the Tn5 insertion and whenever possible determination of the "phenotype" of the Tn5 recombinant plasmid to construct a correlated physical and genetic map. The method is often coupled to the reintroduction of specific Tn5 mutated segments into their original bacterial background followed by replacement of the wild type gene or region with its Tn5 mutated homolog via forced double reciprocal recombination (gene replacement or homogenotization). In the present study random Tn5 insertion was carried out in the genome of *Rhizobium leguminosarum*, to serve as marker tagging of the strain.

MATERIALS AND METHODS

Preparation of packets containing different carriers

Different samples were taken out from different sources for biomanure VC-MA, vermiculite, charcoal (from market), and farm yard manure. These samples were weighed and mixed to give treatments as follow:

T_1. Biomanure VC-MA.
T_2. Farm yard manure (FYM)
T_3. Vermiculite
T_4. Charcoal
T_5. Biomanure VC-MA and Vermiculite (1:1)
T_6. FYM and vermiculite (1:1)
T_7. Biomanure VC-MA and Charcoal (1:1)
T_8. FYM and T Charcoal (1: 1)

Tn5 marked drug resistant *S. freddi* was cultured in Yeast Extract

Table 1. Antibiotic sensitivity of mutant of Sinorhizobium freddi R0132::Tn5 (str^R) developed.

Antibiotics	Antibiotic concentration (µg/ml)							
	200.0	100.0	50.0	25.0	12.5	6.25	3.125	1.56
Gentamycin	-	-	-	-	-	-	-	-
Kanamycin	-	-	-	-	-	-	-	-
Tetracycline	-	-	-	-	-	-	-	-
Streptomycin	+	+	+	+	+	+	+	+
Nalidixic acid	-	-	-	-	-	-	-	+

Table 2. Optimization of incubation period for patch mating to transfer pGS9 plasmid into Sinorhizobium freddi R0132::Tn5.

S/No.	Incubation time (h)	Average titre of cells per ml (TY + str + kan)
1.	0	Nil
2.	4	Nil
3.	8	1.2×10^2
4.	16	6.2×10^3

Mannitol broth with 100 µg/ml streptomycin and 25 µg/ml kanamycin. After 36 h, it was incubated at 28°C on a rotatory shaker, the bacterial culture at ODA$_{600}$= 0.4 was centrifuged at 70000 rpm for 20 min and resuspended in sterile water. The 5 ml of the culture was then mixed with 25 g carrier uniformly. Each inoculated carrier material stored in polythene bag was incubated at 28 and 37°C for 50 to 60 days, respectively. In these experiments the carrier materials was not sterilized. One gram of each carrier sample was taken from the polythene pack and was suspended in 10 ml of sterile water in 30 ml screw cap tubes and vigorously vortexes to mix. The suspension was allowed to stand for 10 min. The 10^2 to 10^6 dilutions were made from suspension and 0.1 ml solution was spread plated on YEMA containing 200 µg/ml of streptomycin and 50 µg/ml kanamycin. The colonies appearing after four days were counted in each plate.

RESULTS

Isolation of antibiotic resistant Sinorhizobium freddi R0132::Tn5

The drug resistance efficiency of S. freddi R0132::Tn5 was characterized by broth assay and Poison agar method. It is clear that the mutant was resistant to streptomycin as it could grow in media supplemented with 200 µg/ml of streptomycin. This strain was designated as R0132::Tn5 and was further used for transposon tagging (Table 1).

Tn5 tagging of rhizobial inoculants

The Tn5 mutagenesis was carried out by the random transposon mutagenesis procedure. The period of mating

on agar had been found to significantly affect the vector transfer and recovery of Tn5 mutants. Therefore to standardize the mating period between WA803 (pGS9) and S. freddi R0132 ::Tn5, the patch from the mixture of their cells on TY agar (str + kan) surface were allowed to grow for 4, 8 and 16 h at 28°C.

Viability of Tn5 tagged drug resistant Rhizobium in biomanure VC-MA was subjected to different treatments. The population of drug resistant S. freddi R0132::Tn5 in untreated control could not be detected after 25 days when kept at 37°C. However, appreciable number of colonies could be observed in autoclaved biomanure VC-MA even after 50 days (Figure 1 and Table 2). The population of drug resistant S. freddi R0132::Tn5 in general, remained higher in various carriers when kept at 28°C. No significant change in population of drug resistant S. freddi R0132:Tn5 was detected in autoclaved biomanure VC-MA (Figure 2 and Table 3).

DISCUSSION

Low cost inert materials like charcoal are quite common earners for bioinoculants. Attempt were made to search and identify new carrier material which apart from being cheap, should also provide a good substrate for enhancing a survival and growth of bioinoculants. One of biomanure produced from agro-wastes generated at farm containing all major nutrients, was tested for suitability as a carrier for drug resistant S. freddi R0132::Tn5. Treatment of isolated strain of S. freddi R0132::Tn5 with sensitivity to common antibiotic streptomycin was tested; the major reason is being lack of suitable carrier which

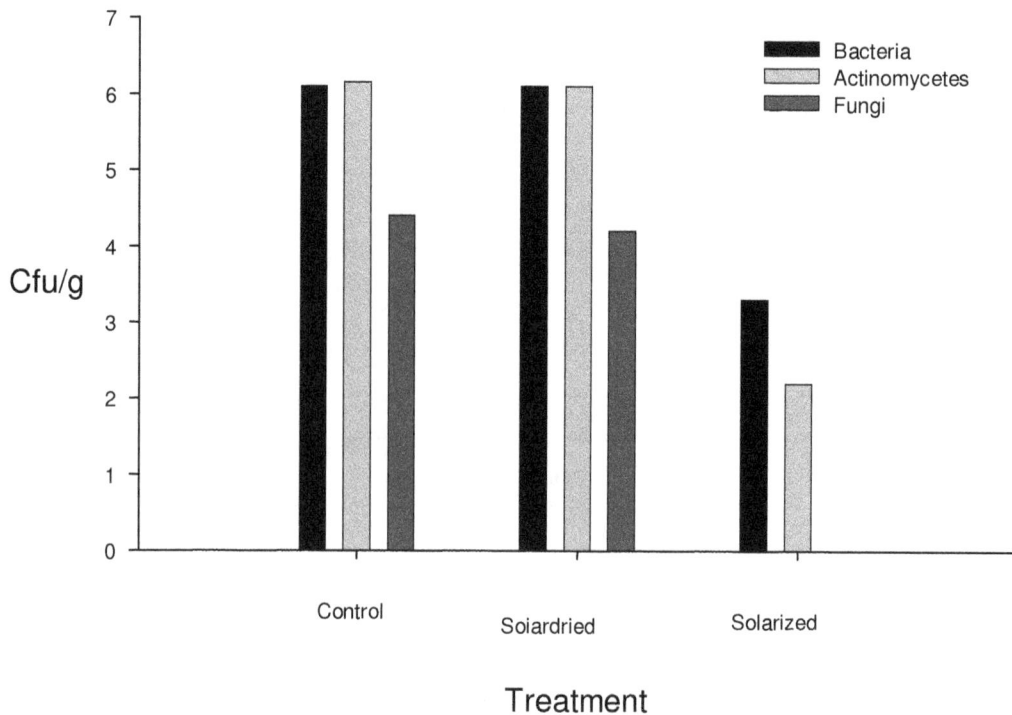

Figure 1. Microbial population of bacteria, actinomycetes and fungi biomanure VC-MA.

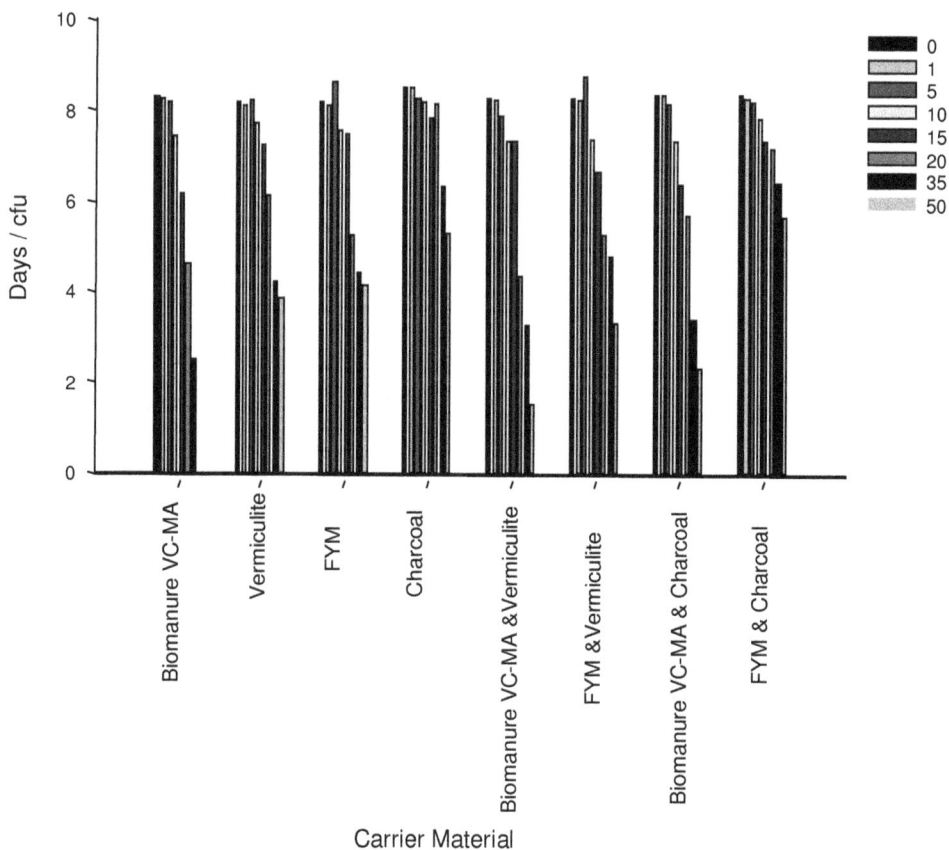

Figure 2. Population of drug resistant *Sinoshizodium freddi* in various carrier materials at 28 °C.

Table 3. Microbial population of bacteria, actinomycetes and fungi in biomanure VC-MA.

S/No.	Treatment	Bacteria (cfu/g)	Actinomycetes (cfu/g)	Fungi (cfu/g)
1.	Control	1.07×10^6	1.52×10^6	4.0×10^4
2.	Solardried	1.0×10^6	1.0×10^6	2×10^4
3.	Solarized	3.0×10^3	2.0×10^2	Not detected
4.	Autoclaved	Not detected	Not detected	Not detected

could not only increase the viability of bacteria but also support its growth and multiplication after application in the fields. So far, charcoal powder has been most commonly and extensively used as carrier for most of the *rhizobial* inoculants, though vermiculite (Sparrow and Ham, 1983), FYM (Chakrapani and Tilak, 1974), compost (Iswaran et al., 1972) has also been tested. The present study indicated that the biomanure VC-MA as such, could not support the growth of drug resistant *S. freddi* R0132::*Tn*5 and sharp decline in population with passage of time was observed. On the other hand high microbial density biomanure *S. freddi* R0132::*Tn*5 VC-MA probably did not allow drug resistant *S. freddi* R0132::*Tn*5 to establish itself. Some of the microorganisms present in biomanure *S. freddi* R0132::*Tn*5 VC-MA antagonistic or their metabolites produced in this biomanure might betoxic to the test bioinoculant (drug resistant *Sinorhizobium freddi* R0132::*Tn*5. Temperature plays an important role in the survival viability and growth of any microorganism. Our results indicated that considerable decline in the population of drug resistant *S. freddi* R0132::*Tn*5 occurs when kept at higher (37 - 40°C) temperature. To improve its sustainability it should be kept in mind that the sample should be kept at cooler (28°C) places.

The major problem is being with the poor survival of bioinoculant in carrier materials. Some material like biomanure VC-MA tested during the present investigations could be a better substitute. However, the post application population dynamics of the test organism *S. freddi* R0132::Tn5 need to be further studied.

ACKNOWLEDGEMENT

The authors are grateful to the Director, Indian Agriculture Research Institutes Pusa New Delhi, for providing facilities and encouragement.

REFERENCES

Chakrapani B, Tilak KVBR (1974). Comparative performance of different carriers of R. japonicum on soybean. Sci. Cult., 40: 433-435.

Selvaraj G, Iyer VN (1984). Transposon Tn5 specifies streptomycin resistance in Rhizobium, In Bacteriol., 158: 580-589.

Khanuja SPS (1987). Genetic dissection of the nodulation process and phase resistance in *Rhizobium meliloti* using Transposon mutagenesis. P.h.D (Genetics) thesis, I.A.R.I New Delhi pp. 114-133.

Iswaran V (1972). Madras Agril. J. 59: 52-53.

Tomer VK, Bhatnagar RK (1998). Palta, In Bhartiya Krishi Anusandhan Patrika, 13: 153-156.

Watsons H, Roberts S, Weiner (1992). "Genetic complementation of rhizobial nod mutants with Frankia DNA: Artifact or reality?" In molecular Biology of the gene, 4: 332-337.

Wiegand TW, Reznikoff WS (1992). Rhizobium- legume symbiosis and nitrogen fixation under sever condition and in an arid climate in J. Bactreiol. pp. 1229-1239.

7

Cry 1Ac levels and biochemical variations in Bt cotton as influenced by tissue maturity and senescence

S. Poongothai[1]*, R. Ilavarasan[2] and C. M. Karrunakaran[3]

[1]Department of Biotechnology, Bharath University, Chennai-600 073, India.
[2]Captain Srinivasa Murthi Drug Research Institute for Ayurvedha and Siddha, Chennai-600 106, India.
[3]Department of Biotechnology, Bharath University, Selaiyur, Chennai-600 073, India.

Quantification of Cry 1Ac protein in two field-grown Bt (*Bacillus thuringiensis*) cotton hybrids (MECH-184, RCH-2) was performed in relation to tissue maturity. Leaves of upper, middle and lower canopies and in bolls and bracts attached to the plant were chosen. Similar measurements were also made in fully mature tagged leaves attached to plants in the upper canopy and excised leaf discs. The leaf discs were incubated in the dark following the ELISA method with commercially available kits. Cry 1Ac levels declined with tissue maturity and senescence in both attached and excised plant parts of the Bt cotton hybrids examined. With advancing maturity in leaves attached to the plants till 21 days, steady increase in the levels of chlorophyll and decrease in the amounts of total protein were observed. Moreover, with increasing maturity, the concentration of reducing sugars rise in contrast to decline in total soluble amino acids. On the other hand, in case of the excised leaf discs, while total soluble amino acids exhibited an increasing trend, chlorophyll, total protein and reducing sugar contents decreased gradually.

Key words: Bt cotton, senescence, ELISA, Cry 1Ac insecticidal protein, amino acids, reducing sugars, chlorophyll.

INTRODUCTION

The transgenic Bt cotton is rapidly dominating the world agriculture (Ismael et al., 2002). With the genetically modified (GM) crops grown on 90 million ha globally in 2005 (James, 2005). Among these, transgenic cotton expressing insecticidal proteins from *Bacillus thuringiensis* (Bt) is one of the most adopted GM crops in the world (Dong et al., 2005). Bt cotton is considerably effective in controlling lepidopteran pests owing to the presence of Cry genes such as Cry 1Ac, Cry 1Ac + Cry 2Ab or Cry 1Ac + Cry 1F. Moreover, they are beneficial to the grower and the environment as they reduce chemical insecticides and preserving population of beneficial arthropods. However, poor performance of the transgenic traits during boll period and variable performance between different regions has been reported (Olsen and Daly, 2000). The loss of efficacy is associated with a

reduction in insecticidal proteins and production of toxin. Those are in turn influenced by plant age, reproductive stage and/or by a variety of environmental factors (Wu et al., 1997). Defoliation is induced by crop senescence which is characterized by loss of chlorophyll and ribonucleic acids (RNA), break down of proteins and complex forms of carbohydrates, decrease in inorganic ion levels and resulting in stimulation or inhibition of enzymes due to changes in hormone levels.

A field investigation indicated that the reduction of the insect-resistant efficacy for Bt cotton was due to senescence of the plant cells (Benedict et al., 1993). Hence to ensure that resistance management strategies designed for use with transgenic cotton is successful, the assessment of the insecticidal protein expression by senescence becomes important. It is indicated that insecticidal protein content in Bt cotton is variable with plant age, plant structure and environmental stresses. Variability in Bt cotton efficacy against target insect pests is mainly attributed to the changes in Bt protein content. Still physiological changes associated with the

*Corresponding author. E-mail: poongothai.sun@gmail.com.

production of secondary compounds in plant tissues may also play an important role. As a part of total protein, the insecticidal protein in plant tissues changes its level through inhibited synthesis, degradation or translocation to developing plant parts particularly under environment stresses (Dong et al., 2006). The synthesis of the Bt protein and its cycle in plant was controlled by several key enzymes such as NR, GPT, GOT, protease and peptidase (Steward, 1965). Control efficacy of Bt cotton is dependent upon the expression of Cry genes through synthesis of insecticidal protein in Bt cotton (Gutierrez et al., 2006). The temporal, spatial and environmental variation of efficacy may lead to insufficient control of targeted pests and evolution of resistance to Bt cotton. In recent years, it is becoming increasingly evident that several natural and induced defense mechanisms operate in host plants against different diseases. One of such defense mechanism is the presence of certain compounds inhibitory to the pathogen. Sometimes, the host plant is induced to synthesize these compounds on infection. In view of present day of Bt cotton, it was thought to analyse the biochemical in Bt cotton and accordingly non-Bt cotton (Hosagoudar et al., 2008). They also reported that the leaf insecticidal protein content of Bt cotton has close correlation with amino acids, chlorophyll and reducing sugars. This indicates that the biochemical aspects of the Bt cotton and the Bt protein content were reduced due to excised and intact senescence. This research aims to: (i) monitor Cry 1Ac protein in upper, middle, lower canopies, bolls and bracts at different ages of the plant, (ii) assess Cry 1Ac levels in relation to chlorophyll, total protein content and soluble amino acids in tagged intact leaves at different ages of their maturity and (iii) time course determination of Cry 1Ac levels in relation to chlorophyll, total protein content and soluble amino acids in detached dark-incubated leaves.

MATERIALS AND METHODS

Plant material and experimental design

The experiment was conducted under field conditions at International Institute of Biotechnology and Toxicology, Paddapai, India (Temperature maintained during season 32°C) during 2005 - 2006 in cotton growing season. Two Bt transgenic varieties (Medium in maturity, *Gossypium hirsutum L)* such as Mahyco cotton hybrid cotton (MECH)-184 (Maharashtra Hybrid seeds company Ltd, Jalna, India) and Rasi cotton hybrid (RCH)-2 (RASI seeds Pvt Ltd, Attur, Tamil Nadu) were grown. The plot size was 40 sq.m laid in Complete Randomized Block Design with four treatments replicated thrice. A spacing of 90 x 60 cm for irrigation was maintained and the seeds were dibbled. The organic manure at 10 to 12.5 tons per hectare was incorporated into soil for 3 to 4 weeks before sowing. The basal dose of 40:60:40 kg of N: P: K per hectare was applied at the time of sowing. Then the first and second top dressing was done, 30 days after sowing (DAS) and 60 DAS @ 40:0:20 kg of N: P: K per hectare. The irrigation was done at the critical stages of crop growth that is, germination, seedling growth, flowering and boll formation. The Bt varieties MECH-184

and RCH-2 flowered at 38 and 45 DAS.

Sample preparation

After square initiation, leaf samples were collected from different parts on the upper, middle, lower canopy, bolls and bracts of the plant on day 50, 70, 110 and 130; flowers were collected on day 38 and 45 from the squares. All the samples were frozen in liquid nitrogen, lyophilized and stored in -20°C for the quantification of the Cry 1Ac protein content.

Intact senescence

Phenotypically similar leaves and bolls were marked in the initial budding stage of the Bt cotton plant. The leaf samples were collected on day 0, 7, 14, 21 and 28. The samples were lyophilized on the same day and stored in -20°C for the estimation of Cry 1Ac protein. Ethanol extraction was done from leaves and bolls for the estimation of chlorophyll, reducing sugars, amino acids and total proteins.

Excised senescence

Phenotyphically similar leaves were collected during initial budding stage and the leaf discs were immersed in petri plates containing 10 ml of sterile water and incubated in dark. The leaf discs were taken from the petri plate on day 0, 2, 4, 6, 8 and 10, lyophilized and stored in -20°C for the estimation of Cry 1Ac protein. Ethanol extraction of the leaves was done for the estimation of chlorophyll, reducing sugars, amino acids and total proteins.

Cry 1Ac protein concentration assay

The Cry 1Ac protein in cotton leaf extracts was determined through immunological analysis by means of the ELISA (Shan et al., 2007). Leaf tissue extract of about 0.5 mg was prepared by homogenizing the lyophilized tissue in 500 μl of the ice cold 1 X sample extraction buffer (The complete recipe contains the following ingredients (in wt%): phosphate-buffered saline containing 0.05% Tween 20 and 1% polyvinyl pyrrolidone and the extract was then treated with trypsin). The lyophilized tissue was macerated at 3000 rpm using mortar driven pestle for 30 s, chilled on ice for 30 s and macerated again for 30 s then centrifuged at 8000 rpm for 15 min. Then the supernatant was collected for the quantification of Cry 1Ac. The antibodies against the Cry 1Ac, that is, goat anti-Cry 1Ac *(Ab2)* was added to each well. Buffer blank, standards, positive and negative controls were added to each well and incubated at 37°C for 1.5 h in a humid environment. AP–Conjugated Affinipure Donkey Anti-Goat IgG (Cat.No.705-0550147, Jackson immunological Research Laboratories Inc., USA) was added next and incubated at 37°C for 45 min in humid environment. Finally, the buffered enzyme substrate pNPP was added and the enzyme reaction was carried in dark at room temperature for 30 min. The absorbance was measured at 405 nm.

Assay of chlorophyll, amino acids, reducing sugars and total proteins

The extraction of the plant tissues in alcohol were prepared by using the following procedure. Four leaf discs of the same size were plunged into hot 80% ethanol (10 ml) and maintained to boil for 10 min at 60°C. Then the tissue was ground in a mortar with pestle and filtered through the Whatman No. 41 filter paper. The

Table 1. Lethal dose of Cry 1Ac level in the leaves of different canopies and in bolls and bracts of two Bt cotton cultivars MECH-184 and RCH-2. Values in μg/g are mean ± SD of three observations

DAS*	Upper canopy		Middle canopy		Lower canopy		Bolls		Bracts	
	MECH-184	RCH-2	MECH-184	RCH-2	MECH-184	RCH-2	MECH-184	RCH-2	MECH-184	RCH-2
50	2.67±0.7[a]	3.77±0.4[a]	2.52±0.8[a]	3.52±0.6[a]	2.42±0.9[a]	3.21±1.8[a]	2.26±0.5[a]	2.24±0.1[a]	1.50±0.1[a]	2.25±1.9[a]
70	1.62±0.5[b]	2.77±0.5[b]	1.28±0.2[b]	2.24±0.2[b]	1.16±0.4[b]	2.40±0.5[a]	1.26±0.2a	1.17±1.2[b]	1.21±0.5[a]	1.16±1.8[bc]
90	1.20±0.2[c]	1.16±1.6[c]	1.14±0.1[b]	1.22±0.5[c]	1.01±0.9[b]	0.94±0.3[b]	1.04±0.9[ab]	1.07±0.2[b]	1.02±1.5[a]	0.11±1.1[c]
110	0.94±0.1[c]	1.06±1.2[c]	0.74±0.3[b]	0.96±0.8[c]	0.88±0.2[b]	0.57±0.0[b]	0.99±0.5[ab]	0.98±0.8[b]	0.94±0.8[a]	0.01±1.5[c]
130	0.21±0.1[d]	0.71±0.6[d]	0.34±0.01[d]	0.01±0.5[c]	0.25±0.01[b]	0.06±0.1[b]	0.07±0.2[b]	0.01±1.5[c]	0.12±0.2[a]	0.01±0.01[bc]

*DAS – Days after sowing. Mean values of different superscript letters (a, b, c) are significantly different (P<0.05) as determined by Duncan's multiple range test.

extract was made up to 5 ml for the estimation of chlorophyll, amino acids and reducing sugars. Finally, the residue was used for the estimation of total proteins. The chlorophyll estimation was done on the same day and the absorbance was measured at 665 nm. The amino acid content was determined by Ninhydrin assay (Moore and Stein, 1948), absorbance readings were converted to fresh tissue (μg/g) of amino acid using a standard curve prepared with glycine. The total protein content was determined by Coomassie Blue Dye–binding assay (Bradford, 1976). The absorbance readings were converted using bovine serum albumin (BSA) as standard curve. Reducing sugar was determined by Nelson's method (Nelson, 1944), absorbance readings were converted to fresh tissue (μg/g) of reducing sugar using glucose as standard curve.

RESULTS

Cry 1Ac levels influenced by the age of Bt cotton hybrids MECH-184 and RCH-2 in different canopies

In upper canopy, the Cry 1Ac levels were initially high in a range 1.20 - 3.77 μg/g during 50 - 90 DAS. The Cry 1Ac level decreases with time in both hybrids. The decline started relatively early in the season in Bollgard MECH-184 and RCH-2 with a range of 0.21 - 1.06 μg/g during 110 - 130 DAS. In middle canopy, the Cry 1Ac levels were initially high in a range of 1.14 - 3.52 μg/g during 50 - 90 DAS. At 110 - 130 DAS, Bollgard MECH-184 and RCH-2 displayed a gradual decline of Cry 1Ac levels and the decline started relatively early in the season with a range of 0.34 - 0.96 μg/g. The Cry 1Ac expression in the lower canopy leaves initially during 50 - 90 DAS ranged between 1.01 - 3.21 μg/g in both hybrids. At 110 - 130 DAS, in both Bollgard MECH-184 and RCH-2, a gradual decline was observed and the rapid rate of decrease started relatively early in the season with a range of 0.025 - 0.57 μg/g. The Cry 1Ac expression in the bolls and bracts initially during 50 - 90 DAS ranged between 1.04 - 2.24 and 0.02 - 2.25 μg/g in both hybrids. At 110 - 130 DAS, in both Bollgard MECH-184 and RCH-2, a gradual decline was observed and rapid decrease was started relatively early in the season with a range of 0.07 - 0.98 and 0.12 - 0.01 μg/g (Table 1).

Biochemical changes influenced by senescence of intact tissues of Bt cotton hybrids

The contents of the leaf Cry 1Ac proteins, chlorophyll, reducing sugars, amino acids and total proteins were different over the duration of the intact senescence of the Bt cotton. The contents of the Cry 1Ac protein decreased from 0 to 28 days for MECH-184, which was 2.11 - 0.74 μg/g of lyophilized tissue, the content of RCH-2 1.62 - 0.87 μg/g of lyophilized tissue. The chlorophyll content was increased for both MECH-184 and RCH-2 varieties with the range of 1.52 - 2.90 and 1.37 - 2.21 absorbance at 665 nm respectively. The results of the reducing sugar decreased between 0 to 28 days for MECH-184, which was 18.66 - 40.26 μg/g of fresh weight and the content of RCH-2 were 14.56 - 62.67 μg/g of fresh weight. The contents of leaf amino acid decreased from 0 to 28 days for MECH-184, which was 7.94 - 1.03 μg/g of fresh weight, the contents of RCH-2 was 8.77 - 1.66 μg/g of fresh weight. The total protein content in the leaf decreased from 0 to 28 days for MECH-184, which was 13.21 - 3.13 μg/g of fresh weight; the contents of RCH-2 were 15.55 - 5.62 μg/g of fresh weight (Figure 1).

Biochemical changes influenced by senescence of excised leaves of Bt cotton hybrids

The contents of the leaf Cry 1Ac proteins, chlorophyll, reducing sugars, amino acids and

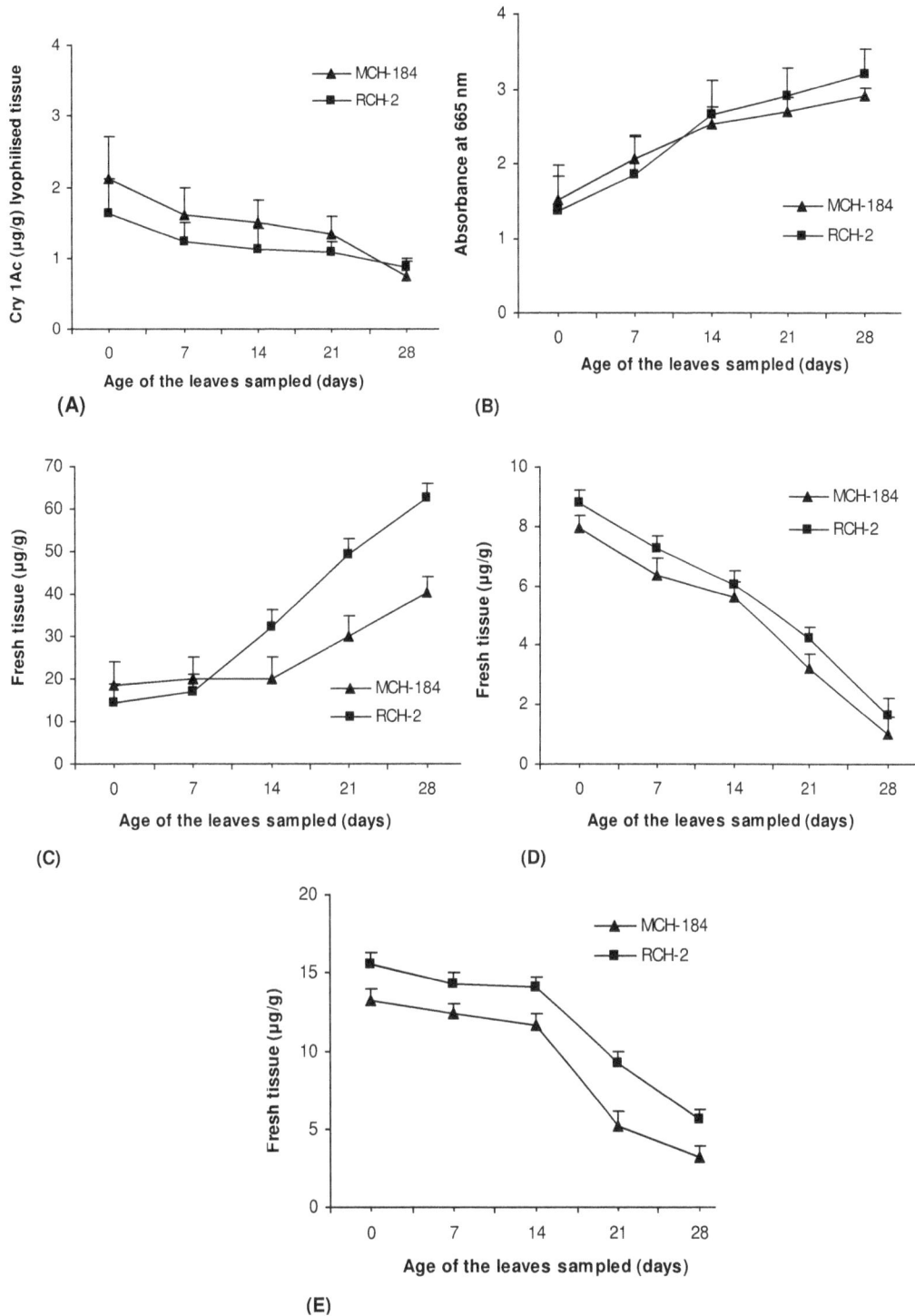

Figure 1. Changes in (A) Cry 1Ac expression, (B) chlorophyll, (C) reducing sugar, (D) amino acids and (E) total proteins of the intact senescence. Symbol MCH-184 and RCH-2 are the name of the two Bt cultivars, vertical bar represent S.E. of the mean (n = 4), when value exceeds the size of the symbol. The value at 0 day was the level of the control.

total proteins were different over the duration of the intact senescence of the Bt cotton. The contents of the Cry 1Ac protein decreased from 0 to 28 days for MECH-

184, which was 2.64 - 1.15 µg/g of lyophilized tissue and the content of RCH-2 was 2.73 - 0.56 µg/g of lyophilized tissue. The chlorophyll content decreased for both

MECH-184 and RCH-2 with the range of 2.58 - 0.52 and 1.92 - 0.16, absorbance at 665nm respectively. The results of the reducing sugar decreased during 0 to 28 days for MECH-184, which was 40.57 - 14.66 µg/g of fresh weight and the content of RCH-2 was 50.25 - 10.09 µg/g of fresh weight. The contents of the leaf amino acid decreased from 0 to 28 days for MECH-184, which was 2.01 - 6.11 µg/g of fresh weight, the content of RCH-2 was 1.20 - 7.05 µg/g of fresh weight. The total protein content in the leaf decreased from 0 to 28 days for MECH-184, which was 7.21 - 1.89 µg/g of fresh weight; the contents of RCH-2 were 8.97 - 2.90 µg/g of freshweight (Figure 2).

DISCUSSION

Cry 1Ac expression in different durations of cotton growth

Dong and Li (2007) reported that the efficacy of trans-genic Bt cotton against target pests varies with plant age, plant part or structure and also due to aging of plants. It is also considerably affected by environmental stresses, such as high temperature, heavy drought, water logging, elevated CO_2 and nitrogen deficiency (Luo et al., 2008). They also reported that the variability in efficacy is mostly attributed to reduction in the amount of endotoxin proteins in plant tissues, but physiological changes accompanied with production of some secondary compounds also play an important role in changes in Bt cotton. Chen et al. (2000) demonstrated that toxin protein content in the fully expanded leaves was significantly higher than those in roots, stems and petioles. During the seedling stage, ovaries at anthesis expressed considerably more toxin protein than pistils and stamens at the flowering stage. Based upon multi-site experiments with 35 transgenic Bt cotton varieties, Greenplate et al. (1999) suggested that differences in field efficacy for less sensitive insect species such as armyworm and cotton bollworm are likely functions of differences in levels of Cry 1Ac as influenced by plant age, field site and variety background. Kranthi et al. (2005) reported that the quantitative levels of Cry 1Ac and the seasonal decline in expression differed significantly among the Bollgard hybrids and also between the different parts of the plant. The Cry 1Ac expression declines progressively over the crop growth with toxin level falling below the critical level of 1.9 µg/g after 110 DAS. The variability in toxin expression and the pest control properties are unlikely to be affected significantly at least, until the crops are 100 - 125 days old (Greenplate et al., 1998).

The Cry 1Ac δ-endotoxin protein content decreased as plant ages during the profiling season–long expression in Bollgard cotton (Adamczyk and Sumerford, 2001). Mahon et al. (2002) commented that salinity stress (200 *mM* NaCl) significantly decreases the insecticidal protein

leaves, but it does not affect the control efficacy against Bollworm in terms of Bioassay results. Changes in efficacy are seen to be mediated through modification of the physiological background of the plant rather than changes in the level of Cry 1Ac expression or in the concentration of Bt toxin. The secondary compounds alter the toxicity of Bt proteins against lepidopteran larvae, either negatively or positively (Olsen and Daly, 2000; Zhang and Guo, 2000). Loss of reproductive organ can be induced by myriad of causes and such losses can elicit many morphological and physiological responses including compensatory growth. They also suggested that fruit loss may enhance photosynthetic rate in cotton. In comparison to the control, the leaf insecticidal protein content was reduced for both the hybrids after 50 DAS and suggests a rapid decline in Cry 1Ac expression in the early stage of the different canopies and in bolls and bracts.

Effect of environmental stress on Bt cotton

Boyer (1982) and Martins et al. (2008) explained that environmental stresses, such as extreme levels of light, temperature, water deficit, salinity or nutrient deficiency, reduce the agricultural production and quality of many crops. In transgenic Bt cotton, the Cry gene expression in terms of concentration of toxin proteins is impacted by environmental factors. Reduction in toxin concentration with the elevated CO_2 was found in a Bt transgenic cotton cultivar Ck-12 (Coviella et al., 2002). The elevated CO_2 enhances water-use efficiency and photosynthetic production of the crop (Samarakoon and Gifford, 2004). Chen et al. (2005) commented that the exposure of the Bt transgenic cotton to high temperature (37°C) significantly reduced the Cry 1Ac protein content at boll-setting stage. Under some circumstances, the amount of insecticidal protein in Bt cotton tissues is considerably reduced, but the toxin level does not fall below the critical level, and still maintains a relative high efficacy against the insect pest. They added that exposure of Bt transgenic cotton plants to high temperature resulted in a significant decline in glutamic-pyruvic transaminase (GPT) activity and soluble protein content, suggesting that high temperature may result in the degradation of soluble protein in the leaf, with a resulting decline in the level of the toxin Cry 1A. Otherwise Abel and Adamczyk (2004) reported that low chlorophyll content of leaf tissue does not fully express Cry 1A and further suggested that photosynthesis regulating factors related to mRNA transcription and translation should have effects on Cry 1A production and insect control.

Olsen and Daly (2000) concluded that there is less Bt protein in older plants and it appears that the protein is either less available or less toxic to neonates. The concentration of Cry 1Ac protein, as a proportion of total protein, also declines during the season.

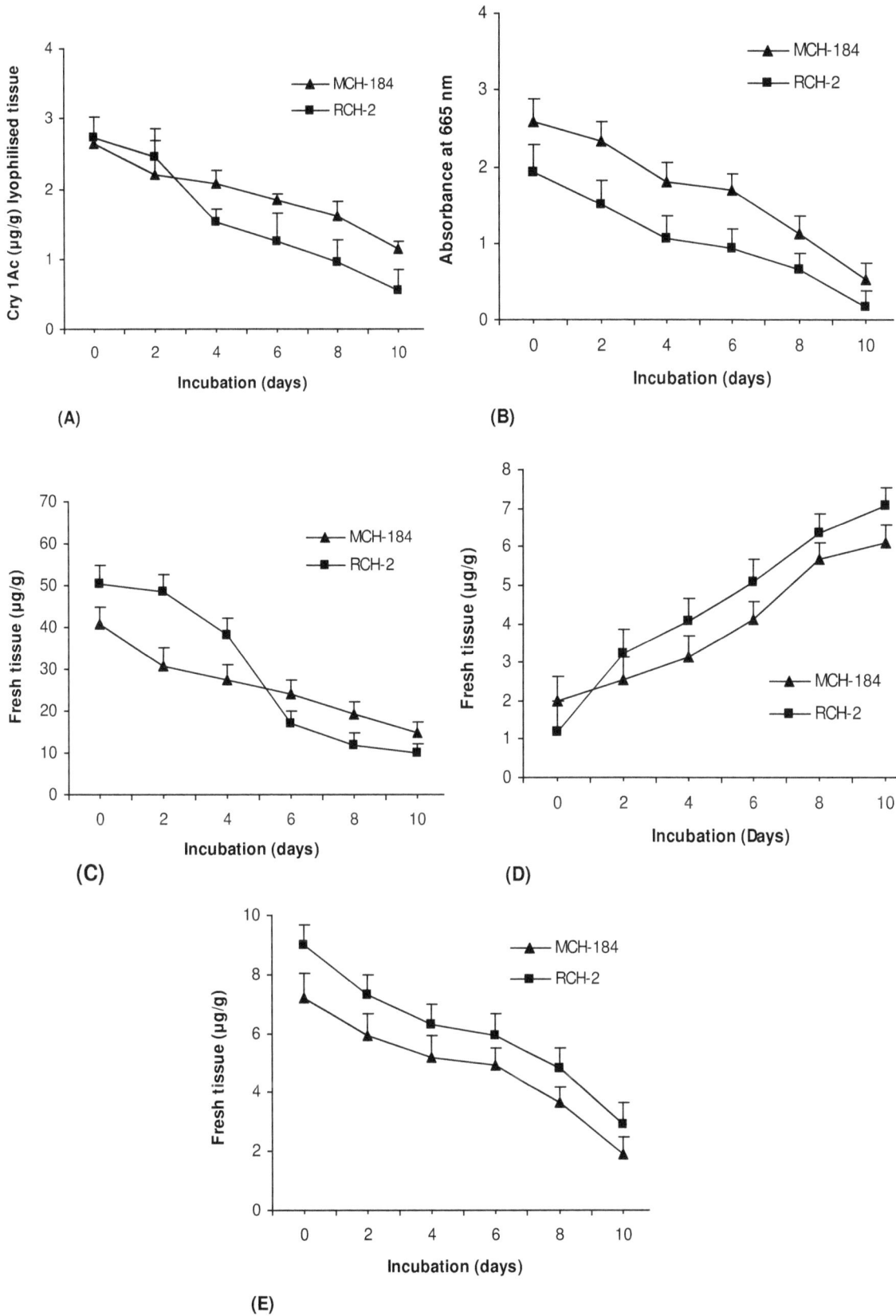

Figure 2. Time course study on the levels of (A) Cry 1Ac expression, (B) chlorophyll, (C) reducing sugar, (D) amino acids and (E) total protein content in excised leaf discs of two Bt cotton hybrids treated on water and incubated in dark. Symbol MCH-184 and RCH-2 are the name of the two Bt cultivars, vertical bar represent S.E. of the mean (n = 4), when value exceeds the size of the symbol. The value at 0 day was the level of the control.

Brown and Oosterhuis (2003) stated that chaperone, a plant growth regulator appears to increase the protein concentration and the efficiency of endotoxin expression even at high temperature stress. Pettigrew and Adamczyk (2006) mentioned that reduced levels of toxin protein in cotton leaves planted early were presumably caused by the remobilization of the leaf to support the larger developing boll load as compared with late planted cotton. There are many physiological changes in plants either as the plant ages or under environmental stresses, which may contribute to the variation in endotoxicity in Bt cotton. Studies on non transgenic cotton indicate that there is a change in the level of secondary compounds, such as phenolics and terpenoids as plants mature.

The results of the senescence induced changes in intact leaves of Bt cotton hybrid MECH-184 and RCH-2 showed that the Cry 1 Ac insecticidal protein, total proteins and reducing sugar decrease due to the aging of plant and the chlorophyll and amino acids increase. Amino acid content increases due to the breakdown of the proteins (Figure 1). In excised leaves of Bt cotton hybrids, that is, MECH-184 and RCH-2, it was observed that the Cry 1Ac, chlorophyll, reducing sugar and total protein content decreases, and the amino acid increases due to aging of plant. This can attributed to the reduction of total protein content under artificially induced senescence incubated in dark. The variation of Cry 1Ac insecticidal protein and other biochemical constituents is not only due to the transformation of the Cry 1Ac protein in the cotton hybrids but also due to the environmental conditions as well as the ageing of the plant. Therefore, it is recommended to spray 5% NSKE (Neem seed kernel extract), HaNPV (*Helicoverpa armigera* nuclear polyhedrosis virus) and farmers should mentally prepare to spray the insecticides such as thiodicarp, quinolphos, chlorpyriphos, navaluron etc., during weekly intervals of fruiting phase of crop as normally bollworm is susceptible to these insecticides (Kranthi et al., 2002). The companies should evaluate their hybrids critically for the highest levels of expression of toxin in late season.

ACKNOWLEDGEMENTS

The authors gratefully acknowledge the support of Dr. P. Balakrishna Murthy, Director of International Institute of Biotechnology and Toxicology (IIBAT). The authors also sincerely appreciate the colleagues of IIBAT for their contribution towards this work.

REFERENCES

Abel CA, Adamczyk JJ (2004). Relative concentration of Cry1A in maize leaves and cotton bolls with diverse chlorophyll content and corresponding larval development of fall armyworm (Lepidoptera: Noctuidae) and southwestern corn borer (Lepidoptera Crambidae) on maize whorl profiles. J. Econ. Entomol. 97: 1737-1744.

Adamczyk JJ, Sumerford DV (2001). Potential factors impacting season-long expression of Cry1Ac in 13 commercial varieties of Bollgard cotton. J. Insect Sci. 1: 1-6.

Benedict JH, Sachs ES, Altman DW, Ring DR, Stone TB, Sims SR (1993). Impact of δ-endotoxin-producing transgenic cotton on insect-plant interactions with Heliothis virescens and Helicoverpa zea (Lepidoptera: Noctuidae). Environ. Entomol. 22: 1-9.

Boyer JS (1982). Plant Productivity and Environment. Science 218: 443-448.

Bradford MM (1976). A rapid and sensitive method for the quantification of microgram quantities of protein utilizing the principle of protein-dye binding. Annal. Biochem. 72: 248-254.

Brown RS, Oosterhuis DM (2003). Effect of foliar Chaperone TM applications under elevated temperatures on the protein concentrations and physiological responses of cotton. Ark. Agric. Exp. Sta. Res. Ser. 521: 108-111.

Chen S, Wu J, Zhou B, Huang J, Zhang R (2000). On the temporal and spatial expression of Bt toxin protein in Bt transgenic cotton. Acta Gossypii Sin. 12: 189-193.

Chen FJ, Wu G, Ge F, Parajulee MN, Shrestha RB (2005). Effects of elevated CO_2 and transgenic Bt cotton on plant chemistry, performance, and feeding of an insect herbivore, the cotton bollworm. Entomol. Exp. Appl. 115: 341-346.

Coviella CE, Stipanovic RD, Trumble JT (2002). Plant allocation to defensive compounds: interactions between elevated CO_2 and nitrogen in transgenic cotton plants. J. Exp. Bot. 53: 323-331.

Dong HZ, Li WJ, Tang W, Zhang DM (2005). Increased yield and revenue with a seedling transplanting system for hybrid seed production in Bt cotton. J. Agron. Crop Sci. 191: 116-124.

Dong HZ, Li WJ, Zhang DM, Tang W, Zhang DM, Li ZH (2006). Effects of genotypes and plant density on yield, yield components and photosynthesis in Bt transgenic cotton. J. Agron. Crop Sci. 192: 132-139.

Dong HZ, Li WJ (2007). Variability of endotoxin expression in Bt transgenic cotton. J. Agron. Crop Sci. 193: 21-29.

Greenplate JT, Head GP, Penn SR, Kabuye VT (1998). Factors potentially influencing the survival of Helicoverpa zea on Bollgard cotton. In: Dugger, P., Richter, D. A., (Eds.), Proceedings, 1998 Beltwide Cotton Conference. National Cotton Council of America. Memphis. TN. 1030-1033.

Greenplate JT (1999). Quantification of Bacillus thuringiensis insect control protein Cry1Ac over time in Bollgard cotton fruit and terminals. J. Econ. Entomol. 92: 1377-1383.

Gutierrez AP, Adamczyk JJ, Ponsard S, Ellis CK (2006). Physiologically based demographics of Bt cotton-pest interactions II. Temporal refuges, natural enemy interactions. Eco. Model. 191: 360-382.

Hosagoudar GN, Chattannavar SN, Srikant Kulkarni (2008). Biochemical studies in Bt and non-Bt cotton genotypes against alternaria blight disease (*Alternaria macrospora* Zimm). Karnataka J. Agric. Sci. 21(1): 70-73.

Ismael Y, Bennett R, Morse S (2002). Benefits from Bt cotton use by small holder farmers in South Africa. Agric. Bio. Forum. J. Agrobio. Mana. Eco. 5(1): 1-5.

James C (2005). Executive Summary of Global Status of Commercialized Biotech/GM Crops: ISAAA Briefs No. 34. ISAAA, Ithaca, NY. pp. 1-12.

Kranthi KR, Jadhav DR, Kranthi S, Wanjari RR, Ali S, Russell D (2002). Insecticide resistance in five major insect pests of cotton in India. Crop Protect. 2: 449-460.

Kranthi KR, Naidu S, Dhawad CS, Tatwawadi A, Mate K, Patil E, Bharose AA, Behere GT, Wadaskar RM, Kranthi S (2005). Temporal and intraplant variability of Cry1Ac expression in Bt-cotton and its influence on the survival of the cotton bollworm, Helicoverpa armigera. Curr. Sci. 89: 291-298.

Luo Z, Dong H, Li W, Ming Z, Zhu Y (2008). Individual and combined effects of salinity and water logging on Cry1Ac expression and insecticidal efficacy of Bt cotton. Crop Prot. 27: 1485-1490.

Mahon R, Finnergan J, Olsen K, Lawrence L (2002). Environmental stress and the efficacy of Bt cotton. Aust. Cotton Grower 22: 18-21.

Martins CM, Beyene G, Hofs J, Kruger K, Van der Vyver C, Schluter U, Kunert KJ (2008). Effect of water-deficit stress on cotton plants expressing the Bacillus thuringiensis toxin. Ann. Appl. Biol. 152: 255-262.

Moore S, Stein WH (1948). Photometric method for use in the chromatography of aminoacids. J. Biol. Chem. 176: 367-388.

Nelson N (1944). A Photometric adaptation of the Somogyi method for the determination of glucose. J. Biol. Chem. 153: 375-380.

Olsen KM, Daly JC (2000). Plant-toxin interactions in transgenic Bt cotton and their effects on mortality of *Helicoverpa armigera*. Entomol. Soc. Am. 93: 1293-1299.

Pettigrew WT, Adamczyk JJ (2006). Nitrogen fertility and planting date effects on lint yield and Cry 1Ac (Bt) endotoxin production. Agron. J. 98: 691-697.

Samarakoon AB, Gifford RM (2004). Water use and growth of cotton in response to elevated CO_2 in wet and drying Soil. Aust. J. Plant Physiol. 23: 63-74.

Shan G, Embrey SK, Schaffer BW (2007). A highly specific enzyme-linked immunosorbent assay for the detection of Cry1Ac insecticidal crystal protein in transgenic Wide Strike cotton. J. Agric. Food Chem. 55: 5974-5979.

Steward FC (1965). Metabolism of nitrogenous compounds. In: Steward, F.C. (Ed.), Plant Physiology : A Treatise [M]. Academic Press. New York pp. 1128-1136.

Wu JY, He XL, Shu C, Chen S, Fu CX, Huang JQ (1997). Influence of water logging on the bollworm resistance of Bt cotton. (In Chinese with English abstract). J. Jiangsu Agric. Sci. 4: 231-233.

Zhang Y, Guo Y (2000). Interactions between condensed tannin and Bt Crystal protein in cotton. Acta Gossypii Sin. 12: 294-297.

8

Analysis of the resistance to *Phytophthora* pod rot within local selections of cacao (*Theobroma cacao* L.) for breeding purpose in Cameroon

M. I. B. Efombagn[1], S. Nyassé[1], D. Bieysse[2] and O. Sounigo[1,2]

[1]Institute of Agricultural Research for Development (IRAD), P. O. Box 2067 or 2123, Yaoundé, Cameroon.
[2]CIRAD-CP, TA 80/02, 34398 Montpellier Cedex 5, France.

The level of resistance against *Phytophthora* pod rot (PPR) was assessed among locally-selected accessions of cacao, the so-called selection of Nkoemvone (SNK), in Cameroon. The evaluation approaches have included four tests: Leaf disc inoculation (LDT); detached pod inoculation (DPT); artificial inoculation of pods and counting of successful infections (%SI); and field observations of the disease incidence (FI). The studied accessions were classified into four groups according to their genetic origin: SNK; SNK600-Tr (SNK*ICS); SNK600-Tr*UA (UPA*ICS); SNK600-Tr*UA (UPA*SNK). In all four testing methods, the groups UPA*ICS and SNK appeared respectively as the least and the most susceptible accession groups. Four resistance classes were defined for each testing method. A great part of the accessions were found in the intermediary resistance classes, based on DPT, LDT and %SI. Few accessions have displayed a significant level of PPR resistance in all testing methods. Rank correlation of the different accessions groups based on their performance in the four tests have shown that all correlation coefficients were positive ($r = 0.37 - 0.87$) and significant ($p \leq 0.06$). The best correlation was registered between %SI and FI with $r = 0.87$ ($p = 0.01$). Subsequently, a unique classification of all accessions was done according to the scores recorded in all the testing methods. The use of locally-selected accessions (SNK) in the varietal development as well as in the cacao seed production is discussed.

Key words: Cacao plant pathology, oomycetes, screening methods, resistant cultivars, selection.

INTRODUCTION

Cacao (*Theobroma cacao* L.) is a perennial crop of high economic importance in several cocoa-producing countries, including Cameroon. However, cacao cultivation is threatened by many constraints, such as the *Phytophthora* pod rot (PPR) disease. In absence of any chemical control, cacao pod losses may reach 90 to 100% (Despréaux et al., 1988), posing to the ongoing research the need to find out resistant cultivars. In 1950s, field surveys were carried out in existing cacao farms of

southern Cameroon in order to identify high yielding cacao trees (Braudeau et al., 1952). The outcomes of these surveys were the selection of two main populations belonging to the two major genetic groups of cacao: the Trinitarios (Tr) and the Lower Amazon Forasteros (LAF). The pod size was the first criterion of selection. Therefore, most of the LAF known to bear pods with limited size (Bartley, 2005) failed to be selected. The selected trees were transferred and multiplied on the research station of Nkoemvone and thus identified using the prefix selection of Nkoemvone ('SNK'). All the SNK accessions were subsequently used to create open-pollinated crosses. As part of the breeding program, all these accessions planted in clonal trials were assessed based on the following criteria: vigor; yield; number of beans per pod; bean size and weight. The best ones were maintained in so-called clone collections. Base on their genetic traits such as yield, some of the SNK cultivars were used as progenitors in seed gardens. Unfortunately, no assessment of the level of tolerance to *Phytophthora* was performed on these clones before their use as progenitors. This resulted in a rather high level of susceptibility of the progenies released to farmers.

Progress in breeding has been facilitated by effective screening methods to identify genotypes with superior abilities to be incorporated into breeding programmes. Generally, there are two types of pathology tests to evaluate PPR resistance: *in vitro* early screening tests on leaves (Nyassé et al., 1995) and on detached pods (Iwaro et al., 2000), *in vivo* screening tests on pods attached to the cacao tree (Blaha and Lotode, 1976) and field evaluation of the natural infection (Ndoumbè et al., 2002). At the end of 1970s, an artificial inoculation test was performed on pods yet attached to the trees, followed by the counting of pods showing symptoms 10 days after inoculation (Blaha and Lotodé, 1976). This test was used in order to assess 112 genotypes, among which 78 clones were locally selected (SNK). The data obtained from this test showed a high level of susceptibility to *Phytophthora* for most of the clones used as progenitors in the commercially released progenies. This test was then used to assess the level of tolerance to *Phytophthora* of trees in local progeny trials and this study resulted in the selection of 42 genotypes showing a very low proportion of infected pods after artificial inoculation. These genotypes were cloned, named as SNK 600 to SNK 641, and grouped based on their different genetic origins: (i) full-sib progenies issued from crosses between two Trinitario parents, (ii) crosses between one Trinitario and one Upper Amazon Forastero parent, and (iii) selfing of SNK 64.

Later on, Berry and Cilas (1994) compared the pod inoculation data obtained from six clones with the level of natural infection observed on these clones in the field, and failed to find any significant positive correlation. On the other hand, they noticed that clones with large pods were more heavily infected in natural conditions than those with small pods and suggested that pod volume might be a major factor in natural infection. During the last 15 years, efforts have been made to develop new tests in order to assess the level of resistance of cacao to PPR. A first test, based on lab artificial inoculation of leaf discs followed by a scoring of the symptoms, was developed in the mid 1990s by Nyassé et al. (1995) and, later on, a test based on pod inoculation in the lab, followed by scoring symptoms was developed by Iwaro et

al. (2000), in order to select progenitors combining high yield and resistance to PPR.

Selection of resistant material in the field and all the known tests for resistance based on pod inoculation necessitates waiting until a tree bears pods before its level of resistance to black pod infection can be determined. These methods, therefore, do not offer a quick means of breeding for resistance. Consequently, tests at the seedling stage that provide information about pod resistance are earnestly needed to save time and cost. Few reliability studies were carried out to confirm the validity of those screening tests have shown positive correlation between pod and leaf tests (Tahi et al., 2000; Nyadanu et al., 2009). However, the correlation between the reactions of detached leaves and pod tests to *Phytophthora* sp. infection is not well understood and utilized. Wheeler (1992) emphasized the need to understand the relationship between pod resistance and leaf resistance to PPR and, the relationship between detached leaves and pod tests infection in the field to expedite the screening of germplasm. With the objectives of verifying the screening power of four assessing methods towards identification of the best PPR-resistant genotypes among the SNK collection, the present paper shows the data obtained on the levels of PPR resistance for the SNK clones, including those selected in local farms (SNK 1 to 599) in the 1950s, and those from local progeny trials in the late 1970s (SNK 600 to SNK 641).

MATERIALS AND METHODS

Cacao accessions

A total of 105 accessions available in the clone collections of the Nkoemvone IRAD Station (Southern Cameroon) was considered in the study. The accessions were classified into two groups, according to their role in the cacao breeding program in Cameroon (Table 1):

(1) The SNK group (72 accessions) labeled from SNK 1 to SNK599, selected in the 1950s at local farms for their yield levels.
(2) The SNK600 group (33 accessions) composed of cloned progenies and subdivided as follows, according to their genetic origin:
(i) The SNK600-Tr group (13 accessions) originated from crosses between a SNK clone (SNK 1 to SNK 599) and an ICS clone (from the 'Trinitario' genotypes introduced from Trinidad and Tobago in the early 1960s).
(ii) The SNK-Tr*UA group (20 accessions) originated from crosses between a Trinitario clone (SNK 1 to SNK 599 or ICS) and an Upper Amazon Forastero clone.

Screening tests for resistance to PPR

Leaf disc inoculation test (LDT)

An early screening method (Nyassé et al., 1995) based on a leaf disc inoculation tests and a disease rating was used in the study. All the plant material was inoculated with a moderately aggressive isolate of *Phytophthora megakarya* collected in farmers' field and

Table 1. List of the locally-selected accessions used in the study and available in cacao genebank at Nkoemvone (Southern Cameroon).

Genetic group	Accession groups	Progenitors	Corresponding accessions
Tr	SNK-Tr[1]	-	SNK 7 10 12 13 15 16 17 27 29 30 31 32 33 35 37 46 48 52 64 67 109 111 141 181 203 213 223 267 269 277 322 332 343 344 348 353 361 376 377 392 404 413 415 416 417 422 443 444 450 451 452 456 459 460 461 467 469 476 478 479 480 485 488 490 496 500 501 502 503 504 505 506
TrxUA	SNK600[2]-Tr	SNK*ICS (SNK13*ICS95; SNK30*ICS95; SNK416*ICS95)	SNK 600 602 603 604 605 606 607 608 609 610
		ICS*SNK (ICS43*SNK37; ICS84*SNK37; SNK416*ICS95)	SNK 611 612 613
		UPA*ICS (UPA134*ICS1; UPA143*ICS95; UPA143*ICS43;UPA134*ICS43)	SNK 614 615 616 618 619
	SNK600-TrxUA	ICS*UPA (ICS84*UPA337)	SNK 620 621 622 623 624
		UPA*SNK (UPA134*SNK64; UPA143*SNK64)	SNK 625 626 627 628 629
		SNK*UPA (SNK30*UPA337; SNK16*UPA143; SNK109*UPA143)	SNK 630 631 632 633 634

[1] Trinitario clones selected from farmers fields in the South of Cameroon in the 1960s and transferred to the genebank of the IRAD Nkoemvone Research Station (Southern Cameroon), [2] So-called SNK600 series of clones, they were selected on-station in UAxTr and in TrxTr crosses.

maintained in the Plant Pathology Laboratory at IRAD Nkolbisson Centre (Cameroon). Prior to their evaluation with the leaf disc inoculation test, all the studied cacao accessions existing in the field as adult trees (ages above 30 to 40 years) were duplicated in the nursery using a vegetative propagation technique (top-grafting on 6-months rootstocks).

Three inoculations series were carried out for each accession tested and the interval between each one was at least 30 days. In each inoculation series (replicate), two or three leaves of approximately two months old from non-lignified twigs as described by Nyassé et al. (1995) were collected early in the morning from relatively flexible branches (green turning to brown) of each accession. Only the synchronized flushes of all the plants were used at the same time for a single replicate. Thirty leaf discs of 15 mm diameter were cut from those two or three collected leaves per accession in the afternoon and placed upside down in three wetted plastic trays, with 10 discs per tray. The covered trays were incubated in the darkness overnight at approximately 25°C. Inoculations were carried out the next morning with a 10 μl suspension of *P. megakarya* calibrated at 300,000 zoospores per ml. Symptoms (disease scores) were observed five days after inoculation by using a 5 point scale, with 0 = no symptoms; 1 = very small localized penetration points; 2 = small penetration spots, sometimes in a network; 3 = coalescing lesions of intermediate size; 4 = large coalescing brown patches; and 5 = uniform large dark brown lesions.

Detached pod test (DPT)

The resistance of cacao accessions was also assessed according to the method developed by Iwaro et al. (2000). Half of the pods

surface was sprayed at a distance of 30 cm, using an atomizer containing a zoospore suspension of the same isolate used with the leaf inoculation test. For each accession, three inoculation rounds were carried out at different dates and ten pods were inoculated during each round. After four days of incubation, the inoculated pods were assessed based on the frequency and size of the lesions formed. The severity of infection was rated on an 8-point scale as follows: 1 = No symptom, highly resistance to penetration; 2 = 1-5 localized lesions, resistant; 3 = 6-15 localized lesions, moderately resistant; 4 = >15 localized lesions, partially resistant/resistant to spread of lesions alone; 5 = 1-5 expanding lesions, partially resistant/resistant to penetration alone; 6 = 6-15 expanding lesions, moderately susceptible, 7 = >15 expanding lesions, susceptible; 8 = coalesced lesions, highly susceptible.

Field tests

Percentage of successful infections (% SI) on attached pods

This method was developed by Blaha and Lotode (1976). It has the advantage to be closer to what happens naturally in the field. The tests are carried out on living pods attached to cacao trees, and therefore differ from the previous testing methods (LDT and DPT) that are undertaken under laboratory conditions. Fruits (pods) are usually tested during heavy rainy seasons, corresponding to the peak of PPR natural infections. Cacao trees in the field experiments are maintained in natural conditions, with no fertilizers and pesticides applications. Attached pods are inoculated with zoospores of *P. megakarya* on any part of the husk. In this work, the evaluation of symptoms was monitored continuously during twelve days. The resistance to the penetration of the pathogen was assessed by counting the number of pods successfully infected. The average number of pods infected per accession varied between 50 and 200. The rate of infection for each accession (% SI) was estimated by the ratio between the numbers of successful infections and the total number of pods inoculated. A total of 61 accessions were tested.

Field incidence (FI) of PPR

The natural infection in the field was measured during the period of PPR incidence. This period corresponds to the start and the end of the rainy seasons within a year. The evaluation of PPR incidence for each accession consists on the counting of both the number of PPR-infected and the healthy pods during the campaign. Pods that were rotten (affected by PPR) wilted (early desiccation of a physiological nature) damaged by rodents, and ripe and healthy, were counted weekly. During the counts, sanitation harvesting was carried out: all pods except unripe healthy pods were removed. Losses caused by PPR were estimated in relation to potential production (excluding rodent-damaged pods) using the formula giving the pod rot rate per tree (Rrot):

$$ \text{Rrot} = \frac{\sum_{\text{rotten pods}}}{\sum_{\text{rotten pods}} + \sum_{\text{ripe pods}} + \sum_{\text{healthy pods on last count}}} $$

Statistical analyses

The data collected were analyzed with SAS software version 8 (2000). ANOVA was computed for the results obtained in all testing methods and the data was tested for adjustment to the major assumptions of ANOVA analysis. The Newman and Keuls (Cochran and Cox, 1957) test at 5% probability level was used to compare means of individual accessions and means of groups of accessions. Four resistance classes were designed in the four testing methods: class A (resistant); class B (moderately resistant); class C (moderately susceptible); class D (susceptible). Each accession was assigned to one of the classes according to their average disease score. Subsequently, the overall level of resistance of each accession was determined by compiling the performance recorded in each of four testing methods as follows: Susceptible ('S') = C or D in all the testing methods; Moderately Susceptible ('MS') = more than 50% of C or D in all the testing methods; Moderately Resistant ('MR') = maximum 50% of C or D in all the testing methods; Resistant ('R') = A or B in all the testing methods. Spearman rank correlation (1904) was used to calculate the correlated coefficients at their associated probabilities among the four different testing methods of the study.

RESULTS

Analysis of PPR resistance for each testing methods

Leaf disc test (LDT)

The average disease score (DS) varied between 2.48 (UPA*ICS) and 3.27 (SNK) (Table 2). Considering the four resistance classes of the LDT, only the accessions group UPA*ICS was found in the class B (moderately resistant), the three other ones were all grouped in class C (moderately susceptible). In the SNK group, about 24.5

and 62.3% accessions were grouped respectively in B and C resistance classes. The same classes B and C were the most representative in SNK*ICS and UPA*SNK groups, with 18.2 and 72.7% (SNK*ICS), and 25.0 and 50.0% (UPA*SNK) respectively. In the most resistant group UPA*ICS, all the accessions were found in one class (B). The lowest DS were recorded in three resistant accessions: SNK478 (DS=1.6) and SNK111 (DS=1.7) in SNK group; SNK630 (DS=1.3) in UPA*SNK group. Even if two of the three resistant accessions were identified in the SNK group using LDT, the five most susceptible ones were also registered in the same group (SNK 376; 267; 52. 416; 10) with DS ranging from 4.1 to 4.5 (class D).

Detached pod test (DPT)

Following the mean score of their corresponding accessions, the four groups displayed different levels of resistance in this test (Table 3). The SNK group was the less resistant against PPR with an average DS of 5.6, followed by the UPA*SNK group (DS = 5.56, not significantly different to SNK AG at α = 5%). The UPA*ICS group was less susceptible (DS=4.60). The accessions were distributed in different resistant classes according to their average disease score. In the SNK group, most of these accessions belong to the moderately susceptible (47.7%) or the susceptible class (40.0%). In the groups SNK*ICS and UPA*SNK, the most susceptible classes have also registered the great number of accessions. In contrast, half of the accessions were found in the moderately resistant class. No accessions were grouped in the resistant class (class A) in UPA*ICS and UPA*SNK groups.

Successful infection rate on attached pods (%SI)

Amplitude of average DS of the four accessions groups was high. The DS values varied from 11.6 (UPA*ICS) to 77.75 (SNK) (Table 4). However, three out of the four accessions groups belonged to the same resistance class (A). Except for SNK 490; 413 and 64 in SNK group, all the other accessions were classified either as moderately susceptible (class C, 28.3%) or susceptible (class D, 65.2%). Despite that not all accessions of the other groups were tested with this method, all the tested SNK600 accessions were found in the same class (A), being thereby classified as resistant (Table 4). The accession SNK616 had the lowest percentage of successful infections (DS = 4.3), whereas 13 accessions (including SNK10 usually known as the most susceptible SNK accession) recorded an infection rate above 90%.

Field incidence (FI) of PPR (% of rotten pods)

Generally, only the limited number of accessions that are

Table 2. Distribution of the accessions in PPR resistance classes based on their average disease scores in leaf disc test (LDT).

Accession group	Mean score	Resistance class	Number of accessions	Amplitude of variation	Ranking of corresponding accessions (*)
SNK	3.27	A	2	1.6-1.7	SNK 478; 111
		B	13	2.2-2.9	SNK 15; 413; 7; 277; 332; 213; 485; 37; 344; 12; 16; 269; 460
		C	33	3.0-3.9	SNK 461; 13; 32; 505; 31; 223; 422; 504; 33; 46; 404; 459; 467; 30; 203; 377; 488; 181; 417; 29; 48; 109; 64; 343; 415; 479; 414; 450; 506; 17; 480; 392; 65
		D	5	4.1-4.5	SNK 376; 267; 52; 416; 10
SNK600-Tr (SNK*ICS)	3.30	A	-	-	-
		B	2	2.2-2.5	SNK 605; 606
		C	8	3.0-3.9	SNK 609; 610; 600; 602; 607; 613; 603; 608
		D	1	4.29	SNK 601
SNK600-TrxUA (UPA*ICS)	2.48	A	0	-	-
		B	7	2.0-2.8	SNK 618; 616; 623; 619; 624; 615; 620
		C	0	-	-
		D	0	-	-
SNK600-TrxUA (UPA*SNK)	3.08	A	1	1.3	SNK 630
		B	2	2.7-2.9	SNK 626; 632
		C	4	3.0-3.9	SNK 634 ; 633; 628; 631; 627
		D	1	-	-

*Accessions are classified in ascending order (from the most to the least resistant) in each class.

planted in the field allowed us to observe the PPR incidence. The values of infection rate (percentage of rotten pods) varies from 15.74 (UPA*ICS) to 33.59 (SNK) (Table 5). No accessions were found in the susceptible class of the SNK*ICS group, and in the moderately susceptible and susceptible classes in the UPA*ICS and UPA*SNK groups. Very low percentages (<10%) of rotten pods were observed in four accessions: SNK620 (4.6%); SNK416 (7%); SNK608 (8.3%); and SNK619 (8.8%). The accession SNK10 was the most susceptible of the study with 87.5% of their pods rotten. About 90% of the UPA*ICS accessions had less than 25% (class A) compared to the groups SNK (20%); SNK*ICS (36.4%) and UPA*SNK (50%).

Analysis of PPR resistance for overall testing methods

After ranking the different groups based on their PPR resistance values in different testing methods (DPT, LDT, %SI, FI) (Table 6), the different ranks obtained were correlated (Table 7). All the correlation coefficients (r-values) were positive but variable (0.37-0.87) and significant (0.01-0.06). The best rank correlation was

registered between %SI and FI (% of rotten pods) with an r=0.87 (p=0.01). The different laboratory screening methods were also positively correlated: r=0.60 (p=0.03) between DPT and LDT; r=0.61 (p=0.03) between DPT and %SI; and r=0.45 (p=0.05) between LDT and %SI. Correlation between FI and DPT recorded the lowest r-value (0.37, p=0.06).

DISCUSSION

The data obtained from the different methods of assessment clearly show a higher level of resistance of the SNK600 genetic group to PPR (accessions labeled from SNK 600 to 641), which confirms the efficiency of the selection of these genotypes in the 1970s. The data also clearly indicate that among these SNK600 accessions, genotypes issued from crosses with one Upper amazon Forastero parent show a higher level of tolerance than the ones issued from crosses between two Trinitario parents. Anyway, even among the SNK genetic group (accessions labeled from SNK 1 to 599), some clones showed a promising level of resistance to PPR, such as SNK 15, 27, 37, 64, 413, 417, 485 and 490, and are or will be included in the local breeding program.

Table 3. Distribution of the accessions in PPR resistance classes based on their average disease scores in detached pod test (DPT).

Accession group	Resistance class	Number of accessions	Amplitude of variation	Ranking of corresponding accessions*
SNK	A	2	2.0	SNK 27; 417
	B	6	2.6-4.0	SNK 15; 267; 64; 450; 461; 490
	C	31	4.2-6.0	SNK 67;478; 479; 52; 485; 500; 460; 353; 37; 17; 181; 322; 361; 416; 443; 332; 377; 452; 269; 505; 501; 343; 480; 30; 213; 422; 7; 12; 29; 33; 46; 496
	D	26	6.2-8.0	SNK 344; 413; 506; 13; 376; 467; 476; 469; 31; 111; 16; 348; 141; 404; 444; 456; 459; 488; 109; 32; 203; 35; 48; 10; 223; 415
SNK600-Tr (SNK*ICS)	A	1	2.0	SNK 604
	B	1	2.25	SNK 605
	C	8	4.1-6.0	SNK 603; 611; 613; 608; 602; 600; 609; 610
	D	2	6.75-7.0	SNK 607; 601
SNK600-TrxUA (UPA*ICS)	A	0	-	-
	B	4	3.5-4.0	SNK 616; 623; 614; 619
	C	3	5.0-5.9	SNK 620; 622; 615; 624
	D	1	-	-
SNK600-TrxUA (UPA*SNK)	A	0	-	-
	B	1	4.0	SNK 626
	C	3	4.8-5.7	SNK 629; 632; 628
	D	3	6.1-6.7	SNK 633; 627; 625

*Accessions are classified in ascending order (from the most to the least resistant) in each class.

The results of the study have shown that there is a strong genetic correlation between the level of PPR resistance and the group of accessions evaluated. The group SNK basically composed of accessions from local Trinitario origin was less resistant compared to the accessions which are hybrids between local SNK and introduced Trinitario or Upper Amazon Forastero (UAF). Low average resistance of pure local Trinitario genotypes compared to the introduced ones such as ICS was already detected by Efombagn et al. (2004) during field tests. Hybrid accessions with UAF genes were averagely more resistant than pure local and introduced Trinitario ones, and averagely more susceptible than the pure UAF accessions. UAF are known to harbor good sources of resistance to PPR (Wood and Lass, 1985) and different breeding strategies aiming at improving PPR resistance in cacao genetic banks have included a great proportion of UAF genotypes. Outstanding UAF derived progenies have always shown low susceptibility to PPR (Zhang 2011). In Cameroon a particular type of SNK accessions (so-called SNK600) was developed after hybridization with UAF (as defined in Table 1). The recent use of some of these accessions as parents of hybrid progenies in field trials already shows good levels of resistance.

Trinitario genotypes are known to perform well regarding agronomic traits such yield components. A great part of Trinitario accessions has good bean size. As consequence for breeding strategy in Cameroon, cacao cultivars deriving from hybrids between UAF and Trinitario were developed and released through seed gardens (Efombagn et al., 2009). However, PPR incidence in farmers' field remains high. This suggests that new resistant progenitors should be detected within the available germplasm in addition to the genotypes recently introduced through international cacao quarantines. The present study has contributed to confirm that some locally-selected accessions may be used as progenitors to developed new cacao hybrid or clonal cultivars with good level of resistance to PPR. Some SNK accessions which were poorly used in the past in varietal and seed production (such as SNK 413 and 490) should be intensively used in variety trials prior to their selection as progenitors in the future seed gardens.

As consequence in the breeding process, cacao genetic bank deriving from farmers' germplasm could enhance the level of genetic diversity as well the sources of resistance to PPR. Field surveys carried out by Pokou

Table 4. Distribution of the accessions in PPR resistance classes based on their average infection rates during the counting of successful infections (%SI) on attached pods.

Accession group	Mean score	Resistance class	Number of accessions	Amplitude of variation	Ranking of corresponding accessions
SNK	77.53	A	1	10.3	SNK 490
		B	2	36.0-45.9	SNK 413; 64
		C	13	55.0-74.0	SNK 267; 416; 32; 180; 213; 223; 461; 30; 67; 16; 15; 181; 450
		D	30	76.5-100	SNK 476; 269; 452; 37; 456; 48; 46; 12; 376; 478; 111; 7; 422; 344; 460; 332; 488; 52; 27; 10; 459; 13; 109; 343; 203; 277; 467; 33; 348; 415
SNK600-Tr (SNK*ICS)	14.72	A	4	6.3-19.2	SNK 600; 610; 611; 601
		B	0	-	-
		C	0	-	-
		D	0	-	-
SNK600-TrxUA (UPA*ICS)	11.6	A	4	4.3-16.7	SNK 616; 614; 618; 615
		B	0	-	-
		C	0	-	-
		D	0	-	-
SNK600-TrxUA (UPA*SNK)	12.66	A	7	4.5-25.0	SNK 625; 633; 629; 626; 628; 631; 627
		B	0	-	-
		C	0	-	-
		D	0	-	-

*Accessions are classified in ascending order (from the most to the least resistant) in each class.

et al. (2008) in Côte d'Ivoire, Opoku et al. (2011) in Ghana, Aikpokpodion et al. (2010) in Nigeria and Efombagn et al. (2007) in Cameroon have enable the detection of great sources of resistance against *Phytophthora* sp., causal agent of PPR. The exploitation of the local cacao genetic bank such as SNK accessions is rather supported by the fact the introduction of cacao elite clones through appropriate international quarantines become more and more difficult, due to the costs linked to management of that quarantine facilities. In addition, direct exchanges of cacao genetic material may harbor serious risks of disease transmission, given the fact all the five major diseases known in cacao cultivation worldwide are not distributed in all producing countries. The direct use of the best locally-selected accessions as released clonal variety might speed up the breeding achievements of this genetic material, regarding the constant increase in demand for high performing cacao cultivars in the country. The best PPR-resistant SNK accessions could be immediately distributed to the cacao farmers through budwood gardens or available cacao genetic banks, which therefore could shorten the time needed to produce planting material. However, PPR is not the only constraint of cacao production in Cameroon as well as in the others producing countries. In the

selection process of PPR-resistant clonal material, emphasis should also be placed on the local selections with good yield performance (bean size, pods index, number of pods per tree).

Ranking of accessions groups were correlated to confirm the stability of the resistance patterns independently of the testing methods used. All the correlations were found positive and significant, confirming the reliability among laboratory and field screening approaches as identified earlier by Tahi et al. (2000). Other earlier studies including some SNK accessions have shown good correlation between leaf discs tests and attached pod resistance tests and among between leaf disc tests, detached pod tests and field observations of PPR incidence (Efombagn et al., 2011). The use of additional SNK accessions as potential progenitors raises the need to combine the results obtained with leaves and pods using screening tests or field observations. Correlations between pod testing methods (DPT and %SI) were positive but lower compared to the other r-values of the study. This due to the fact that the method of counting the number of successful infections developed in seventies (Blaha and Lotode, 1976), were improved by Iwaro et al. (2000) by describing the levels of the PPR evolution on pods

Table 5. Distribution of the accessions in PPR resistance classes based on their average infection rates in field observations (FI).

Accession group	Mean score	Resistance class	Accessions number (%)	Amplitude of variation	Ranking of corresponding accessions
SNK	33.59	A	2 (20.0)	7.0-11.6	SNK 37; 64; 416
		B	3 (30.0)	23.4-38.9	SNK 13; 30; 413
		C	3 (30.0)	41.0-47.6	SNK 109; 16
		D	2 (20.0)	74.6-87.5	SNK 48; 10
SNK600-Tr (SNK*ICS)	27.02	A	4 (36.4)	8.3-18.0	SNK 600; 602; 607; 608
		B	4 (36.4)	20.0-32.6	SNK 610; 609; 613; 611
		C	3 (27.2)	40.5-50.3	SNK 604; 605; 603
		D	0	-	-
SNK600-TrxUA (UPA*ICS)	15.74	A	9 (90.0)	4.6-23.6	SNK 620; 619; 614; 622; 615; 621; 624; 623; 616
		B	1 (10.0)	27.5	SNK 618
		C	0	-	-
		D	0	-	-
SNK600-TrxUA (UPA*SNK)	17.04	A	3 (50.0)	4.7-14.4	SNK 625; 633; 627
		B	3 (50.0)	23.2-2701	SNK 632; 628; 630
		C	0	-	-
		D	0	-	-

*Accessions are classified in ascending order (from the most to the least resistant) in each class.

Table 6. Ranking of genetic and accessions groups following their average disease scores or infection rates in all the four testing methods.

Genetic group	Accession group	Leaf disc test (LDT)	Detached pod test (DPT)	Infected pod (%SI)	Incidence of PPR in the field (FI)
SNK	-	3.27^{1a3} $(53)^2$	5.60b (66)	77.53a (46)	37.58^a (10)
SNK600	All	2.99^b (26)	4.96^b(27)	13.8b (15)	20.62^b (27)
	SNK*ICS	3.30^a (11)	4.86^a (12)	14.72b (4)	27.02^{ab} (11)
	UPA*SNK	3.07^a (7)	5.56^a (7)	14.47b (7)	15.74^b (10)
	UPA*ICS	2.48^b(8)	4.60^a (8)	11.60b (4)	17.04^b (6)

[1],PPR score [2]number of accessions; [3]different letters in the same column indicate significant differences at 5% probability among accessions groups tested with the same method (Student-Newman-Keuls test).

(symptoms) classified in eight stages in the PPR detached pod test rating scale. However, rankings of accessions were different from testing method to another. This may due to the fact that screening methods that use detached structures from the plant tend to miss any possible systemic signaling or reaction by the plant, which would be a good reason why *in vivo* (e.g pods or leaves attached to the plants) tests tend to be more reliable. Moreover, the overall interaction of the plant and the pathogen with the environment and its variations is not considered in detached-structures tests.

Besides screening methods and estimation of pod rot rates, other factors should be taken into account as PPR resistance components during the selection process of SNK genetic material. Berry and Cilas (1994) found that the length of fruiting cycle of Trinitario material such as SNK may contribute to their poor performance in the field because the fruit development and maturation stages coincide with the peak of PPR severity. Escape mechanisms developed by cacao genotypes such as upper Amazon accessions could have an influence on the level of field resistance to PPR as shown by Efombagn et al. (2004) in Cameroon in a study which included very few SNK accessions. Spatial distribution of pods on a cacao tree (Ndoumbè et al., 2002) should also been investigated in locally-selected cacao accessions regarding the high PPR pressure in cacao plantations in Cameroon.

Table 7. Rank correlations among accessions groups in different testing methods.

Groups	LDT	DPT	%SI	FI
LDT	-	0.60[1](0.03)[2]	0.61 (0.03)	0.37(0.06)
DPT		-	0.45 (0.05)	0.65[1] (0.03)
%SI			-	0.87[1] (0.01)
FI				-

[1]Correlation coefficient, [2]Probability.

The composition of groups of accessions tested with different screening methods do not interfere on the results founds in resistance classes. During the assessment of resistance to *Ceratocystis cacaofunesta* in cacao genotypes, Sanches et al. (2008) found that the clones from the CEPEC collection in Brazil were grouped as resistant, moderately resistant and susceptible. This was also observed in our study with the SNK and SNK600 groups of accessions which displayed different levels of resistance.

ACKNOWLEDGEMENTS

The authors thank all the field and laboratory technicians of the Institute of Agricultural Research for Development, IRAD, who actively participate in data collection, maintenance of field trials and genetic banks, and artificial screening tests.

REFERENCES

Aikpokpodion PO, Kolesnikova-Allen M, Adetimirin VO, Guiltinan MJ, Eskes AB, Motamayor JC, Schnell RC (2010). Population structure and molecular characterization of Nigerian field genebank collections of cacao, *Theobroma cacao* L. Silvae Genet. 59(6):273-285.

Bartley BGD (2005). The genetic diversity of cocoa and its utilization. CABI Publishing (ed), Wallingford, UK. P. 341.

Berry D, Cilas C (1994). Etude génétique de la réaction à la pourriture brune des cabosses chez des cacaoyers issus d'un plan de croisements diallèle. Agronomie 14:599-609.

Blaha G, Lotodé R (1976). Un critère primordial de sélection du cacaoyer au Cameroun : la résistance à la pourriture brune des cabosses (*Phytophthora palmivora*). Variations des réactions à la maladie en liaison avec les données écologiques et l'état physiologique des fruits. Café, Cacao, Thé 20(2):97-116.

Braudeau J, Grimaldi E, Lavabre E (1952). Rapport Annuel de la Station du Cacaoyer de Nkoemvone (IRA). In : Rapport Annuel du Service de l'Agriculture, Territoire du Cameroun. pp. 56-112.

Cochran WG, Cox GM (1957). Experimental Designs. 2nd edn. London: John Wiley.

Despréaux D, Clément D, Partiot M (1989). La pourriture brune des cabosses du cacaoyer au Cameroun: mise en évidence d'un caractère de résistance au champ. Agronomie 9:683-691.

Efombagn MIB, Bieysse D, Nyassé S, Eskes AB (2011). Selection for resistance to *Phytophthora* pod rot of cocoa (*Theobroma cacao* L.) in Cameroon: Repeatability and reliability of screening tests and field observations. Crop Prot. J. 30:105-110.

Efombagn MIB, Marelli JP, Ducamp M, Cilas C, Nyassé S, Vefonge D (2004). Effect of some fruiting traits on the field resistance of several cocoa (*Theobroma cacao* L.) clones to *Phytophthora megakarya*. J.

Phytopathol 152:557-562.

Efombagn MIB, Nyassé S, Sounigo O, Kolesnikova-Allen A, Eskes AB (2007). Participatory cocoa Selection in Cameroon: *Phytophthora* pod rot resistant accessions identified in farmers' field. Crop Prot. J. 26(10):1467-1473.

Efombagn MIB, Sounigo O, Eskes AB, Motamayor JC, Manzanares-Dauleux MJ, Schnell R, Nyasse S (2009). Parentage analysis and outcrossing patterns in cacao (*Theobroma cacao* L.) farms in Cameroon. Heredity 103(1):46-53.

Iwaro AD, Sreenivasan TN, Butler DR, Umaharan P (2000). Rapid screening for *Phytophthora* pod rot resistance by means of detached pod inoculation. In: Working procedures for cocoa germplasm evaluation and selection. IPGRI, Rome. pp. 109-113.

Ndoumbè NM (2002). Incidence des facteurs ago-écologiques sur la pourriture brune des fruits du cacaoyer au Cameroun : contribution à la mise en place d'un modèle d'avertissement agricole. Thèse de doctorat, INAPG, Paris, France.

Nyadanu D, Assuah M, Adomako B, Asiama Y, Apoku I, Adu-Ampomah Y (2009). Repeatability of leaf disc test in assessing resistance levels of international clone trial selections of Phytophthora megakarya and Phytophthora palmivora infection in Ghana. Int. J. Trop. Agric. Food Syst. 3:20-26.

Nyassé S, Cilas C, Hérail C, Blaha G (1995). Leaf inoculation as early screening test for cocoa (*Theobroma cacao* L.) resistance to *Phytophthora* black pod disease. Crop Prot. 14:657-663.

Opoku SY, Dadzie MA, Padi FK, Opoku IY, Adu-Ampomah Y, Adomako B (2011). Selection of new varieties on-farm and on-station in Ghana. In: CFC Technical Report N° 59. pp. 73-79.

Pokou ND, N'Goran JAK, Kébé I, Eskes A, Tahi M, Sangaré A (2008). Levels of resistance to *Phytophthora* pod rot in cocoa accessions selected on-farm in Côte d'Ivoire. Crop Prot. 27(3-5):302-305.

Sanches CLG, Pinto LRM, Pomella AWV, Silva DVM., Loguercio LL (2008). Assessment of resistance to *Ceratocystis cacaofunesta* in cacao genotypes. Eur. J. Plant Pathol. 122:517-528.

SAS (2000). Statistical Analysis System. SAS/STAT Software Changes and Enhancements through release 6.12. SAS Institute Inc., Cary.N.C., USA.

Spearman C (1904). "General intelligence" objectively determined and measured. Am. J. Psychol. 15:201-293.

Tahi M, Kebe I, Eskes AB, Ouattara S, Sangare A, Mondeil F (2000). Rapid screening of cocoa genotypes for field resistance to *Phytophthora palmivora* using leaves, twigs and roots. Eur. J. Plant Pathol. 106:87-94.

Wheeler BE (1992) Assessment of resistance to major cocoa diseases. In: Proceedings of the International Workshop on Conservation, Characterization and Cocoa Genetic Resources in the 21st Century. Cocoa Res. Unit, University of the West Indies, Trinidad and Tobago. pp. 139-145.

Wood GAR, Lass RA (1985). Cocoa. Longman. 4th Ed. Tropical Agriculture series, London, UK, P. 620.

Zhang D (2011). Cocoa breeding. In: W ild Crop Relative: Genome and Breeding Resources. Plantation and ornamental crops. Chittarajan Kole (Ed). P. 303.

9

Growth and yield of cucumber (*Cucumis sativus* L.) as influenced by farmyard manure and inorganic fertilizer

E. K. Eifediyi* and S. U. Remison

Department of Crop Science, Ambrose Alli University, P. M. B. 14, Ekpoma, Edo State, Nigeria.

The growth and yield of Ashley variety of cucumber in response to the effect of farmyard manure and inorganic fertilizer NPK 20:10:10 was evaluated at the Teaching and Research Farm of the Ambrose Alli University, Ekpoma, Nigeria Lat 6°45'N and long 6°08'E. The farmyard manure was applied at the rates of 0, 5 and 10 t/ha and the inorganic fertilizer at 0, 100, 200, 300 and 400 kg/ha. The layout was a 3 × 5 factorial scheme with three replicates. The combined rates of farmyard manure at 10 t/ha × 400 kg/ha fertilizer increased the growth characters such as the vine length and the number of leaves. At 8 weeks after planting (WAP), the application of 10 t/ha of farmyard manure × 400 kg/ha of fertilizer gave the longest vine length of 276.93 cm and the highest number of leaves. The fruit length, fruit girth, fruit weight per plant and fruit weight per hectare were significantly influenced by the application of farmyard manure × fertilizer. The highest weight of 2.43 kg per plant and yield per hectare of 43,259 kg/ha were obtained with 10 t/ha farmyard manure and 400 kg/ha of fertilizer which were 166.42% higher than the control.

Key words: Cucumber, farmyard manure, inorganic fertilizer, growth and yield.

INTRODUCTION

Cucumber *(Cucumis sativus* L.) is an important vegetable and one of the most popular members of the Cucurbitaceae family (Lower and Edwards, 1986; Thoa, 1998). It is thought to be one of the oldest vegetables cultivated by man with historical records dating back 5,000 years (Wehner and Guner, 2004). The crop is the fourth most important vegetable after tomato, cabbage and onion in Asia (Tatlioglu, 1997), the second most important vegetable crop after tomato in Western Europe (Phu, 1997). In tropical Africa, its place has not been ranked because of limited use.

Fertile soils are used for the cultivation of cucumber; infertile soils result in bitter and misshapen fruits which are often rejected by consumers. Bush fallowing has been an efficient, balanced and sustainable agricultural system for soil productivity and fertility restoration in the tropics (Ayoola and Adeniran, 2006), but as a result of increase in the population, the fallowing periods have decreased from ten years to three years and this has had an adverse effect on the fertility restoration leading to poor yields of crops. Therefore, the use of external inputs

in the form of farmyard manures and fertilizer has become imperative. Farmyard manure has been used as a soil conditioner since ancient times and its benefit have not been fully harnessed due to large quantities required in order to satisfy the nutritional needs of crops (Makinde et al., 2007). The need for renewable forms of energy and reduced cost of fertilizing crops, have revived the use of organic manures worldwide (Ayoola and Adeniran, 2006). Improvement in environmental conditions and public health are important reasons for advocating increased use of organic materials (Ojeniyi, 2000; Maritus and Vleic, 2001). However, because it is bulky, the cost of transportation and handling constitute a constraint to its use by peasant farmers.

Farmyard manure release nutrients slowly and steadily and activates soil microbial biomass (Ayuso et al., 1996; Belay et al., 2001). Organic manures can sustain cropping systems through better nutrient recycling and improvement of soil physical attributes (El-Shakweer et al., 1998). The use of inorganic fertilizer has not been helpful under intensive agriculture because of its high cost and it is often associated with reduced crop yields, soil degradation, nutrient imbalance and acidity (Kang and Juo, 1980; Obi and Ebo, 1995).

The complementary use of organic and inorganic fertilizers has been recommended for sustenance of long

*Corresponding author. E-mail: keveifediyi@yahoo.com.

Table 1. Chemical analysis of soil and farmyard manure.

Parameter	Soil sample	Farmyard manure
pH (in 2: 1 water)	5.80	6.0
Organic matter content	24.15 g/kg	53.73 g/kg
Organic carbon	14.0 g/kg	31.13 g/kg
Nitrogen	1.290 g/kg	2.23 g/kg
Ca	8.80 mg/kg	39.08 cmol/kg
Mg	0.96 mg/kg	4.32 cmol/kg
Available P	10.40 cmol/kg	61.29 mg/kg
Exchangeable K	0.29 cmol/kg	2.53 cmol/kg

term cropping in the tropics (Ipimoroti et al., 2002). Fuchs et al. (1970) reported that nutrients from mineral fertilizers enhance the establishment of crops while those from mineralization of organic manures promoted yield when both fertilizers were combined. Titiloye (1982) reported that the most satisfactory method of increasing maize yield was by judicious combination of organic wastes and inorganic fertilizers. It has been observed that addition of manure increases the soil water holding capacity and this means that nutrients would be made more available to crops where manures have been added to the soil (Costa et al., 1991). Murwira and Kirchman (1993) observed that nutrient use efficiency might be increased through the combination of manure and inorganic fertilizer. This study was therefore conducted to investigate the effects of varying rates of farmyard manure and inorganic fertilizers on the growth and yield of cucumber.

MATERIALS AND METHODS

Field experiment was conducted at the Teaching and Research Farm of the Ambrose Alli University Ekpoma on Lat. 6°45'N and Long.6° 08' E in a forest, savanna transition zone of Nigeria. The area is characterised by a bimodal rainfall pattern with a long rainy season which starts in late March and the short rainy period extends from September to late October after a dry spell in August. The soil order is an ultisol and the site is classified locally as kulfo series (Moss, 1957).

The site was left to fallow for three years after it was cropped to maize (Zea mays), yam (Dioscorea sp.) and cassava (Manihot sp.) for two years prior to the establishment of the experiment. A composite soil sample was collected from 0 - 30 cm depth prior to planting of cucumber before the farmyard manure incorporation to determine the pH and the nutrient status of the soil. Soil pH was analyzed by 1:2 in H_2O, total N content was determined by Kjeldahl method (Bremner, 1965); available phosphorus was analyzed using the modified Walkley and Black (Nelson and Sommers, 1982). The farmyard manure was collected from a deep litter pen of the Poultry Unit of the Teaching and Research Farm of the Ambrose Alli University Ekpoma and left to decompose for three months. The NPK fertilizer was bought from the Edo State Ministry of Agriculture and Natural Resources. Chemical analyses of the soil and farmyard manure used are presented in Table 1. The result indicated a soil pH of 5.86 (slightly acidic), 24.15 g/kg organic matter (medium), organic carbon 14.00 g/kg, nitrogen 1.29 g/kg (low) available P

(Bray Pi) 10.40cmol/kg and exchangeable K 1.12 cmol/kg.

The experiment commenced on the 23rd of August, 2007 by planting of two seeds of "Ashley" variety of cucumber at a spacing of 75 by 75 cm and later thinned after two weeks to one seedling per stand to give a population of 17,777.8 plants per ha. This is the current recommended density of planting cucumber in Nigeria. The treatments used were three levels of farmyard manure (0, 5 and 10 t/ha) and five of NPK 20: 10:10 (0,100, 200, 300 and 400 kg/ha). The experiment was laid out in a 3 × 5 factorial scheme with three replicates. The plot size was 3.75 m × 3.75 m with 2 m pathways. The farmyard manure was uniformly spread on the plots and a West Indian hoe was used to turn the manure into the soil two weeks before planting. Two weeks after planting, NPK 20:10:10 was applied at the rate of 0, 100, 200, 300 and 400 kg/ha to the plots; this is the period recommended for the application of NPK fertilizer in this zone. Manual weeding was carried out at 3 and 5 weeks after planting. Insect pests were controlled with lamdacyahalothrin as Karate (2 L per ha.) at biweekly intervals for effective insect control.

Growth parameters were assessed at 4, 6 and 8 weeks after planting. Cucumber vine length was measured by using a flexible tape rule. Number of leaves was assessed by visual count of the green leaves. At every harvest the fruit girth was assessed by using a Vernier calliper, the fruit length was measured by using a flexible tape before the fruits were weighed using a 10 kg scale. The cumulative weights of the entire harvests (10 times) were summed up for data analysis using SAS version 17 software.

RESULTS AND DISCUSSION

Vine length and number of leaves

Cucumber vine length increased significantly ($P < 0.05$) with the application of farmyard manure and fertilizer at 4 weeks after planting (WAP). The application of a combination of 10 t/ha of farmyard manure and fertilizer at the rate of 400 kg/ha gave the longest vine of (28.98 cm) and the control produced the shortest vine length (20.56 cm).

At 6WAP, the longest vine of 142.56 cm was observed in the plot treated with only 400 kg/ha NPK and the control had the shortest vine of (119.20 cm).

The differences between the treatment means at this stage of growth were not significant. The application of fertilizer on cucumber had significant ($P < 0.01$) effect on the vine length but no significant FYM and FYM × fertilizer

Table 2. Effect of farmyard manure and inorganic fertilizer rates on the vine length of cucumber at 4, 6 and 8 WAP.

FYM rates (t/ha)	4 WAP Fertilizer rates (kg/ha)						6 WAP Fertilizer rates (kg/ha)						8 WAP Fertilizer rates (kg/ha)					
	0	100	200	300	400	Mean	0	100	200	300	400	Mean	0	100	200	300	400	Mean
0	20.56	24.21	25.74	26.99	27.36	24.97	119.20	126.03	131.19	136.45	142.56	131.08	187.21	219.33	237.48	262.84	269.29	235.23
5	22.00	26.31	26.40	27.19	27.87	25.96	129.06	130.31	125.94	124.53	128.06	127.59	230.79	241.72	246.72	246.17	261.13	239.28
10	25.54	28.17	28.49	28.65	28.98	27.96	129.01	130.82	129.11	127.96	133.16	130.16	225.15	239.01	249.86	256.66	276.93	249.52
Mean	22.70	26.23	26.88	27.61	28.07		125.75	129.05	130.26	130.12	133.66		209.65	229.71	243.02	225.22	269.12	

LSD: FYM – 0.541* Fertilizer 0.703* FYM × Fert 1.212**; FYM – 4.531[NS], Fertilizer 0.707[NS], FYM × Fert. 1.213[NS], Fertilizer 0.707[NS]; FYM – 13.761[NS] Fertilizer 17.774** FYM × Fert 7.950[NS].

Table 3. Effects of farmyard manure and inorganic fertilizer rates on the number of leaves of cucumber at 4, 6 and 8 WAP.

FYM rates (t/ha)	4WAP Fertilizer rates (kg/ha)						6WAP Fertilizer rates (kg/ha)						8WAP Fertilizer rates (kg/ha)					
	0	100	200	300	400	Mean	0	100	200	300	400	Mean	0	100	200	300	400	Mean
0	5.67	5.98	6.07	6.30	6.43	6.06	30.03	31.08	31.57	34.33	30.39	31.50	42.31	47.58	49.55	59.13	60.12	51.74
5	5.84	6.57	7.02	7.43	7.68	6.91	33.68	33.32	32.19	31.26	32.13	32.52	58.55	59.44	60.32	60.57	61.26	60.03
10	6.28	6.61	7.25	7.73	8.43	7.26	31.03	34.84	35.47	34.75	35.33	31.58	61.26	58.02	59.50	61.22	61.73	60.33
Mean	5.92	6.39	6.78	7.15	7.51		31.53	33.08	33.08	33.45	32.65		54.04	55.01	56.46	60.31	61.04	

LSD: FYM 6.727*, Fertilizer 0.934**, FYM × Fert NS, FYM NS, Fertilizer NS, FYM × Fert. NS, Fertilizer NS; FYM 2.671* Fertilizer 3.444**, FYM × Fert NS.

interaction at 8 WAP was observed (Table 2). The mean number of leaves per plant was significantly influenced by farmyard manure (P < 0.05) and fertilizer (P < 0.01) but no significant FYM × Fertilizer interaction at 4 WAP was observed. The highest number of leaves (8.43) was produced by a combination of farmyard manure at 10 t/ha and 400 kg/ha of fertilizer while the least number of leaves (5.68) was produced by the untreated plots (Table 3). At 6 WAP, the influence of farmyard manure, fertilizer and their interactions was not significant but the highest number of leaves (35.33) was produced by the 10 t/ha farmyard manure × 400 kg/ha fertilizer treated plots and the control produced the least number of leaves (30.03).

At 8 WAP, the mean number of leaves was influenced by farmyard manure (P < 0.05) and fertilizer (P < 0.01) but no significant FYM × fertilizer interaction. The highest number of leaves was produced by the treatment in which a combination of 10 t/ha of farmyard manure and 400 kg/ha of fertilizer was used and the control produced the least number of leaves (Table 3). The cucumber growth parameters were strongly influenced by the combined application of farmyard manure and fertilizer and yield was highest with the combination. The cucumber plant had enough nutrients for rapid growth and development considering the composition of the farmyard manure which was incorporated into the soil during land preparation. It was observed that the higher the nutrients applied, the higher the values of the vine length and number of leaves produced per plant. The vigorous growth in cucumber which was experienced during the growing period as evidenced in the vine length and number of leaves produced per plant (Tables 2 and 3) was in agreement with Fuchs et al. (1970) who reported that nutrients from mineral fertilizers enhanced the establishment of crops while those from the mineralization of organic matter promoted yield when manures and fertilizers were combined.

Yield and yield components

The fruit length of cucumber was significantly

Table 4. Effects of farmyard manure and inorganic fertilizer rates on the fruit length and girth of cucumber.

FYM rates (t/ha)	Fruit length (cm) Fertilizer rates (kg/ha)						Fruit girth (cm) Fertilizer rates (kg/ha)				
	0	100	200	300	400	Mean	100	200	300	400	Mean
0	14.20	14.27	14.54	15.18	15.39	14.72	4.28	4.42	4.49	4.62	4.39
5	14.76	14.93	15.30	15.40	15.42	15.16	4.97	5.02	5.06	5.48	5.08
10	15.40	15.55	15.53	15.55	15.71	15.55	5.59	5.64	5.70	5.59	5.60
Mean	14.79	14.91	15.12	15.38	15.51		4.95	5.03	5.08	5.23	

LSD: FYM 0.282*, Fertilizer 0.331**, FYM × Fert NS; FYM 0.171*, Fertilizer NS FYM × Fert. NS.

Table 5. Effects of farmyard manure and inorganic fertilizer rates on fruit weight per plant and yield/ha of cucumber.

FYM rates (t/ha)	Fruit wt. (kg/plant) Fertilizer rates (kg/ha)						Fruit yield (kg/ha) Fertilizer rates (kg/ha)					
	0	100	200	300	400	Mean	0	100	200	300	400	Mean
0	0.91	1.17	1.47	1.65	1.62	1.36	16237.01	20740.73	26133.32	29392.57	28859.24	24272.57
5	1.21	1.47	1.61	2.22	2.23	1.75	21511.10	226192.58	28681.46	39466.65	39644.43	31099.24
10	1.19	2.06	2.27	2.30	2.43	2.05	21155.54	36681.46	40414.79	40888.87	43259.24	36479.24
Mean	1.10	1.57	1.78	2.30	2.09		19634.55	27871.59	31743.19	36582.70	37254.30	

LSD FYM 0.234 LSD FYM 633.252* Fertilizer 0.251** Fertilizer 817.520* Fertilizer 0.408** FYM × Fertilizer 817.520* FYM × Fertilizer 0.408** FYM × Fert 1415.981**.

enhanced by the application of farmyard manure (P < 0.05) and fertilizer (P < 0.01) but no significant farmyard manure × fertilizer interaction. The longest fruit of 15.71 cm was observed in the 10 t/ha farmyard manure × the 400 kg/ha fertilizer combination and the shortest fruit of 14.20 cm was observed in the control (Table 4).

The fruit weight per plant was significantly influenced by farmyard manure (P < 0.05), fertilizer (P < 0.01) and their interaction (P < 0.01). The highest fruit weight (2.43 kg) per plant was observed in the treatments receiving farmyard manure at the rate of 10 t/ha and 400 kg/ha of fertilizer and the least value of 0.90 kg was observed in the control.

The cucumber fruit weight per hectare was significantly influenced by the application of farmyard manure (P < 0.05), fertilizer (P < 0.05) and their interaction (P < 0.01). The highest fruit weight per hectare of 43,259.24 kg/ha was observed in the 10 t/ha farmyard manure and 400 kg/ha fertilizer combination and the least yield of 16237.01 kg/ha was observed in the untreated control. The combination of 10 t/ha farmyard manure and 400 kg/ha fertilizer gave a yield of 166.42% over the control. The interaction thus showed that response to FYM was more at higher rates of NPK application. The farmyard manure applied at 10 t/ha combined with 400 kg/ha fertilizer was just enough to satisfy the nutritional requirements of cucumber plant. This was evident in the significant yield experienced in the treatment over the other treatments. The yield (Table 5) from the experiment was in agreement with the report of Murwira and Kirchman (1993) who found increased yield of crops through the combination of farmyard manure and inorganic fertilizer and the findings of Titiloye (1982) who reported that the best way to increase maize yield was by the combination of organic wastes and inorganic fertilizer. The addition of farmyard manure increased the water holding capacity and reduced the incidence of erosion thereby making more nutrients available to the soil (Costa et al., 1991). The highest grain yield of rice has been obtained when farmyard manure was applied at 10 t/ha combined with 120:60:45 N: P_2O_5 and K_2O ha^{-1} (Satyanaraya et al., 2002). Bayu et al. (2006) also reported that sorghum yield increased when 5 t/ha of farmyard manure was combined with 20 kg N + 10 kg P ha^{-1}. Makinde et al. (2007) reported increased melon growth and optimum yield with organo-mineral fertilizer at 4 t/ha or the application inorganic fertilizer at 41 kg N+ 20kg P. The combination of farmyard manure × inorganic fertilizer significantly influenced cucumber yields compared to farmyard manure and fertilizer alone especially at higher rates of application (Table 5). The increase in yield of cucumber could be attributed to the fact that nutrients were more readily available when organic and inorganic fertilizers were combined.

This study has clearly shown that cucumber growth can be promoted by the combined application of farmyard manure and inorganic fertilizer and farmers will be encouraged to apply the combination in their cropping practices.

REFERENCES

Ayoola OT, Adeniran ON (2006). Influence of poultry manure and NPK fertilizer on yield and yield components of crops under different cropping systems in South West Nigeria. Afr. J. Biotechnol., 5 1336-1392.

Ayuso MA, Pascal. JA, Garcia C, Hernandez T (1996). Evaluation of urban wastes for urban agricultural use. Soil Sci. Plant Nutr., 42:105-111.

Bayu W, Bethman NFG, Hammes PS, Alemu G (2006). Effects of farmyard manure and inorganic fertilizer on sorghum growth, yield and nitrogen use in semi-arid area of Eithiopia. J. Plant Nutr., 29:391-407.

Belay A, Classens AS, Wehner FC, De Beer JM (2001). Influence of residual manure on selected nutrient elements and microbial composition of soil under long term crop rotation. S. Afr. J. Plant Soil, 18:1-6.

Costa FC, Hernadez GC, Polo A (1991). Residuos organicos urbanicos in manejoy utilizacion CSIC Munica p. 181.

El-Shakweer MHA, El-Sayad EA, Ewees MS (1998). Soil and Plant analysis as a guide for interpretation of the improvement efficiency of organic conditioners added to different soils in Egypt. Communicationin soil science and plant analysis. 29:2067-2088.

Fuchs W, Rauch K, Wiche HJ (1970). Effect of organic fertilizer and organo mineral fertilizing on development and yield of cereals. Abrecht- Thaer. Arch., 14:359-366.

Ipimoroti RR, Daniel MA, Obatolu CR (2002). Effect of organic mineral fertilizer on tea growth at Kusuku Mabila Plateau Nigeria. Moor J. Agric. Res., 3: 180-183.

Kang BT, Juo ASR (1980). Management of low activity clay soils in tropical Africa for food crops production pp129-133 In: Terry ER, KA Oduro and S Caveness (eds.) Tropical Roots crops. Research strategies for the 1980s.Ottawa, Ontario IDRC.

Lower RL, Edwards MD (1986). Cucumber breeding In: M J Basset (ed.). Breeding vegetables crops. Westport, Connecticut USA: AVI Publishing Co. pp. 173-203.

Makinde EA, Ayoola OT, Akande MO (2007). Effects of organo-mineral fertilizer application on the growth and yield of egusi melon. Australian J. Basic Appl. Sci., 1:15-19.

Maritus CHT, Vleic PLG (2001). The management of organic matter in tropical soils. What are the priorities? Nutrient cycling in Agro-ecosystems. 61:1-6.

Moss RP (1957). Report on the classification of soil found over sedimentary rocks in Western Nigeria. Departmental Report. Research Div., MANR, Ibadan, Nigeria. p. 88.

Murwira HK, Kirchman AK (1993). Carbon and nitrogen mineralization of cattle manures subjected to different treatment in Zimbabwean and Swedish soils: In Mulongoy K and Merckr KR (editors) Soil organic matter dynamics and sustainability of tropical agriculture. pp. 189-198.

Nelson DW, Sommers LE (1982). Total carbon and organic matter. In Page A.L (editor).Methods of soil analysis. Part 2. 2nd edition. Chemical and Microbiological properties. Agronomy monograph 9, Madison, WI, USA, ASA and SSSA. pp. 149 – 157.

Obi ME, Ebo PO (1995).The effect of different management practices n the soil physical properties and maize production in severely degraded soil in /Southern Nigeria. Biol. Res. Technol., 51: 117-123.

Ojeniyi, S.O. (2000). Effect of goat manure on soil nutrients and okra yield in a rain forest area of Nigeria. Appl. Tropical Agric., 5:20-23.

Phu NT (1997). Nitrogen and potassium effect on cucumber yield. AVI 1996 report, ARC/AVRDC Training Thailand.

Tatlioglu T (1997). Cucumber (Cucumis sativus L.) In: Kailov, G and Bo Bergn,(eds.). Genetic improvement of vegetable crops. Oxford Pergamon Press. pp. 197-227.

Thoa DK (1998). Cucumber seed multiplication and characterisation. AVRDC/ARC Training Thailand.

Titiloye EO (1982). The chemical composition of different sources of organic wastes and effects on growth and yield of maize. Ph.D. thesis University of Ibadan, Nigeria. p. 316.

Wehner TC, Guner N (2004). Growth stage, flowering pattern, yield and harvest date prediction of four types of cucumber tested at 10 planting dates. Proc. xxvi IHC. Advances in Vegetable Breeding (Eds) J.D McCreight and E. J Ryder Acta. Hort., 637, ISHS 2004.

10

Variation in streptomycin-induced bleaching and dark induced senescence of rice (*Oryza sativa*) genotypes and their relationship with yield and adaptability

S. Das*, R. C. Misra, S. K. Sinha and M. C. Pattanaik

Department of Plant Breeding and Genetics, College of Agriculture, Orissa University of Agriculture and Technology, Bhubaneswar, Orissa, India.

Variation in streptomycin sensitivity and dark sensitivity of 36 rice genotypes of 3 different maturity groups was studied. Streptomycin sensitivity and dark sensitivity of rice genotypes were expressed in terms of bleaching index (BI) and senescence index (SI) respectively. Genotypes of each maturity group showed wide variation in their BI / SI value. The objective of this investigation was to find relationship of BI / SI parameter with yield, adaptability and stability in yield performance of rice genotypes. Yield performance of rice genotypes were evaluated over 12 environments. Adaptability and stability analysis were done following linear regression model of Eberhart and Russell and AMMI Stability Value (ASV) of purchase. BI parameter showed positive correlation with yielding ability and deviation from regression and negative correlation with adaptability parameter (b) for all the 3 maturity groups. But SI parameter showed negative correlation with yielding ability and positive correlation with adaptability parameter (b) for all the 3 maturity groups. This experimental study revealed that sensitivity of rice genotypes to SM in terms of BI could be used to predict yielding ability of genotypes and dark sensitivity (SI) could be used to indicate adaptability to rich environments or poor environments. This novel approach may help the breeder in indirect selection of high yielding genotypes and genotypes well adapted to rich or poor environments at an early seedling stage before going for multilocation trials.

Key words: Streptomycin, dark treatment, *Oryza sativa,* adaptability.

INTRODUCTION

Rice (*Oryza sativa*) occupies a pivotal place in Indian agriculture. It is grown in an area of 44.6 million hectares in India with a production of about 90 million tonnes and productivity of 2.07 ton/ha (Economic Survey, 2007). Rice accounts for 43% of food grain production and 55% of cereal production in the country. It is the staple food of more than two-thirds of the population of India and occupies a key position in national food security. All the high yielding varieties of rice released so far for cultivation have not gained equal popularity due to their unstable performance over wide range of environmental conditions. Thus, multilocation testing of genotypes under diverse

agro-ecological conditions for evaluation of yield potential, adaptability and stability is essential before recommending a genotype for release as variety. But multilocation trials need large quantity of seeds, more money, manpower, land, labour and most important of all, time. Though apparently a non-monetary input, time contributes to cost in diverse ways. It would be of great help and immense value if some method (s) could be developed to screen genotypes for their adaptability and stability before taking them to multilocation trials.

The present study aimed at developing some simple, rapid and inexpensive laboratory methods for evaluating adaptability and stability of performance of crop varieties that could be used for a preliminary selection of breeding lines before going for the more expensive multi-location trials. The study took the cue from works of Sinha and Satpathy (1979), Das and Sinha (1992) and Mohapatra

*Corresponding author. E-mail: swarnalata1967@rediffmail.com.

Table 1. Parentage of mid early, mid-late and late rice genotypes.

Name of genotype	Parentage
Mid-early (115-125 days)	
1. OR 1739-47	Sankar/IR 72
2. OR 1916-19	Lalat/Ratna
3. OR 1929-4	OR 929-3-2/RP 2423-108-97
4. OR 1976-11	TRC 87-125//IR 49517/Prana
5. OR 2006-12	Sarathi/IR 36
6. OR 2168-1	IR 36/UPRI 3
7. OR 2172-7	IR 64///IR 72//Jagannath/NCJ 10
8. OR 2200-5	RP 2423-108-97/ORS 199-2
9. Konark	Lalat/OR 135-3-4
10. Lalat	Obs 677/IR 2071// Vikram/W1263
11. Bhoi	Gouri/RP 825-45-1-3

Mid-late (126-140 days)

1. OR 1681-11	Bhoi/Surendra
2. OR 1912-25	Swarna/Lalat
3. OR 1914-8	Swarna/IR 36
4. OR 1964-8	RTN 14-1-1//IR 72
5. OR 1967-15	RTN 14-1-1//IR 49517/OR 1301-32
6. OR 2156-15	Swarna/IR 72
7. OR 2310-12	Swarna/Birupa
8. Pratikshya	Swarna/IR 64
9. Gouri	Rajeswari/Vikram
10. Surendra	OR 158-5/Rasi
11. Gajapati	OR 136-3/IR13429-196-1-20
12. Kharavela	Daya/IR 13240-108-2-2-3
13. MTU 1001	MTU 5249/MTU 7014

Late (145-165 days)

1. OR 1885-16-34	IR 72/Kanchan
2. OR 1898-2-15	Mahalaxmi/OR 633-7
3. OR 1898-3-16	Mahalaxmi/OR 633-7
4. OR 1901-14-32	Manika/IR 72
5. OR 2001-1	RP 1125-606-32/Rambha
6. OR 2109-2	Indravati//IR 72/Salivahan
7. OR 2119-13	Manika/Manasarovar
8. Savitri	Pankaj/Jagannath
9. Salivahan	RP 5-32/Pankaj
10. Mahanadi	IR 19661/Savitri
11. Kanchan	Jajati/Mahsuri
12. Jagabandhu	Savitri/IR 4819-77-3-2//IR 27301-154-3

(1997) who found certain relationship between sensitivity to streptomycin (SM) and dark treatment and adaptation pattern of rice and ragi varieties and indicated the possibility of SM and dark sensitivity serving as aids in preliminary laboratory evaluation of broad adaptation of crop varieties.

So in this investigation, an attempt has been made to find indicators of adaptability and stability of yield performance of rice genotypes of three different maturity groups in terms of their response to streptomycin (SM) induced bleaching and dark induced senescence at early seedling stage.

MATERIALS AND METHODS

Materials of the present study comprised of 36 rice genotypes of 3 different maturity groups (11 mid-early, 13 mid-late and 12 late group). The list of genotypes along with their parentage was presented in Table 1.

Streptomycin (SM) treatment

Streptomycin (SM) is an amino-glycoside antibiotic and it induces bleaching of leaves in seedlings by inhibiting chlorophyll synthesis. For SM treatment, 50 seeds of each genotype were soaked in 5 ml of 500 ppm SM solution for 48 h along with control (seeds soaked only in distilled water).

After 48 h, the treated and control seeds were washed and put on moist blotting paper in petridishes for germination at room temperature. Observations were recorded on SM-induced seedling bleaching on the 9th day using a random sample of 30 seedlings per genotype. Bleaching of seedlings in each genotype was scored in a 0 - 2 scale where 0 for green/normal, 1 for partially bleached and 2 for fully bleached seedlings (Sinha and Satapathy, 1977).

Estimation of bleaching index

SM sensitivity of each genotype was measured in terms of bleaching index and it was calculated following Sinha and Satapathy (1977) as follows.

$$BI = \frac{n_1 \times 0 + n_2 \times 1 + n_3 \times 2}{2N}$$

where:
n_1, n_2 and n_3 are numbers of green, partially bleached and fully bleached seedlings, respectively, in a genotype and
N is total number of seedlings scored for bleaching.

Dark treatment

One hundred seeds of each genotype were soaked in distilled water for 48 h. The soaked seeds were then transferred to blotting papers in petri dishes for germination at room temperature. Forty fully expanded first leaves were excised from 9 days old seedlings and floated on nutrient solution. The nutrient solution was prepared as: 2 ml each of 1 M solutions of KNO_3, KH_2PO_4 and ($MgSO_4$, $7H_2O$) + 3 ml of 1 M solution of $CaCl_2$ + 1 ml of 1 M solution of $FeSO_4$ + 1 ml of micronutrient solution, made to 1 L by adding distilled water. Finally, the excised leaves were incubated in dark at room temperature for 48 h.

For control, the excised leaves were floated on nutrient solution and kept in sun light. Observations were recorded on the senescence (yellowing) pattern of the 40 excised leaves in treated and control samples of each genotype. The leaves in the dark-exposed samples were scored for degree of senescence in a 0 - 2 scale where 0 for green/normal, 1 for partially senesced and 2 for fully senesced excised leaves, was measured by eye estimation. All the excised leaves in control were normal green.

Estimation of dark-induced senescence index (SI)

As a measure of the effect of dark-exposure on chlorophyll development, an index called dark-induced senescence index (SI) based on the senescence scores of the dark-exposed leaves, was calculated for each genotype (following Sinha and Satpathy, 1977). The SI was computed as follows:

$$SI = \frac{n_1 \times 0 + n_2 \times 1 + n_3 \times 2}{2N}$$

Where:

n_1, n_2 and n_3 are numbers of green, partially senesced and fully senesced excised leaves, respectively, in a treatment and

N is total number of excised leaves scored for senescence.

The experiments on streptomycin and dark were repeated four times at intervals of 15 -20 days during the period July 2004 to April 2005. Analysis of variance was carried out on SI and BI in completely randomized design using repetitions of experiments as replications. The significance of differences among genotypes for this parameter was tested by F-test and t-test through critical difference (CD). Means of genotypes for SI and BI were the averages over the four replications.

Field evaluation

Three multi-location-year trials were conducted for the three duration groups of rice. The genotypes were evaluated at four different locations of Orissa (Bhubaneswar, Chiplima, Jeypore and Ranital), over three years during 2003 - 2005 in 'kharif' season using a randomized block design with three replications. For all trials, nursery sowing was done during last week of June to 1st week of July. Twenty-five to thirty days old seedlings were transplanted with 20 × 15 cm spacing and 2 seedlings per hill. In each trial, the plot size was 2 × 3 m containing 10 rows of 3 m length each. Normal cultural practices and plant protection measures were followed in each trial to raise the crop. In all trials, data were recorded on net plot grain yield. Stability analysis was performed following the linear regression model of Eberhart and Russell (1966) and AMMI Stability Value (ASV) of Purchase (1997).

Analysis of relationship of BI and SI with yield, adaptability and stability

Correlation study was done for each maturity groups separately to find out the relationship of SI/BI parameter with yield, adaptability and stability parameters. Correlation study by combining each maturity group is not possible as the yield level of each maturity group is different. Therefore a 2 x 2 contingency classification method was followed to analyse the relationship of SI/BI parameter with yield, adaptability and stability of all the rice genotypes irrespective of their maturity duration and done as follows. For SM and dark sensitivity, the genotypes in a trial were classified into two classes, those having above average BI/SI value as highly sensitive (HS) and those having below average BI/SI value as less sensitive (LS). Similarly, for yield, adaptability and each stability parameter, the genotypes in a trial were classified into 2 classes, those having above average value for the parameter as 'above average' and those having below average value as 'below average'. The frequencies of genotypes in the four contingency classes, were determined for each maturity group. Then the frequencies of genotypes in the 4 contingency classes of each maturity group were combined to get a single 2 x 2 contingency Table. Means of HS and LS classes for yield, adaptability and stability parameters

were also computed. The relationship of BI / SI parameter with yield, adaptability and stability parameters was inferred from χ^2 test. A significant χ^2 value indicated the presence of relationship.

RESULTS

Analysis of variance for G x E interaction in mid-early, mid-late and late groups over 12 environments (Table 2) showed significant differences due to genotypes, environments and genotype x environment interactions. Highly significant interaction component indicated differential response of the genotypes to environmental changes.

Mean yield, adaptability parameter that is regression coefficient (b) and stability parameters (S_d^2 and ASV) along with BI and SI of mid-early, mid-late and late group genotypes are presented in Table 3, 4 and 5 respectively. In case of mid-early group, average yield of the genotypes ranged from 34.41 to 38.42 q/ha with a grand mean of 36.55 q/ha and six genotypes were found to be high yielder. Seven genotypes had b-values less than one indicating their adaptation to poor environments like less fertile soils and the rest four had b-values more than one indicating their adaptation to rich environments like highly fertile soils. S_d^2 of genotypes Lalat, OR 2200-5, Konark, OR 1929-4 , OR 1916-19 and Bhoi were not significantly different from zero, indicating stability of performance over environments. S_d^2 of the remaining five genotypes were significantly different from zero, indicating that the genotypes were not stable.

In case of mid-late group, the genotypes OR 1912-25, Pratikshya, MTU 1001, OR 2156-15, Surendra, OR 2310-12 and OR 1964-8 gave above average yield and the rest gave below average yield. The regression coefficient/adaptability parameter (b-values) of the genotypes varied from 0.59 to 1.49. Seven genotypes had b-values less than one indicating their adaptation to poor environments and the rest six had b-values more than one indicating their adaptation to rich environments. On the basis of S_d^2 values, the genotypes OR 1912-25, OR 1914-8, OR 2310-12, Gajapati and MTU 1001 were classified as stable ($S_d^2 \approx 0$). The remaining 8 genotypes showed high deviation from regression (S_d^2 significantly different from zero), indicating that these genotypes lacked stability in yield performance.

In late group, seven genotypes were found to be high yielder and five were low yielder. The regression coefficient (b-values) of the genotypes ranged from 0.39 to 1.64. On the basis of the magnitude of b-values, five genotypes were found to have b-values greater than unity (b > 1) and seven genotypes had b-values less than unity (b < 1). Based on S_d^2 values, four genotypes were found

Table 2. Pooleanalysis of variance for grain yield (q/ha) in mid-early rice genotypes

Source	Df	MS	F
Genotypes (G)	10	24.94	3.85**
Environments (E)	11	395.70	61.16**
G x E	110	17.03	2.63**
E + G x E	121		
Environment (linear)	1	4352.71	296.65**
G x E (linear)	10	28.90	1.97*
Pooled deviation	10	14.67	2.26**
Pooled error	240	6.47	

Pooled analysis of variance for grain yield (q/ha) in mid-late rice genotypes

Genotypes (G)	12	83.19	10.57**
Environments (E)	11	900.88	114.50**
G x E	132	24.19	3.08**
E + G x E	143		
Environment (linear)	1	9909.73	483.94**
G x E (linear)	12	44.29	2.16*
Pooled deviation	130	20.48	2.60**
Pooled error	288	7.87	

Pooled analysis of variance for grain yield (q/ha) in late rice genotypes

Genotypes (G)	11	167.12	17.20**
Environments (E)	11	404.01	41.60**
G x E	121	39.93	4.11**
E + G x E	132		
Environment (linear)	1	4444.08	125.30**
G x E (linear)	11	52.26	1.47
Pooled deviation	120	35.47	3.65**
Pooled error	264	9.71	

Table 3. BI, SI, mean yield, S_d^2, ASV and b values of mid-early rice genotypes.

Genotype	SM-BI	Dark-SI	Mean yield (q/ha)	S_d^2	ASV	b
1.OR 1739-47	0.23	0.17	36.04	10.21*	1.26	0.78
2.OR 1916-19	0.36	0.14	34.95	6.15	2.08	0.94
3.OR 1929-4	0.43	0.24	35.24	3.09	0.61	0.76
4.OR 1976-11	0.56	0.20	37.63	19.50**	0.38	0.99
5.OR 2006-12	0.54	0.28	37.72	14.46**	2.44	0.92
6.OR 2168-1	0.51	0.40	35.07	24.95**	0.93	0.69
7.OR 2172-7	0.55	0.13	37.82	9.28*	2.10	1.23
8.OR 2200-5	0.53	0.59	37.87	0.30	1.23	1.47
9.Konark	0.45	0.68	36.87	0.34	1.38	1.35
10.Lalat	0.37	0.47	38.42	-3.95	0.40	1.08
11.Bhoi	0.37	0.75	34.41	7.16	1.95	0.79
Average	0.44	0.38	36.55	8.20	1.34	1.00

* and ** implies significant at 5 and 1% level respectively.

Table 4. BI, SI, mean yield, S_d^2, ASV and b values of mid-late rice genotypes.

Genotype	SM-BI	Dark-SI	Mean yield (q/ha)	S_d^2	ASV	b
1.OR 1681-11	0.45	0.22	41.30	16.59**	0.60	0.58
2.OR 1912-25	0.68	0.35	47.45	6.29	1.57	0.83
3.OR 1914-8	0.31	0.68	40.84	4.26	0.65	1.08
4.OR 1964-8	0.63	0.40	42.41	19.69**	1.55	1.01
5.OR 1967-15	0.46	0.52	40.11	7.81*	0.94	1.17
6.OR 2156-15	0.31	0.39	43.05	15.68**	1.31	0.59
7.OR 2310-12	0.42	0.39	42.66	6.00	0.56	0.97
8.Pratikshya	0.83	0.37	46.05	38.26**	2.61	0.93
9.Gouri	0.43	0.16	39.60	12.55**	0.90	0.93
10.Surendra	0.57	0.82	42.71	19.01**	1.05	0.93
11.Gajapati	0.63	0.58	39.61	4.45	1.26	1.49
12.Kharavela	0.57	0.61	38.27	15.59**	1.50	1.31
13.MTU 1001	0.63	0.36	44.08	-2.26	0.92	1.08
Average	0.53	0.45	42.16	12.61	1.19	1.0

* and ** implies significant at 5% and 1% level respectively

Table 5. BI, SI, mean yield, S_d^2, ASV and b values of late rice genotypes

Genotype	SM-BI	Dark-SI	Mean yield (q/ha)	S_d^2	ASV	b
1.OR 1885-16-34	0.44	0.85	32.15	52.42**	2.33	0.71
2.OR 1898-2-15	0.52	0.30	37.02	11.09*	0.88	1.06
3.OR 1898-3-16	0.68	0.37	43.94	3.83	0.75	1.12
4.OR 1901-14-32	0.61	0.46	44.45	8.11	2.43	0.68
5.OR 2001-1	0.55	0.42	40.72	11.46*	2.56	0.93
6.OR 2109-2	0.74	0.32	42.99	22.39**	1.22	0.92
7.OR 2119-13	0.65	0.54	41.46	72.37**	6.47	0.39
8.Savitri	0.45	0.77	38.59	36.52**	2.94	1.61
9.Salivahan	0.52	0.61	35.87	38.50**	3.32	1.33
10.Mahanadi	0.36	0.27	41.50	0.46	0.74	0.84
11.Kanchan	0.55	0.69	37.26	49.40**	4.37	1.64
12.Jagabandhu	0.42	0.37	42.65	2.47	0.88	0.77
Average	0.54	0.43	39.88	15.60	2.41	1.0

* and ** implies significant at 5% and 1% level respectively

to be stable and rest eight were unstable. Data pooled over the three maturity groups showed that of the 36 rice genotypes (11mid-early + 13 mid-late + 12 late), 20 (6 mid-early + 7 mid-late + 7 late) were found to be HY, 15 (4 mid-early + 6 mid-late + 5 late) had b>1 and 15 (6mid-early + 5 mid-late + 4 late) genotypes were found to be stable.

Bleaching response of rice genotypes

Rice genotypes treated with streptomycin showed varying degree of bleaching in seedlings, which was measured in terms of bleaching index (BI). From Table 3 it was observed that the genotypes OR 1739-47, OR 1916-19, OR 1929-4, Lalat and Bhoi of mid-early group had BI values less than group average (that is < 0.44) and considered as lowly sensitive (LS); the rest six genotypes had BI values greater than group average (> 0.44) and considered as highly sensitive (HS). The genotypes OR 1681-11, OR 1914-8, OR 1967-15, OR 2156-15, OR 2310-12 and Gouri of mid-late group (Table 4) showed low degree of bleaching (BI < 0.53), while OR 1912-25, OR 1964-8, Pratikshya, Surendra, Gajapati, Kharavela

Table 6. Correlation of BI and SI parameters with yield, S_d^2, ASV and b.

	Mid-early				Mid-late				Late			
	Yield	S_d^2	ASV	b	Yield	S_d^2	ASV	b	Yield	S_d^2	ASV	b
BI	0.445	0.351	-0.062	-0.524	0.480	0.401	0.755*	-0.395	0.443	0.135	0.210	-0.367
SI	-0.072	-0.373	-0.039	0.301	-0.235	-0.071	-0.024	0.493	-0.683*	0.738*	0.557	0.335

* Significant at 5% probability

Table 7. 2 x 2 contingency Tables of BI and SI parameters and yield and adaptability parameters.

Sensitivity parameter	Class	No. of genotypes	Yield class				b class			
			LY	HY	c2	Av. Yield (q/ha)	b < 1	b > 1	c2	Av.b-value
SM-BI	HS	19	4	15		40.8	10	9		1.04
	LS	17	12	5	8.92**	38.5	11	6	0.54	0.94
Dark-SI	HS	16	9	7		38.4	6	10		1.10
	LS	20	7	13	1.62	40.7	15	5	5.14*	0.92

HS: Highly sensitive, LS: Lowly sensitive, LY: Low yielder, HY: High yielder.

and MTU 1001 showed high degree of bleaching (BI > 0.53). Six genotypes of late-group (Table 5) were found to be less sensitive (LS) to the bleaching action of SM, while other six were highly sensitive (HS). This result revealed that rice genotypes showed wide variation in their bleaching response (Tables 3, 4 and 5). Mean BI value of the mid-early genotypes was 0.44, while those of mid-late and late groups were 0.53 and 0.54, respectively, indicating that the bleaching effect was generally low on mid-early genotypes and higher on those of mid-late and late genotypes. Pooled data indicated that out of 36 rice genotypes, 17 were LS and 19 were HS to the bleaching action of SM.

Dark-response

Dark response of a genotype was measured in terms of senescence index (SI), which was calculated from scores for the proportion of the leaves turning yellow. The genotypes of different maturity groups showed differences in their SI values (Tables 3, 4 and 5). The genotypes OR 2172-7, OR 1916-19 and OR 1739-47, OR 1976-11, OR 1929- 4 and OR 2006-12 of the mid-early group showed low degree of senescence (SI < 0.38) and considered as less sensitive (LS) to dark treatment, while the rest five had high response to dark treatment. Eight genotypes of the mid-late group were less sensitive to dark treatment (SI < 0.45), while the rest five were highly sensitive (SI > 0.45). The SI values of the late group genotypes ranged from 0.27 to 0.85 with a mean of 0.43. Six genotypes of the late group showed low response to dark treatment (SI < 0.43), and the rest six genotypes had high response to

dark treatment (SI > 0.43). Mean SI values of the three maturity groups ranging between 0.38 and 0.45, were quite similar, indicating that genotypes of high and low dark sensitivity occurred evenly in all maturity groups. Pooled data revealed that out of 36 rice genotypes, 20 were LS and 16 were HS to dark treatment.

Relationship of BI parameter with yield, adaptability and stability parameter

Correlation study (Table 6) indicated that BI parameter showed positive correlation (though high but non significant) with yielding ability and deviation from regression (S_d^2) and negative correlation with adaptability parameter (b) for all the three maturity groups. It showed a significant positive correlation with the ASV parameter of mid-late group.

The 2 × 2 contingency Table (Table 7) revealed that on the basis of SM-BI values, 19 rice genotypes of the 36 were HS and 15 of the 19 HS genotypes were high yielder (HY). In contrast, 12 genotypes of the 17 lowly sensitive (LS) genotypes were low yielder (LY). Contingency chi-square value was found to be significant (Table 7); indicating that the distribution was non-random and HS class had high frequency of HY genotypes, while LS class had high frequency of LY genotypes. In conformity, the HS class had higher average yield of 40.8 q/ha as against 38.5 q/ha of the LS class. Nine genotypes of HS class had b > 1 and 10 had b < 1. The LS class contained 11 genotypes with b < 1 and 6 genotypes with b > 1 (Table 7). The contingency

Table 8. 2 x 2 contingency Tables of BI and dark-SI parameters and stability parameters (S_d^2 and ASV) in rice.

Sensitivity parameter	Class	No. of genotypes	S_d^2 Class				ASV Class			
			S	U	χ^2	Av. S_d^2	S	U	χ^2	Av.ASV
Rice										
SM-BI	HS	19	7	12		15.34	7	12		2.02
	LS	17	8	9	0.39	15.97	12	5	3.54	1.22
Dark-SI	HS	16	7	9		21.69	7	9		2.08
	LS	20	9	11	0.01	11.36	12	8	0.94	1.36

chi-square was non significant, indicating the distribution to be random. The HS class had average b-value of 1.04 as against 0.94 of the LS class. Moreover, both HS and LS classes of SM-BI had very similar class means for b, both close to 1. Thus, sensitivity to SM does not appear to have any significant relationship with b-values of genotypes. The HS class also had higher average values for S_d^2 and ASV than LS class (Table 8). However, contingency chi-square values in all three cases were not significant at 5% level, though it was quite high (chi-square = 3.54) in case of ASV, indicating almost random distribution of genotypes in SM - BI classes.

Relationship of SI parameter with yield, adaptability and stability parameter

Table 6 showed correlation of SI parameter with yield, stability and adaptability parameters. For all the 3 maturity groups the SI parameter showed negative correlation with yield indicating that highly dark sensitive genotypes may be low yielder. It showed positive correlation with adaptability parameter (b) for all the 3 maturity groups and a significant positive correlation with the S_d^2 parameter only for late maturity group. Sensitivity to dark treatment was measured as senescence index (SI) and 16 of the 36 genotypes were HS of which 7 were HY and 9 were LY (Table 7). The remaining 20 genotypes were lowly sensitive (LS) to dark exposure and 13 of these were HY and 7 were LY. The contingency chi-square was non significant. But the LS group had higher average yield (as it included more number of high yielders) than HS group genotypes. This study indicates that the SI parameter has some relationship with yield.

Ten genotypes of dark HS class had b > 1 and 6 had b < 1. The remaining 20 genotypes of dark LS class included 15 genotypes with b < 1 and 5 with b > 1 (Table 7). The contingency chi-square was significant, indicating the distribution to be non- random. In addition, the HS class had high average b-value of 1.10 as against 0.92 of

the LS class. Thus, most rice genotypes showing high sensitivity to dark (Dark-SI) had b > 1, indicating that genotypes have better adaptation to rich environments. Similarly, most LS genotypes had b < 1, indicating that they could have better adaptation to poor environmental conditions.

The HS and LS classes of Dark-SI included similar number of genotypes with stable and unstable performance as assessed by S_d^2 and ASV (Table 8). The contingency chi-square value was non-significant indicating the distributions to be random. This implies that Dark-SI parameter does not have any significant relationship with stability of performance of rice genotypes.

DISCUSSION

Photosynthesis is the cornerstone of crop production. It serves as the primary source of all energy for mankind. The yield of agricultural plants depends on the size and efficiency of the photosynthetic system. The green plastids or chloroplasts that constitute the photosynthetic apparatus is likely to contribute to photosynthetic efficiency and ultimately to crop productivity. Therefore, variation in chloroplast behaviour either due to chemical or physical stress might help the breeder in preliminary selection of superior genotypes.

Streptomycin (SM) is an amino-glycoside antibiotic and acts as a protein synthesis inhibitor. In plants, it induces bleaching of leaves in seedlings by inhibiting chlorophyll synthesis. The bleaching effect of SM was first reported by Von Euler (1947) in barley seedlings. Similar bleaching effect of SM on seedling leaves due to inhibition of plastid development has been reported by Khudairi (1961), Babayan et al. (1975), Mancinelli et al. (1975), Pretova and Anna (1980) and Zubko and Dey (1998, 2002) in various plant species. Kinoshita and Reiko (2001) studied SM sensitivity of 103 rice varieties of *Japonica* and *Indica* types and 17 isogenic lines of cv. Shiokari in terms of seedling bleaching. Varieties of both *Japonica* and *Indica* groups showed wide range of variation in SM-sensitivity and the near isogenic lines of Shiokari also showed wide variation in SM-sensitivity.

In the present study, seeds of 36 rice genotypes of three duration groups were treated with 500 ppm SM solution for 48 h and seedling-bleaching index (BI) was estimated. The BI of genotypes of all duration groups showed wide variation ranging from 0.23 to 0.83, indicating differences in SM-sensitivity of genotypes in terms of bleaching. Similar differences in bleaching effect of SM treatment in different genotypes have been reported by Sinha and Satapathy (1979), Das (2001) and Kinoshita and Reiko (2001) in rice; Sinha and Swain (1978), Das and Sinha (1986), Sinha et al. (1996) in ragi, Rath (1977) in wheat, Sinha and Satapathy (1977) in maize and Singh and Nanda (1997) in green gram.

In the present investigation, an attempt was made to find out if sensitivity of genotypes to SM treatment has any relationship with their yield potential, adaptability and stability of performance. So the genotypes in each duration group of rice was classified as highly sensitive (HS) and lowly sensitive (LS) on the basis of SM-BI value. Similarly, genotypes in each group were classified as high yielder (HY) and low yielder (LY), b > 1 and b < 1 and stable and unstable. Of the 36 rice genotypes, 19 were HS on the basis of SM-BI and 15 of them were high yielder, while 12 of the 17 LS genotypes were low yielder. The HS class of genotypes had higher average yield than LS class. Thus, higher degree of seedling bleaching due to SM treatment would be an indicator of high yield potential of genotypes in rice. Sinha and Swain (1978) observed that ragi mutant lines showing SM-sensitivity in terms of bleaching were generally earlier in maturity, shorter in height and higher yielding than SM-resistant lines. Sinha and Satapathy (1979) reported that semi-dwarf high-yielding rice varieties showed more SM bleaching than low-yielding tall *indica* varieties. Das and Sinha (1986) suggested that selection of ragi genotypes showing more SM-bleaching could lead to identification of lines with early maturity, short height and more tillers per plant. SM-response could be used as a criterion in preliminary laboratory evaluation of broad adaptation pattern of new high-yielding rice varieties and for germplasm screening for drought tolerance (Das and Sinha, 1992).

SM-sensitivity in terms of bleaching in the present study does not appear to have any definite relationship with adaptability of genotypes to poorer or rich environments and also with stability. However, Sinha et al. (1996) reported that ragi genotypes showing low SM-sensitivity would show general adaptability.

Senescence in green plants is a complex and highly regulated process that occurs as a part of growth and development. Senescence reduces leaf area duration adversely affecting photosynthesis. Exposure of leaves to dark induces senescence by inhibiting the expression of light regulated genes, responsible for chloroplast development and causes etiolation which may be termed as dark-induced senescence. It reduces leaf area duration (LAD) adversely affecting photosynthesis, respiration, transpiration and also

translocation of nutrients. So in plants early senescence would affect supply of photosynthates leading to inadequate filling of the sink. Thus, plant breeders often look for genotypes showing slow and late senescence of leaves. Saulescu et al. (2001) observed significant correlation between rate of chlorophyll loss following exposure to dark and chloroplast loss during aging in stress free environments. They concluded that seedling test for dark-induced senescence is a potential tool in breeding of wheat for optimum senescence pattern. Dark-induced senescence of seedling leaves has also been used as parameter of aging senescence by Grover et al. (1986), Annamalainathan et al. (1995), Saulescu and Kronstad (1998), Saulescu et al. (1998) and Spano et al. (2003).

The genotypes of each duration group of rice were classified as highly sensitive (HS) and lowly sensitive (LS) to dark treatment on the basis of senescence index (Dark-SI) values. The LS class included more number of genotypes with higher yield and the HS class included more number of genotypes with low yield. The LS class also had high average yield than the HS class. Thus, it appears that low SI of genotypes, which can be attributed to slow or late senescence of leaves give some indication about high yield potential of genotypes in rice. Mohapatra (1997) evaluated senescence response of ragi genotypes to dark treatment in terms of etiolation index and found that in case of early duration group, moderately resistant and resistant classes included more number of higher yielding varieties.

Rice genotypes falling in HS class for high Dark-SI included greater number of genotypes with b > 1 and the LS class included greater number of genotypes with b < 1. Moreover, class mean for b-values of HS class was 1.10 and LS class was 0.92. Thus, it appears that, genotypes showing low senescence under dark treatment would be better adapted to poorer environments and those showing high senescence would be better adapted to rich environments. Pattanaik (1994) working on dark response of ragi genotypes reported that genotypes showing moderate response to dark treatment in terms of etiolation index would show most desirable pattern of adaptation. Saulescu and Mustatea (2002) reported that slow senescence under dark treatment seemed to be a characteristic of wheat cultivars adapted to more favourable environments, whereas fast senescence was found in cultivars adapted to stress environments. However, for stability parameter S_d^2, Dark-SI would not give any lead for identifying stable genotypes.

The present study indicated that the rice genotypes showed wide differences in sensitivity to streptomycin induced bleaching (SM-BI) and dark induced senescence (Dark-SI). High sensitivity of genotypes to streptomycin in terms of BI could be used for laboratory screening of rice genotypes for their yielding ability at an early seedling stage. Use of other chemicals like maleic hydrazide in predicting yielding ability was also suggested by Das et al. (2008). Rice genotypes showing high sensitivity to dark treatment in terms of senescence (SI) would show better adaptation to rich environments and those showing low sensitivity would be better adapted to poorer environments. Selection of SM-HS class may help the breeder in indirect selection of high yielding genotypes at an early seedling stage and makes the multilocation trials more economic.

REFERENCES

Annamalainathan K, Pathmanshan G, Manian K, Nagarajan M, Kanivara-dharaju (1995). Influence of light and regulators on senescence related changes in detached soybean leaves. Madras Agric. J., 82: 201-204.

Babayan RS, Gevorkyan AM, Soakyan NA (1975). Effect of streptomycin on chlorophyll biosynthesis in wheat seedlings. Fiziol. Rast. (Muscow), 22: 484-489.

Das S (2001). Laboratory screening of high yielding rice varieties for drought tolerance. Env. Eco., 19: 14-18.

Das PK, Sinha SK (1986). Relation of a seedling character to yield and yield components in finger millet (Eleusine coracana). J. Orissa Bot. Soc., 8: 49-50.

Das S, Sinha SK (1992). Use of streptomycin as a chemical aid in the laboratory evaluation of broad adaptation patterns in rice. Pl. Ski. Res., 14: 17-20

Das S, Sinha SK, Misra RC (2008). Variation in seedling growth inhibition due to maleic hydrazide treatment of rice (O. sativa) and ragi (E. coracana) genotypes and its relationship with yield and adaptability. J. Crop Sci. Biotechnol., 3: 215-222.

Eberhart SA, Russell WA (1966). Stability parameters for comparing varieties. Crop Sci., 6: 36-40.

Economic Survey (2007). Ministry of Finance, Economic Division, Government of India, New Delhi.

Grover A, Sabat SC, Mohanty P (1986). Effect of temperature on photosynthetic activities of senescing detached wheat leaves. Plant and Cell Physiol., 27: 117-126.

Khudairi AJ (1961). Effect of streptomycin on Xanthium. Biochem. Biophys. Acta., 46: 344-354.

Kinoshita T, Reiko TS (2001). Streptomycin resistance of rice varieties and dwarf lines. Rice Genet. Newslett., 5: 13-14.

Mancinelli AL, Yang CH, Lindquist P, Anderson OR, Rabino I (1975). Photocontrol of anthocyanin synthesis III. The action of streptomycin on the synthesis of chlorophyll and anthocyanin. Plant Physiol., 55: 251-257.

Mohapatra B (1997). Variation on plastid behaviour and chemical induction of plastid mutation on ragi. M.Sc. (Ag.) thesis, O.U.A.T.

Pretova A, Anna S (1980). Effect of streptomycin on growth of immature flax (Linum usitatissimum). Biologia., 35: 413-416.

Purchase JL (1997). Parametric analysis to describe genotype x environment interaction and yield stability in winter wheat. Parametric analysis to describe genotype x environment interaction and yield stability in winter wheat. Ph. D. Thesis. Department of Agronomy, Faculty of Agriculture of the University of the Free State, Bloemfontein, South Africa.

Saulescu NN, Ciocazan L, Lazar C (1998). Genotypic differences in dark-induced senescence are correlated with field "stay green" scores in maize (Zea mays L.). Rom Agric. Res. In Press.

Saulescu NN, Ittu G, Mustatea P (2001). Dark induced senescence as a tool in breeding wheat for optimum senescence pattern. Euphytica., 119: 205-209..

Saulescu NN, Kronstad WE (1998). Genotypic differences in leaf chlorophyll loss during dark induced senescence in seedlings of winter wheat. (Triticum aestivum L.). Rom Agric. Res. In press.

Saulescu NN, Mustatea P (2002). Dark induced senescence in seedlings is correlated with flag leaf senescence in the field. Annual wheat Newsletterlett., 45: 1-3

Singh B, Nanda PR (1997). Prediction of adaptation pattern of some mungbean genotypes based on streptomycin response. Env. Eco., 15: 559-561.

Sinha SK, Satpathy MB (1979). An evidence for relationship between streptomycin sensitivity of plastids and high yield potential in rice. Science Sci. Cult., 45: 373-374.

Sinha SK, Prusty N, Das S (1996). Laboratory evaluation of adaptation on ragi (Eleusine coracana Gaerotn). Numerical classification of seedling response to streptomycin. Pl. Sci. Res., 18: 57-61.

Sinha SK, Satpathy MB (1977). Streptomycin induced variation in albinism in inbred maize. Sci. Cult., 43: 567-568.

Sinha SK, Swain MN (1978). Chemically aided selection in two EMS derived population of ragi. Proc. 65th session of Indian Sci. Congress, pp. 3.

Spano G, Di Funzo N, Perrotta C, Platani C, Ronga G, Lawlor DW, Napier JA, Shewry PR (2003). Physiological characterization of 'Stay green' mutants in durum wheat. J. Exp. Bot., 54: 1415-1420

Von Euler M (1947). Einfluss des streptomycin and dischloro phyllibildung. Kem. Arb., 9: 1-3.

Zubko M, Dey A (2002). Differential regulation of genes transcribed by nucleus encoded plastid RNA polymerase and DNA amplification, within ribosome – deficient plastids in stable phenocopies of cereal albino mutants. Mol. Genet. Genom., 267: 27-37.

11

Assessment of intervarietal differences in drought tolerance in chickpea using both nodule and plant traits as indicators

Nehla Labidi*, Henda Mahmoudi, Messedi Dorsaf, Ines Slama and Chedly Abdelly

The Laboratry of Plant Adaptation to Abiotic Stress (LAPSA), Biotechnology Center at the Technopark of Borj-Cedria (CBBC), BP 901, Hammam-Lif 2050, Tunisia.

5 lines of Tunisian varieties of chickpea (*Cicer arietinum* L.) inoculated with *Mesorhizobium ciceri* UPMCa7 were monitored during the vegetative stage on sterilized sandy soil. 2 levels of soil moisture were compared (100 and 33% of field capacity). The work was aimed at assessing the relative tolerance of these lines to drought and then, to research relationships between the level of sensitivity of plant growth and N content to drought and nodule, leaf and root traits. Drought limited plant growth of Amdoun and Neyer and decreased N content of Chetoui, Amdoun and Neyer. The latter N shortage was associated with increase in nodule mortality and restriction of nodule growth. In view of their minimal decrease in plant biomass and N content, Beja and Kesseb were the most tolerant varieties. Inter-varietal differences for water stress effects on nodule, root and leaf traits were limited to (i) change in root to shoot ratio (ii) loss of chlorophylls and (iii) nodule mortality. Each of these traits was considered as an indicator of stress tolerance. These indicators predict that the most tolerant variety was Beja based on higher increase in root to shoot ratio and Beja and Kesseb based on lower nodule mortality. When choice of the varieties should depend on the likelihood of water stress during the culture, Amdoun which presented the higher biomass per plant and the higher nitrogen content in control condition and ranked similarly to all other varieties in stress condition might represent a reasonable trade-off between high growth and stress tolerance when the probability of water stress to occur during the culture period is low.

Key words: *Cicer arietinum*, drought stress, nodule characteristics, tolerance.

INTRODUCTION

Cicer arietinum L. (chickpea) is an important food legume crop of Mediterranean populations. In Tunisia, it shares the first rank with faba bean and it is a winter-spring crop grown in semi-arid regions. Generally, legumes are highly sensitive to water deficit stress (Mahieu et al., 2009). Drought conditions may limit production of legumes by affecting nodule functioning (Ashraf and Iram, 2005; Clement et al., 2008). The effect can be due to (i) restriction of carbohydrate transport from leaves to nodule (Singh and Singh, 2006) (ii) reduced in shoot N demand reflected by the Rubisco depletion affected negatively malate dehydrogenase and glutamate-oxalate

transaminase nodule's activities (Aranjuelo et al., 2009) (iii) less water which hinders the transport of N-products away from the nodules (Ramos et al., 2003) (iv) direct effects on nodule gas permeability (Ramos et al., 2003) and/or (v) the alteration of nodule metabolic activity (Clement et al., 2008).

Water deficit can also induce premature senescence of nodules (Puppo et al., 2005). During this process, many changes occur in nodules, for example the external colour of the N-fixing tissues of the nodule changes from red (due to functional leghaemoglobin) to green (indicating alteration of this protein) (Swaraj and Bishnoi, 1996), decrease in leghaemoglobin content (Garg and Manchanda, 2008), decrease in nodule membrane integrity (Mhadhbi et al., 2009), degradation of bacteroids (Herder et al., 2008), increase of proteinase activities in nodules (Groten et al., 2006) and loss of N-fixation acti-

*Corresponding author. E-mail: hmidanehla@yahoo.com.

vity regardless of physiological and biochemical mechanisms of N_2 fixation inhibition by water deficit stress, there is evidence that legume species have significant genetic variation in their ability to fix N_2 under drought conditions (Ashraf and Iram, 2005; Charlson et al., 2009). However, there is no available information on the variability of drought tolerance in chickpea cultivated in symbiosis condition. The aim of this work was to establish easy use of indicators of chickpea tolerance to drought based on simple traits of plants and nodules. 5 chickpea varieties were compared. Their relative tolerance was assessed from plant biomass and N content and a variety of traits of nodules, roots and leaves were used as indicators.

MATERIAL AND METHODS

Growth conditions and experimental procedures

The experiment was conducted in a greenhouse of the Biotechnological Center at Borj Cedria (35 km south-east of Tunis), during June 2005. Sterilized seeds of 5 local chickpea (C. arietinum L.) lines named Beja, Neyer, Amdoun, Kesseb and Chetoui were germinated in plastic Petri dishes. 4-day-old seedlings were individually transplanted in sterilized plastic pots (16.5 cm diameter and 15.5 cm height) filled with loam sandy soil (3.1 kg). Prior to transplantation, soil was abundantly washed with distilled water to remove nutrients content and it was heat sterilized in metal buckets at 380°C for 4 h. After transplantation, the seedlings were inoculated with rhizobial suspension (strain Mesorhizobium ciceri, UPMCa7). They were irrigated with nutrient solution (Vadez et al., 1996) at field capacity for 21 days. The average day and night temperatures were $35 \pm 5°C$ and $20 \pm 2°C$, respectively. The day length was 14 h and the relative humidity was $55 \pm 5\%$ by day and $80 \pm 5\%$ at night. Functional nodules were established during this period. After 21 days, plants were divided into 2 lots of 6 plants each, one irrigated with tap water at 100% field capacity(control plants) and the second one at only 33% field capacity (stressed plants)

The water volume necessary to reach 100% field capacity was determined by measuring soil water content after cession of drainage. Regular wetting (every 2 days) enabled to restore the moisture of soil at its nominal value.

Measurements

Plants were harvested after 14 days of treatment (that is, 39 days after germination). The midday leaf water potential of the 3 younger leaves of the main stem was determined using a pressure chamber (Soil Moisture Equipments Corp., Santa Barbara, CA, USA) according to Scholander et al. (1965). Harvested plants were divided into leaves, stems, roots and nodules and the fresh weight (FW) of plant parts was determined. Leaves were counted and their surface area was measured using a portable area meter (LI-3000A). Electrolyte leakage of leaves was measured on the third youngest (fully expanded) leaf as described by Dionisio-Sese and Tobita (1998). Relative water content (RWC) was calculated in the same leaves using the as RWC (%) = 100 FW - DW / (TW - DW) (Schonfeld et al., 1988). Fresh weight (FW) was determined within 2 h after harvest. Turgid weight (TW) was obtained after soaking leaves in distilled water in test tubes for 12 h at room temperature (about 20°C) under laboratory room ceiling light. After soaking, leaves were quickly and carefully blotted dry with tissue paper in preparation for determination turgid weight. Dry weight was measured after oven drying samples at 60°C for 48 h. Chlorophyll content of leaves was determined using the method of Bruinsma

(1963). Fully expanded and mature leaves were randomly selected for this purpose. Nodules were removed, counted and photographed with digital camera. Leghaemoglobin content was measured according to Becana et al. (1986). Shoots, roots and nodules dry weight was determined after oven drying for 72 h at 80°C. Total nitrogen was determined by Kjeldahl method. Statistical analysis was performed using a computer program (Statistica™ software). Data were analysed by 2 way ANOVA (treatment and chickpea lines as factors, Table 1). Least significant difference (LSD) post hoc tests were used for mean comparison within varieties when ANOVA indicated significant effect of the considered factor. Significance differences between the 2 treatments in each variety was examined accordingto student's t- test at p = 0.05. Data are represented by means of 6 replicates ± SEM.

RESULTS

Plant growth

No significant variability appeared among varieties for whole plant biomass, but significant effect of treatment was observed for whole plant biomass (Table 1). In control condition, the plant biomass (dry weight, DW) differed slightly among varieties, with the higher level for Amdoun and Neyer (Figure 1). In these 2 varieties growth was significantly affected by water stress (p = 0.05). However, the absolute values of the plant biomass under stress did not significantly differ between the five varieties. The root biomass of control plants did not differed between varieties (Table 1). It was augmented by the water stress only in Chetoui (p = 0.05, Figure 2). Values in stressed plants of all varieties were statistically similar. However, the root to shoot ratio and its response to water stress significantly differed between varieties (Table 1), it was significantly augmented in water stressed Beja, but not the other varieties (p = 0.05, Figure 2).

Nodules

The N content of the whole plants was significantly diminished by water stress in Chetoui, amdoun and Neyer (p = 0.05, Figure 1) varieties, but this response did not discriminate between varieties (Table 1). The water stress significantly increased nodule defects or mortality (Figure 4), especially in Chetoui and Amdoun. In these varieties, the proportion of empty nodules with dark coloration reached 50 - 60% at the harvest, in contrast to the 3 other varieties in which it was only 10 to 26% (Table 2). In both control and stress condition, Kesseb presented less nodules per plant than the other varieties. No effect of the water stress could be detected for this parameter (p = 0.05, Figure 3). On the contrary, the biomass of non-empty nodules per plant presented the same pattern as observed for N content (Figure 3), with a clear decrease upon water stress and no significant inter-varietal difference (Table 1). Only leghaemoglobin concentration in non-empty nodules depended on both varieties (highest values in Chetoui and Neyer) and treatments, with a large decrease in Amdoun and Neyer upon water stress (p =

Table 1. Results of variance analysis (ANOVA). Five chickpea varieties and two treatments (control and water stress) were compared.

	p values								
	Whole plant root and nodules								
Variables	Whole plant biomass	Whole plant N content	Root system biomass	Root /shoot ratio	Nodule biomass per plant	Nodule number	proportion of empty nodules	legheamoglobin	Mean Specific N fixation
Factor "variety"	0.145	0.121	0.036	$2\ 10^{-04}$	0.326	0,028	$2\ 10^{-05}$	0.005	0.004
Factor "treatment "	0.027	$9\ 10^{-06}$	0.789	0,034	$5\ 10^{-08}$	0.138	$6\ 10^{-10}$	0.036	0.006
Interaction	0.033	0.184	0.085	0.035	0.202	0.992	$2\ 10^{-05}$	0.060	0.277

	Leaves					
Variables	Leaf water potential	leaf number	chlorophyll concentration	Leaf surfacic mass	Electrolyte leakage	Relative water content
Factor "variety"	$3\ 10^{-07}$	0.415	$8\ 10^{-06}$	0.000	0.008	0.093
Factor "treatment "	$3\ 10^{-13}$	0.004	$1\ 10^{-14}$	0.002	0.159	0.347
Interaction	$2\ 10^{-07}$	0.940	$2\ 10^{-07}$	0.704	0.244	0.702

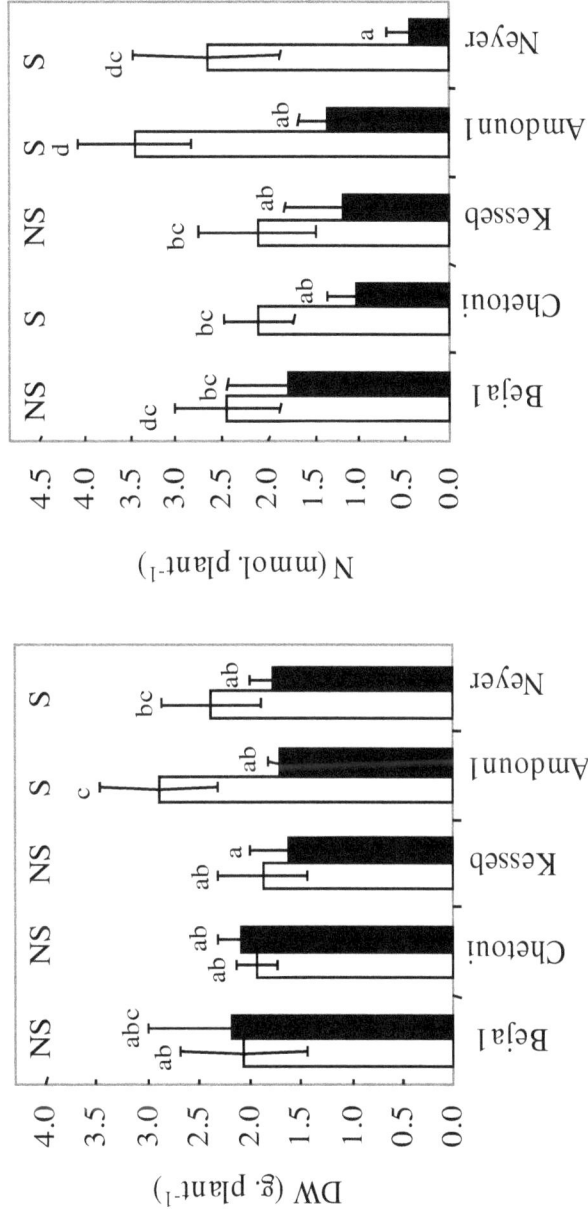

Figure 1. Effect of water stress on growth and nitrogen fixation. Growth was estimated as whole plant dry weight (DW) at the final harvest. Nitrogen fixation was estimated as N content of whole plants grown in the absence of N in the culture solution. Means ± SE (n = 6). Differences among chickpea lines were analyzed using ANOVA. Different letters indicate significant difference (p = 0.05). S, NS: significant, respectively non significant mean difference within each variety, according to Student t test at p = 0.05. Open bars: control. Full bars: water stress.

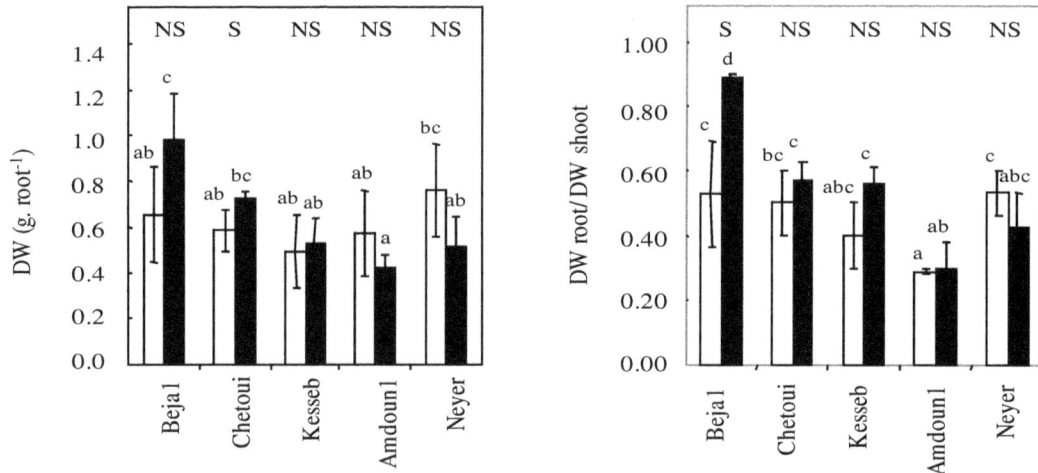

Figure 2. Effect of water stress on root growth. Root growth was estimated as root system dry weight (DW) at the final harvest (left panel). The right panel shows the root to shoot DW ratio. Five chickpea lines are compared. Means ± SE (n = 6). Differences among chickpea lines were analyzed using ANOVA. Different letters indicate significant difference (*p* = 0.05). S, NS: significant, respectively non significant mean difference within each variety, according to Student t test at *p* = 0.05. Open bars: control. Full bars: water stress.

Table 2. Nodule characteristics of five chickpea varieties and effect of water stress. The plants were cultivated for for 14 days either in control condition or at 33% field capacity. Values are means ± standard errors (n = 6). For each parameter different letters indicate significant differences (p< 0.05).

Parameters	Treatments	Varieties				
		Beja	Chetoui	Kesseb	Amdoun	Neyer
Individual nodule	control	6.6± 3.6abc	6.1 ± 2.3abc	7.3 ± 0.8bc	5.1 ± 0.5abc	7.8 ± 2.2c
biomass (mg DW) [1]	water stress	4.5 ± 1.4ab	4.2 ± 0.1ab	7.1 ± 2.2bc	3.4 ± 0.1a	4.1 ± 1.3ab
Specific N fixation	control	15.5± 2.8dc	11.7 ± 3.2bc	14.1 ± 2.3c	18.7 ± 0.8d	16.3 ± 0.6dc
(mmol N. g⁻¹ nodule DW) [1]	water stress	15.8± 5.9dc	9.3 ± 3.0ab	11.9± 4.5bc	14.1 ± 2.2c	7.2 ± 4.6a
Empty nodule number	control	0	0	0	0	0
(% of total nodules)	water stress	10 ± 6 a	60 ± 10 b	12 ± 13 a	51 ± 13 b	26 ± 4 a
Leghaemoglobin	control	3.9± 1.9abc	8.7 ± 3.1d	4.3 ± 2.6bc	5.1 ± 1.4bc	6.3 ± 1.8dc
(mg g⁻¹ nodule FW) [1]	water stress	5.2 ± 1.4bc	6.4 ± 3.0dc	5.0 ± 3.2bc	1.4 ± 0.7a	3.2 ± 1.2ab

Note: Specific N fixation units are $(\text{mmol N. g}^{-1}\text{ nodule DW})$; Leghaemoglobin units are $(\text{mg g}^{-1}\text{ nodule FW})$.

(1) Empty nodules excluded

0.05, Table 2).

The mean specific N fixation provides a crude measure of the nodule efficiency. It was estimated by rationing the whole plant N content to the (non-empty) nodule mass (DW) per plant. Although decrease was observed upon stress treatment in Amdoun and Neyer varieties (p=0.05, Tab. 2), the interaction between treatment and variety factors was not significant (Tab. 1).

Leaves

Water stress lowered the leaf number, but this effect was only significant in Chetoui (Table 3). In both control and stress conditions, the 5 varieties presented statistically similar leaf mean thickness (estimated as the leaf surfacic mass, that is, leaf fresh weight rationed to leaf surface area). The latter parameter was augmented by water stress, but this behavior was significant only in Kesseb

and Amdoun varieties (p = 0.05, Table 3). Leaf tissue hydration (RWC) did not present significant difference between varieties nor between treatments and no interaction between these factors could be observed for this parameter (Table 1). In contrast to RWC, the leaf water potential (Table 1) was strongly dependent on variety, Kesseb presenting a higher value (-0.42 Mpa) compared to the other ones in control condition (ca. -1.15 MPa). It was lowered in water stressed plants, to similar values in all varieties. Decrease in concentration of total chlorophylls in leaf tissues upon water stress occurred for values in Chetoui and Neyer) and treatments, with a large all varieties and the magnitude of this effect differed significantly between them (p = 0.05, Table 3). In stressed plants of Chetoui, Amdoun and Neyer the chlorophyll content per g leaf FW was only 10 to 20% that of control plants, but this value was maintained at ca. 30 - 50% in Beja and Kesseb. Finally, the electrolyte leakage from

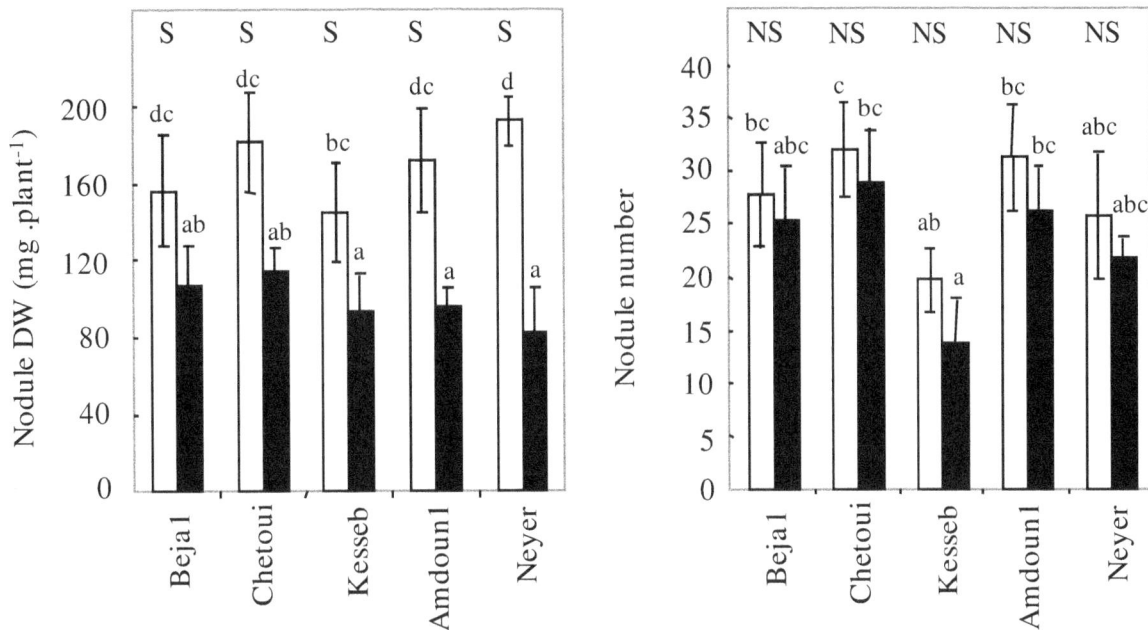

Figure 3. Effect of water stress on nodule development. Empty nodules were excluded. Left panel: total biomass of nodules. Right panel: number of nodules per plant. Five chickpea lines are compared. Means ± SE (n = 6). Differences among chickpea lines were analyzed using ANOVA. Different letters indicate significant difference (p = 0.05). S, NS: significant, respectively non significant mean difference within each variety, according to Student t test at p = 0.05. Open bars: control. Full bars: water stress.

Table 3. Leaf characteristics of five chickpea varieties and effect of water stress. The plants were cultivated for 14 days either in control condition or at 33% field capacity. Means ± standard errors (n = 6). For each parameter different letters indicate significant differences (p< 0.05).

Parameters	Treatments	Varieties				
		Beja	Chetoui	Kesseb	Amdou	Neyer
Leaf number	control	50.3 ± 15c	41.0 ± 1.1abc	46.7 ± 17bc	46.7 ± 15bc	43.3 ± 8abc
plant^{-1})	water stress	41.0 ± 13abc	25,0 ± 2,3a	31.0 ± 14ab	37.7 ± 12abc	30.3 ± 14ab
Total chlorophylls	control	680 ± 170c	1483 ± 95f	1080 ± 192e	910 ± 70d	838 ± 218cd
µg.g^{-1} leaf FW)	water stress	447 ± 82b	176 ± 44a	504 ± 40b	212 ± 9a	141 ± 22a
Leaf surfacic mass	control	(41 ± 8) 10^{-4}abc	(59 ± 21) 10^{-4}ed	(32 ± 1) 10^{-4}ab	(34 ± 5) 10^{-4}abc	(29 ± 14) 10^{-4}a
g.cm^{-2})	water stress	(51 ± 10) 10^{-4}cd	(76 ± 9) 10^{-4}e	(38 ± 4) 10^{-4}abc	(47 ± 5) 10^{-4}bcd	(51 ± 21) 10^{-4}cd
Electrolyte	control	40 ± 12b	20 ± 12a	28 ± 8ab	20 ± 9a	24 ± 6a
leakage (%)	water stress	39 ± 11b	20 ± 2a	30 ± 12ab	38 ± 12b	26 ± 7ab
Relative water	control	85.7 ± 5,0ab	80.6 ± 8,4ab	74.3 ± 5.9a	76.2 ± 7,3ab	84.0 ± 5,0ab
content (RWC, %)	water stress	87.4 ± 12,4ab	83.3 ± 15,1ab	75.2 ± 11.7a	74.0 ± 8,7a	90.1 ± 4,0b
Leaf water	control	-1.15 ± 0.00b	-1.13 ± 0.13b	-0.42 ± 0.21a	-1.17 ± 0.03b	-1.15 ± 0.06b
potential (%)	water stress	-1.55 ± 0.00cd	-1.67 ± 0.13d	-1.53 ± 0.07cd	-1.43± 0.07c	-1.43 ± 0.03c

leaf tissues differed significantly between the varieties (Table 1) that of Beja being twice that of Chetoui and 1.6 fold that of Neyer. However, significant effect of water stress could be detected only in Amdoun (p = 0.05, Table 3).

DISCUSSION

The results show that water deficit significantly limited plant growth (Amdoun and Neyer) and N fixation (Chetoui, Amdoun and Neyer). Inter-varietal variability was observed for plant biomass in control condition, but not under water stress. Probably, this constraint generated some limiting factor for plant growth. Considering the change in whole plant biomass due to water stress leads to distinguish 2 sets of varieties, namely Beja, Chetoui and Kesseb in which there was no significant response and Neyer and Amdoun in which growth was inhibited.

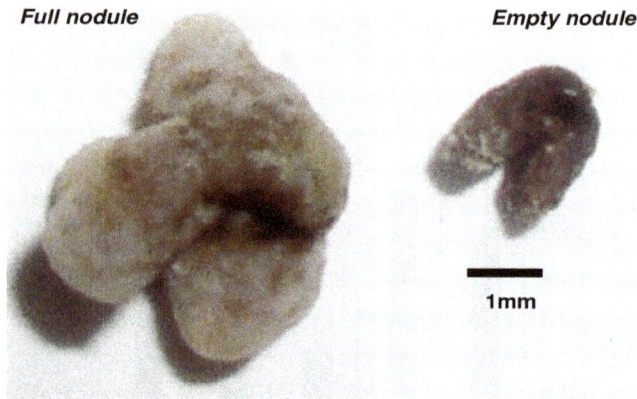

Figure 4. Nodule morphology. Left: full nodule; right: empty nodule. These examples are for Chetoui. The same morphology was observed for the other studied varieties.

However, since the latter plants were the taller in control condition, their biomass under stress was reduced to the same value as that of the other varieties. Differences between varieties appeared for plant development under stress, evidenced as an increase in root to shoot ratio specifically in Beja. Such drought (Shao et al., 2008) and to mineral deficiency, for instance N shortage (Hessini et al., 2009). Drought significantly limited the N content of the whole plants of Cheoui, Amdoun and Neyer, suggesing that N provision for plant growth was limited in these varieties. Since nodule number was poorly dependent on water stress, the putative N shortage of stressed plants would result from nodule functioning rather than nodule initiation. This hypothesis is likely because drought strongly decreased total nodule biomass. This phenomenon could result from adverse effect of drought on individual nodule growth, and from nodule abortion, as indicated by the significant effect of stress on the proportion of empty, dark nodules. Nodule blackening and/or emptying are indicators of degeneration (Gross et al., 2002). The black colour is due presumably to the accumulation of a dark-staining material within the cortex cells (Ramos et al., 2003). It has been reported that leghaemoglobin content declined in dehydrated nodules subjected to severe drought (Figueiredo et al., 2008). In our investigation, this parameter decreased in Amdoun and Neyer, the only varieties which showed a decrease in the nitrogen fixation per unit mass of non-empty nodule. The reduction of nodule leghaemoglobin content can be attributed to early nodules degeneration related probably to the production of O_2^- radicals (Mhadhbi et al., 2009). However, reports on the relationship between symbiotic nitrogen fixation and nodule leghaemoglobin content are controversial (Irigoyen et al., 1992; Gonzalez et al., 2001).

According to Ashraf and Iram (2005) drought do not seem to influence the colonization of roots by rhizobia but it suppresses the growth of nodules. The high sensitivity of chickpea nodule development as compared to other plant parts suggests that water deficit specifically affected nodule development. Inhibition of nodule development in stressed plants has been suggested to be due to restricttion of carbohydrate transport from leaves to nodule (Singh and Singh, 2006). Leaf water relations were not dramatically affected by water stress. Their water potential was lowered, and their thickness was augmented, 2 classical mechanisms for avoidance of tissue desiccation. Indeed, their relative water content was maintained. The cell membrane integrity was preserved as indicated by the insensitivity of electrolyte leakage to water stress. However, leaf chlorophyll content was significantly lower in stressed plants. These observations suggest that limitation of the whole plant photosynthetic capacity rather than hydro unbalance might have limited assimilate provision to nodules.

In conclusion, this work permitted to purpose several indices to predict relative tolerance to drought of chickpea varieties. In view of the minimal decrease in plant biomass and N content, Beja and Kesseb were the most tolerant varieties. Beja was predicted as the most tolerant variety when the increase in root to shoot ratio was used as an indicator of tolerance. Minimal chlorophyll loss and minimal nodule mortality predicted Beja and Kesseb as the most tolerant varieties. However, these 2 varieties do not present the higher performance in the absence of stress. Amdoun which presented the higher biomass per plant in control condition and ranked similarly to all other varieties in stress condition might represent a reasonable trade-off between high growth and stress tolerance when the probability of water stress to occur during the culture period is low. Of course, the agronomic value of these predictions should be evaluated from seed yield rather than vegetative growth, in realistic agricultural condition.

REFERENCES

Aranjuelo I, Irigoyen JJ, Nogués S, Sánchez-Díazb M (2009). Elevated CO_2 and water-availability effect on gas exchange and nodule development in N_2-fixing alfalfa plants. Environ Exp. Botany 65: 18–26.

Ashraf M, Iram AT (2005). Drought stress induced changes in some organic substances in nodules and other plant parts of two potential legumes differing in salt tolerance. Flora. 200: 535–546.

Becana M, Gorgocena Y, Aparicio-Tejo PM, Sánchez-Días M (1986). Nitrogen fixation and leghaemoglobin content during vegetative growth of Alfalfa. J. Plant Phyiol. 123: 117-125.

Bruinsma J (1963). The quantitative analysis of chlorophylls a and b in plant extracts. Photochem. Photobiol. 2: 241-249.

Charlson DV, Korth KL, Purcell LC (2009). Allantoate amidohydrolase transcript expression is independent of drought tolerance in soybean. J. Exp. Botany. 60(3): 84-851.

Clement M, Lambert A, Herouart D, Boncompagni E (2008). Identification of new up-regulated genes under drought stress in soybean nodules. Gene Gene. 426: 15-22.

Dionisio-Sese ML, Tobita S (1998). Antioxidant response of rice seedling to salinity stress. Plant SCI. 135: 1-9.

Figueiredo MVB, Burity HLA, Martïnez CR, Chanway CP (2008). Alleviation of drought stress in the common bean (*Phaseolus vulgaris* L.) by co-inoculation with *Paenibacillus polymyxa* and *Rhizobium tropici*. Appl. Soil Ecol. 40: 182- 188.

Garg N , Manchanda G (2008). Effect of Arbuscular Mycorrhizal Inoculation on Salt-induced Nodule Senescence in *Cajanus cajan* (Pigeonpea). J. Plant Growth Regul. 27: 115-124.

Gonzalez EM, Gálvez L, Arrese-Igor C (2001). Abscisic acid induces a decline in nitrogen fixation that involves leghaemoglobin, but is independent of sucrose synthase activity. J. Exp. Bot. 52: 285-293.

Gross E, Cordeiro L, Caetano FH (2002). Nodule Ultrastructure and Initial Growth of *Anadenanthera peregrina* (L.) Speg. var. falcata (Benth.) Altschul Plants Infected with Rhizobia. Annals Botany 90: 175-183.

Groten K, Dutilleul C, van Heerden PDR, Vanacke H, Bernard S, Finkemeier I, Dietz KJ, Foyer CH (2006). Redox regulation of peroxiredoxin and proteinases by ascorbate and thiols during pea root nodule senescence. FEBS lett. 580: 1269-1276.

Herder GD, Keyser AD, Rycke DR, Rombauts S, Van de Velde W, Clemente MR, Verplancke C, Mergaert P, Kondorosi E, Holsters M, Goormachtig S (2008). Seven in Absentia Proteins Affect Plant Growth and Nodulation in *Medicago truncatula*. Plant Physiol. 148:369-382.

Hessini K, Lacha I M, Cruz C, Soltani A (2009). Role of Ammonium to Limit Nitrate Accumulation and to Increase Water Economy in Wild Swiss Chard. J. Plant Nutr. 32: 821-836.

Irigoyen JJ, Emerich DW, Sánchez-Díaz M (1992). Phosphoenol-pyruvate carboxylase, malate and alcohol dehydrogenase in alfalfa (*Medicago sativa*) nodules under water stress. Physiol. Plantarum. 84: 61-66.

Mahieu S, Germon F, Aveline A, Hauggaard-Nielsen H, Ambus P, Jensen E.S (2009). The influence of water stress on biomass and N accumulation, N partitioning between above and below ground parts and on N rhizodeposition during reproductive growth of pea *(Pisum sativum L.)*. Soil Biol. Biochem. 41:380-387.

Mhadhbi H, Fotopoulos V, Djebali N, Polidoros AN, Aouani ME (2009). Behaviours of *Medicago truncatula*-Sinorhizobium meliloti Symbioses Under Osmotic Stress in Relation with the Symbiotic Partner Input: Effects on Nodule Functioning and Protection. J. Agron. Crop Sci. In Press.

Puppo A, Groten K, Bastian F, Carzaniga R, Soussi M, Lucas M, de Felipe MR, Harrison J, Vanacker H, Foyer CH (2005). Legume nodule senescence: roles for redox and hormone signalling in the orchestration of the natural aging process. New Phytol. 165: 683-701.

Ramos MLG, Parsons R, Sprent JI, James EK (2003). Effect of water stress on nitrogen fixation and nodule structure of common bean. Pesq. Agropec. Bras., Brasilia. 38: 339-347.

Scholander PF, Hammel HT, Bradstreet ED, Hemmingsen EA (1965). Sap pressure in vascular plants. Science 148: 339-346.

Schonfeld MA, Johnson RC, Carver, BF, Mornhinweg, DW (1988). Water relation in winter wheat as drought resistance indicator. Crop Sci. 28: 526-531.

Shao HB, Chu LY, Jaleel CA, Zhao CX (2008). Water-deficit stress-unduced anatomical changes in higher plants. C.R. Biologies 331: 215-225.

Singh B, Singh G (2006). Effects of controlled irrigation on water potential, nitrogen uptake and biomass production in *Dalbergia sissoo* seedlings. Environ. Exp. Bot. 55: 209-219.

Swaraj K, Bishnoi NR (1996). Physiological and biochemical basis of nodule senescence in legumes. Plant Physiol Biochem. 23: 105-116.

Vadez V, Rodier F, Payre H, Drevon JJ (1996). Nodule permeability and nitrogenase-linked respiration in bean genotypes varying in the tolerance to P deficiency. Plant Physiol Biochem. 34: 971-878.

12

Evaluation of eucalyptus essential oil against some plant pathogenic fungi

Nafiseh Katooli[1]*, Raheleh Maghsodlo[2] and Seyed Esmaeil Razavi[2]

[1]Islamic Azad University, Mashhad Branch, Young Researchers Club, Iran.
[2]Department of Plant Protection, University of Agricultural Science and Natural Resources, Gorgan, Iran.

Antifungal activity of eucalyptus *(Eucalyptus camaldulensis* Dehnh.) essential oil was evaluated and suppressed the mycelial growth of postharvest pathogenic fungi, *Penicillium digitatum, Aspergillus flavus, Colletotrichum gloeosporioides* and soilborne pathogenic fungi, *Pythium ultimum, Rhizoctonia solani, Bipolaris sorokiniana* pathogenic fungi. The experiment was carried out with Whatman paper disc method in 25, 50, 75 and 100% concentration of essential oil on PDA culture at 25°C and mycelial growth measured daily for 30 days. The antifungal activity was evaluated under a randomized completely factorial design with three replications. The result showed eucalyptus essential oil in all concentration had completely inhibition of mycelial growth only in *P. ultimum* and *R. solani.*

Key words: Antifungal activity, essential oil, eucalyptus.

INTRODUCTION

Synthetic fungicides are currently used as primary means for the control of plant disease. However, the alternative control methods are needed because of the negative public perceptions about the use of synthetic chemicals, resistance to fungicide among fungal pathogens, and high development cost of new chemicals. The uses of plant-derived products as disease control agents have been studied, since they tend to have low mammalian toxicity, less environmental effects and wide public acceptance (Lee et al., 2007). Essential oils are concentrated, hydrophobic liquid containing volatile aromatic compounds extracted from plants (Isman, 2000). They were previously known to have biological activities such as antifungal (Soliman and Badeaa, 2002), antibacterial (Dorman et al., 2000), insecticidal and nematicidal effects (Pandey et al., 2000).

In this study, the inhibitory effect of eucalyptus *(Eucalyptus camaldulensis* Dehnh.) essential oil was determined against three postharvest pathogenic fungi, *Penicillium digitatum, Aspergillus flavus, Colletotrichum gloeosporioides* and three soilborne pathogenic fungi, *Pythium ultimum, Rhizoctonia solani* and *Bipolaris*

sorokiniana.

MATERIALS AND METHODS

Fungal cultures

The six phytopathogenic fungi such as *C. gloeosporioides* (extracted from citrus), *R. solani* and *B. sorokiniana* (extracted from rice), *P. ultimum* (extracted from Bastard saffron), *P. digitatum* and *A. flavus* (extracted from musty bread) were maintained and grown on potato dextrose agar (PDA). In this experiment, 24 h culture was used.

Extraction of essential oil

Leaves of *(E. camaldulensis* Dehnh.) was collected and washed; then extracted for 4 h by distilled water, using a Clevenger type apparatus established by Montes et al. (2001).

Antifungal activity test

Essential oil (10 µl) was directly assayed to each fungus with 25, 50 and 75% dilution with acetone and undiluted 100%. The control was used for each case by not exposing the fungus to any extract and addition of acetone. Qualitative and quantitative analysis of fungicidal activity was done using Woodward and De Groot (1999) by adding a 5 mm diameter of each fungal growth (three replicates) in the center of Petri dish containing culture PDA. Paper filter discs

*Corresponding author. E-mail: n.katooli@gmail.com.

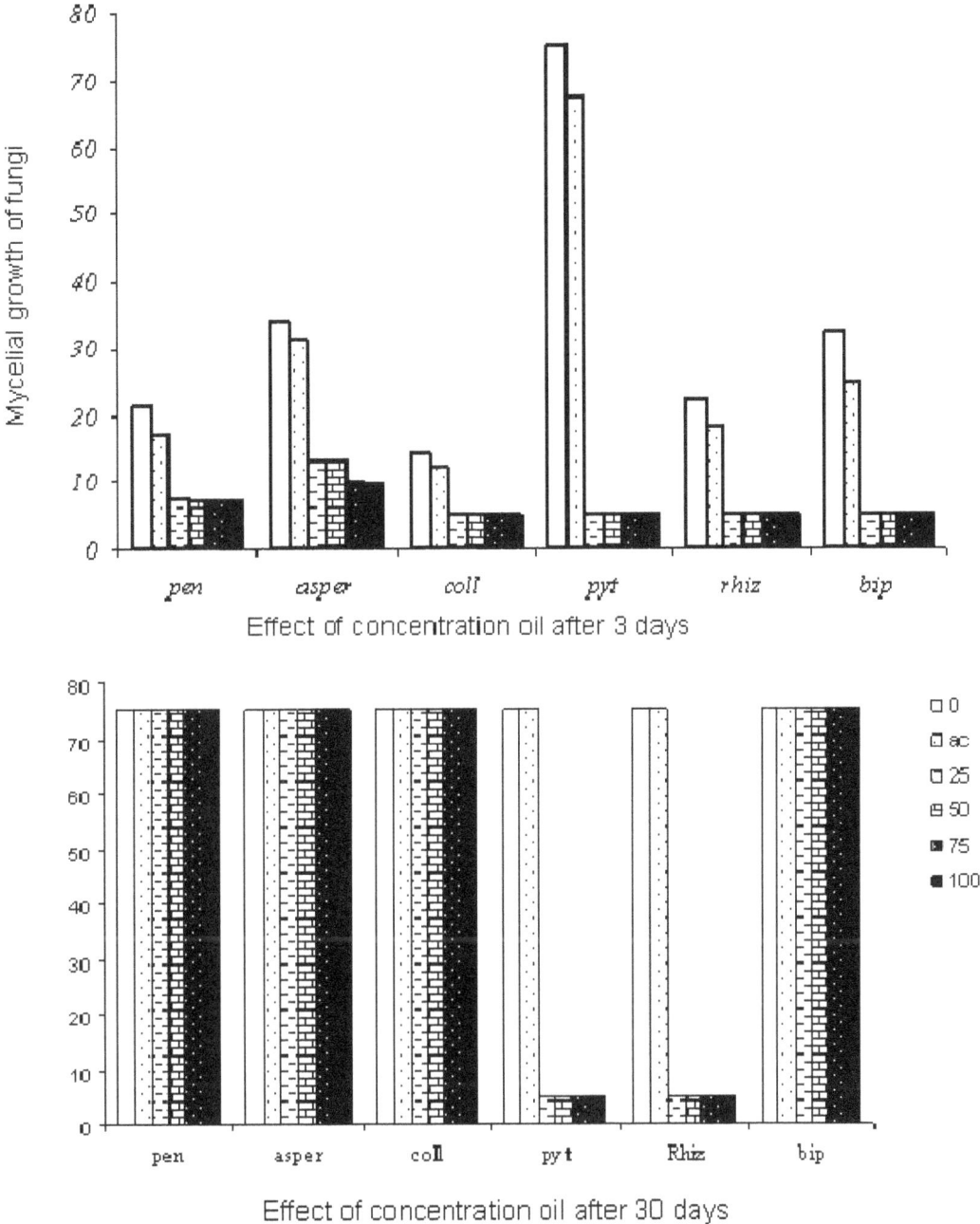

Figure 1. The comparison of mycelial growth (mm) after 3 and 30 days in essential oil present.

(Whatman No. 4 mm diameter) impregnated with increasing dilutions (0, 25, 50, 75, 100%) of essential oil were placed around the fungi and allowed only volatile compounds to be the causative agents for mycelial growth inhibition. The plate was sealed with parafilm immediately after adding each essential oil and incubated for 30 days at 25°C. The diameter of mycelial growth was measured every day.

Statistical analysis

Data were analyzed statistically using analysis of variance (ANOVA) and differences among the means were determined for significance at p≤ 0.01 using LSD test by SASS.

RESULTS

Antifungal activity of eucalyptus (E. camaldulensis Dehnh.) essential oil were tested against six phytopathogenic fungi such as P. digitatum, A. flavus, C. gloeosporioides, P. ultimum, R. solani and B. sorokiniana. According to the results, the effects of four concentration and times of exposure showed significant difference at p ≤ 0.01. The results showed complete inhibition of mycelial growth in P. ultimum and R. solani on all concentration of essential oil after 30 days (Figure 1). B. sorokiniana and C. gloeosporioides showed complete inhibition until 5 days, but after that there was fungi mycelium growth and non inhibition. This essential oil in *P. digitatum* and *A. flavus* had no inhibition. The average of mycelial growth showed significant difference after 5 days.

DISCUSSION AND CONCLUSION

To develop environment-friendly alternatives to synthetic fungicides for the control of fungal plant disease, the interest on essential oils has been increased. In this study we investigated the antifungal activity of eucalyptus essential oil against six soilborne and postharvest disease pathogens. As a result, essential oil of eucalyptus inhibited the mycelial growth in all of the experiment fungi after 3 days. The most important soilborne of fungi, *P. ultimum* and *R. solani*, had 100% complete inhibition of mycelial growth, and agreement with those obtained by Huv et al. (2000), that showed *Eucalyptus unigera* oil inhibited mycelial growth *of* three phytopathogenic fungi such as *C. gloeosporioides* , *R. solani* and *Pythium* spp. This study demonstrated the *in vitro* antifungal activity of essential oil of Eucalyptus against phytopathogenic fungi. However, for the development of essential oils as alternative of synthetic

fungicides, further studies are required to evaluate essential oils for application on plants and sensory quality of treated fruits and vegetables.

REFERENCES

Dorman HJD, Deans SG (2000). Antimicrobial agents from plants: antibacterial activity of plant volatile oils. J. Appl. Microbiol., 88: 308-316

Huv JS, Ahn SY, Koh YJ, Lee CI (2000). Antimicrobial Properties of Cold-Tolerant *Eucalyptus* Species against Phytopathogenic Fungi and Food-Borne Bacterial Pathogens. Plant Pathol. J., 16(5): 286-289.

Isman MB (2000). Plant essential oils for pestand disease management. Crop Prot. 19: 603-608.

Lee SO, Choi GJ, Jang KS, Kim JC (2007). Antifungal Activity of Five Plant Essential Oils as Fumigant Against Postharvest and Soilborne Plant Pathogenic Fungi. Plant Pathol. J., 23(2): 97-102.

Montes MO, Munoz L, Wilkomirsky YT (2001). Plants medicinales de uso en Chile. Quimicay farmacologia. Editorial Universitaria, Santagiago, Chile, 330 p.

Pandey R, Kalra A, Tandon S, Mehrotra N, Singh HN, Kumar S (2000). Essential iols as potent sources of nematicidal compounds. J. Phtopathol., 148: 501-502.

Soliman KM, Badeaa RI (2002). Effect of oil extracted from some medicinal plants on different mycotoxigenic fungi. Food Chem. Toxicol., 40: 1669-1675.

Woodward B, De Groot R (1999). Tolerance of *Wolfiporia cocoa* isolates to copper in agar media. Forest Prod. J., 49(4): 87-94.

13

Host selection behaviour of aphid parasitoids (Aphidiidae: Hymenoptera)

Abdul Rehman* and Wilf Powell

[1]Rice Program, Crop Sciences Institute, National Agricultural Research Centre
Islamabad, Pakistan.
[2]IACR-Rothamsted, Harpenden, Herts., AL5 2JQ, UK.

Biological control is the central stone of Integrated Pest Management (IPM) paradigm and natural enemies are becoming an increasingly desirable prospect. Parasitoids are a widely used group of invertebrate natural enemies as biological control agents and several species are being used to control various aphid pests. In recent years, an increasing emphasis is being given to the conservation and manipulation of naturally-occurring populations of parasitoids in agricultural ecosystems over traditional approaches to biological control. But these approaches must be underpinned by basic knowledge in host preference behaviour and ecology of the parasitoid species being manipulated. Three aspects of host preference behaviour, namely host recognition, host acceptance and host suitability have been discussed in this paper. Parasitoids' host selection strategy is based on using long-range and short-range cues. Parasitoids respond to both semiochemical and physical stimuli to locate and recognise their hosts. These responses are either due to aphid sex pheromones acting as kairomones, or due to aphid-induced plant volatiles, acting as synomones. Various interactions like genetic, learning and conditioning factors, which play an important role in host selection behaviour of foraging parasitoids, have been discussed. The learning ability provides the parasitoid with behavioural plasticity to adapt its responses to suit prevailing foraging opportunities and the maintenance of genetic variability within natural populations of parasitoids may promote long-term population stability and help conserving genetic diversity by ensuring flexibility in host selection.

Key words: Aphid parasitoids, host preference, host selection, host recognition, host acceptance, host suitability, conservation of aphid parasitoids, manipulation of parasitoid behaviour, habitat.

INTRODUCTION

The term 'parasitoid' was introduced by Reuter (1913), but became universally accepted during the last three decades. Godfray (1994) defined a parasitoid based on its larval feeding habits; it exclusively feeds on one host and eventually kills it. Parasitoids are intermediate between predators and true parasites. Like predators, they always kill the host they attack and can have profound effects on host population dynamics (Quicke, 1997). Like many parasites, they require just a single host to develop and often have a short period when they are acting as true parasites. The adult parasitoid is free living only larval stage kills the host.

The aphid parasitoids were regarded as member of separate family Aphidiidae, whereas many authors now consider them as a subfamily, Aphidiinae within the Braconidae (O'Donnell, 1989; Reed et al., 1995). More than 400 parasitoid species have been recorded (Starý, 1988). In Europe, several parasitoid species have been recorded from cereal aphids (Starý, 1976; Carter et al., 1980; Powell, 1982; Dedryver et al., 1991). Alam and Hafiz (1963) listed 23 aphid parasitoid species and 14 insect predator species

*Corresponding author. E-mail: abdul258@gmail.com.

attacking 43 aphid pests in Pakistan. Hamid (1983) studied aphids and their natural enemies on cereal crops in Pakistan and concluded that the natural enemies play a significant role in maintaining a natural balance throughout the country. In this study, 8 aphid parasitoids, including *Praon pakistanum* (Kirkland), were recorded from 8 aphid species.

Aphid parasitoids are important components of the natural enemy guild which helps to control pest aphid populations in a variety of crops. Starý (1987) and Hågvar and Hofsvang (1991) reviewed the impact of aphidiines on aphid populations in some major ecosystems, and in different geographical regions.

Sometimes parasitoids are not as efficient as they could be in the field due to the influence of farming practices such as pesticide use, climatic and other environmental factors which disrupt their synchrony with target pests, cause dispersal away from crops or adversely affect parasitoid populations. However, the parasitoids appear to be more effective if both parasitoids and their host coincide in the crop early in spring (Powell, 1983; Powell et al., 1983, 1986). Early season synchrony depends on parasitoids successfully over wintering near the early sown crops, and grassland may serve as a reservoir for over wintering parasitoids (Vickerman, 1982; Vorley and Wratten, 1987). This paper reviews the host selection behaviour of aphid parasitoids and discusses the opportunities to manipulate their behaviour for better control of aphids.

HOST SELECTION PROCESS

The behaviour of parasitoids in selecting their hosts for oviposition is fascinating (Mackauer et al., 1996). According to Godfray (1994), host preference may be either a rationalised attitude of female that determines host acceptance or rejection, and is influenced by female fitness affecting oviposition behaviour. Host selection may be the female response to the selected attributes that distinguish hosts from non-hosts (Mackauer et al., 1996).

Upon emergence, the female parasitoid needs to locate suitable hosts in order to propagate. The female parasitoids' ability to find suitable hosts is vital as they may be emerging away from suitable aphid populations, or may be emerging in an unsuitable environment, such as within a crop from which the aphids have dispersed (Starý, 1988). Parasitoids sometimes need to disperse from unsuitable habitats (Vinson, 1981). Parasitoids use a variety of chemical and physical cues during the habitat location, host location and host examination phases of host selection (Vinson, 1984; Schmidt, 1991; Vet and Dicke, 1992; Turlings et al., 1993; Powell et al., 1998; Rehman, 1999). The behavioural responses expressed by a foraging parasitoid at any one time are largely determined by its genotype, its physiological state, and its previous environmental adaptability (Vet et al., 1990). Parasitoids have to search for their hosts in

a highly complex environment. Vet (1995) argued that parasitoids search non-randomly, learn cues from different trophic levels during foraging and alter their decisions accordingly. Several parasitoid species respond to stimuli associated with the hosts or their host plants before the host itself is encountered. They also respond to chemical stimuli present during successful foraging bouts by changing their searching behaviour, which improves their chances of finding hosts. Many studies on parasitoids have been conducted considering aspects of their behaviour in the presence of chemical cues. For example, reduced walking speed, stopping and increased turning, have been reported (Hood-Henderson and Forbes, 1988; Bouchard and Cloutier, 1984; van Alphen and Vet, 1986).

Parasitoid host selection is deterministic and focuses on the proximate mechanisms by which a female locates and selects a potential host for oviposition. This assumes a hierarchy of discrete steps that include habitat location, host location, host acceptance, host suitability (Doutt, 1959; Vinson, 1976), and host regulation (Vinson and Iwantsch, 1980). Host selection ultimately results from a sequence of behaviours that guide foraging females to suitable hosts by the elimination of unsuitable habitats and non-hosts. It is thought that the host selection process depends both on environmental and host factors and that the parasitoid is guided to a host habitat and to the host itself by chemical and physical parameters. These cues elicit a series of direct behavioural responses by the female that serve to reduce and restrict the area and habitats searched, leading to host location. Hagvar and Hofsvang (1991) give detail of the host selection processes specifically for aphidiines. Weseloh (1981) and Arthur (1971) review host location and host acceptance, respectively. Michaud and Mackauer (1994) distinguish three discrete steps involved in host selection; host recognition, host evaluation, and host acceptance (oviposition). According to them the entire host selection process may be described as:

1. Host habitat location: The female searches for habitats where suitable host plants and hosts occur.
2. Host location: The female searches for the host, on or very close to the plants.
3. Host recognition: The female encounters the potential host, evaluates it with antennae and ovipositor probing.
4. Host acceptance: The parasitoid examines the host and decides to oviposit and deposit an egg.
5. Host suitability: The deposition of an egg and its subsequent development dependent on the host's physiological state.
6. Host regulation: The parasitoid development may affect its host development, behaviour, physiology and biochemistry.

The first five steps can be combined as aspects of the host selection process which involve the use of olfactory, visual and tactile cues to locate and assess the host. In this review, host habitat location, host location, host recognition, host acceptance and host suitability are discussed.

Host habitat and location

These are the initial steps to locate food and oviposition sources by female parasitoids. Parasitoids use long-range cues including electromagnetic radiation, sound or odour at this step. Chemical cues appear to play a major role at almost every level of the host selection process. Semiochemicals emanating from the host, from the host's food plant, from organisms associated with the host or from a combination of these have been shown to be important cues in the host habitat location (Vinson, 1976, 1984). Parasitoids respond to the aphids' host plants, usually attracted by plant-produced synomones, and sometimes also by visual cues. Olfactory responses to volatiles from aphid host plants are probably more important than vision in host habitat location in aphidiines. Attraction to odour of the host plants has been demonstrated in some species (Read et al., 1970; Singh and Sinha, 1982; Powell and Zhang, 1983; Powell et al., 1998; Rehman, 1999; Storeck et al., 2000; Hatano et al., 2008). Wickremasinghe and van Emden (1992) reported a strong host plant odour and aphid odour attraction in several parasitoids and aphid combinations.

The alteration of a food source by the injury of herbivores may result in the release of different odour. *Cardiochiles nigriceps* Viereck appears to cue first on plant factors, but once in the proper habitat, it may cue on injured plant tissue (Vinson, 1975). In other cases, odour from the host provides the necessary cues for host habitat and host location.

Some aphid parasitoids did not appear to respond to odour from their host plants. Parasitoids that attack polyphagous hosts are probably less likely to use plant volatiles in their host habitat location process. *Aphidius nigripes* Ashmead parasitises polyphagous aphids, and showed no response to plant odour (Bouchard and Cloutier, 1985). Several studies on host-parasitoid interactions of the specialist parasitoid *Diaeretiella rapae* and its host *Brevicoryne brassicae* indicate that this species, which shows a great degree of host and habitat specificity, uses odours of its host food plant rather than from its host in host habitat location. Allyl isothiocyanate, the major chemical constituent of cruciferous plants, is the source of attraction, shown in wind tunnel experiments (Sheehan and Shelton, 1989) and in olfactometer experiments (Read et al., 1970). Gently and Barbosa (2006) reported that leaf epicuticular wax plays an important role on the movement, foraging behaviour and attack efficiency of *D. rapae*. Van Emden (1978) also has demonstrated this interaction and observed differences in parasitization of cabbage aphids on two cultivars of Brussels sprout in greenhouse trials.

In another olfactometer test, Reed et al. (1995) found no response of *D. rapae* to cabbage leaves. However, females were attracted to *B. brassicae* infested leaves and the response was greater than to wheat leaves infested with Russian wheat aphid, *Diuraphis noxia*. In another wind tunnel study, Sheehan and Shelton (1989) found that *D. rapae* reared on collards showed increased flight responses to these plants than to potato. This suggests that *D. rapae* has an innate preference for the crucifer feeding aphid system. Although these experiments permit several conclusions, few data exist from field experiments, and a significant response in an olfactometer does not necessarily imply long range attraction in the field.

There is mixed evidence concerning the attraction of other parasitoid species to volatiles from plants or plant-host complexes. In an olfactometer test, the cereal aphid parasitoids *Aphidius uzbekistanicus* and *Aphidius ervi* responded to uninfested leaves of their host plants (Powell and Zhang, 1983; Powell et al., 1998). In Y-tube olfactometer tests, Wickremasinghe and van Emden (1992) also recorded greater responses by the aphid parasitoids *A. ervi* and *A. rhopalosiphi* to the plants on which the females were reared than to their host aphids, but they responded towards even more to the plant-host system. *A. rhopalosiphi* also showed a greater response towards the particular variety of wheat on which it had been reared.

Herbivore-induced synomones are also involved in habitat location by aphid parasitoids. Guerrieri et al. (1993) recorded a greater upwind flight response by *A. ervi* to a plant-host system than to either the aphid or plant alone in wind tunnel tests. The parasitoid also responded to a host-damaged plant from which aphids had been removed. Similar responses by *A. ervi* to a plant-host complex were demonstrated by Du et al. (1997) and Powell et al. (1998). The female response to broad bean plants damaged by *Acyrthosiphon pisum* was greater than to undamaged plants or mechanically damaged plants. The parasitoid also responded more to *A. pisum* damaged bean plants than to, *Aphis fabae* infested plants. This indicates that the response of parasitoids to herbivore-induced synomones is host specific. Kris and Heimpel (2007) recorded responses of naïve and experienced *Binodoxys communis* (Gahan) females to odours from both target and non-target host plant complexes by using Y-tube olfactometer assays. The study indicated that *B. communis* females respond to a broad array of olfactory stimuli, exhibit low fidelity for any particular odour, and employ some behavioral plasticity in their response to volatile cues.

Once the parasitoid has reached a potential host habitat, it begins to search for the host on or near the host plant. The females respond to physical, chemical and visual stimuli associated with their hosts before the host itself is encountered. Most of the chemical stimuli act as kairomones, being produced by the host itself or arising from host products. Such kairomones are either volatile, perceived by olfaction, or non- volatile, contact kairomones. These chemicals affect female parasitoid behaviour by changing their searching time, reducing walking speed, and increasing frequency of turning, and ameliorate their chances of finding host (Bouchard and Cloutier, 1984).

In a Y-tube olfactometer, Wickremasinghe and van Emden (1992) examined the responses of *A. rhopalosophi, Lysiphlebus fabarum* (Marsh) and species of *Trioxys* and *Praon* and found all were attracted to their respective hosts. van Emden (1995) reported that *A. rhopalosiphi* preferred wheat varieties on which its aphid host had developed, proposing that females became 'conditioned' during immature development, rather than by post-eclosion experience, to volatiles associated with the aphid's host plant. *A. ervi* was attracted to the nettle aphid, *Microlophium carnosum* (Buckton), on which the *A. ervi* was originally collected from field. In another study, Powell and Zhang (1983) found no response to nettle aphid by *A. ervi* collected from *A. pisum*, suggesting the existence of specialised races in the field.

The use of aphid honeydew as a host finding kairomone is common amongst the aphid parasitoids. Several studies have now proved that aphidiines use honeydew as a kairomone for host location (Singh and Sinha, 1982; Powell and Zhang, 1983; Bouchard and Cloutier, 1984, 1985). Contact with honeydew in a petri dish stimulated abdominal protraction in *A. nigripes* (Bouchard and Cloutier, 1984). Ayal (1987) showed *D. rapae* on a crucifer plant searched contaminated lower leaves followed by an upward flight if no host was encountered. She suggested that the parasitoid uses honeydew on the leaves as a cue for evaluating the number of aphids on the plants.

Although the response mechanisms involved in host location of aphidiines vary between species, chemo-orientation apparently dominates over the use of physical cues. Probably both olfaction and chemotactile responses play a part and, at least in some species, may be complemented by vision. Olfactory cues may originate from host plants (Read et al., 1970; Wickremasinghe and van Emden, 1992; Guerreri et al., 1993; Braimah and van Emden, 1994; van Emden et al., 1996; Du et al., 1996; Blande et al., 2008) or from aphid-produced substances such as sex pheromones (Hardie et al., 1991, 1994; Powell et al., 1993) or may be from host and non host plants (Kris and Heimpel, 2007). These cues play significant role in host location of aphid parasitoids, and possibly also in host recognition.

Host recognition and host acceptance

Once the host has been located and contact has been made, the next step for the parasitoid is to accept or reject the host for oviposition. In this paper host acceptance is considered as two behavioural steps: Host recognition

(oviposition attack), and host acceptance (egg deposition). 'Oviposition attack' refers to the visible oviposition behaviour of the aphidiine female until her ovipisitor has penetrated the host cuticle. 'Egg deposition' refers to the release of a parasitoid egg into the host's haemolymph after oviposition insertion. This distinction seems appropriate because oviposition attack behaviour, but not necessarily egg release, is probably induced by aphid external factor that is, physical or chemical stimuli and the parasitoid internal status. The release of a parasitoid egg into the aphid haemolymph, however, may well be affected by the host's internal physiological conditions as detected by receptors on the ovipositor. Antennal sensoria are involved in odour perception as well as in the evaluation of contact chemicals on the aphid cuticle. Sensoria on the ovipositor probably aid in the evaluation of host quality during ovipositor probing. Reviews of host recognition and acceptance are given by Vinson (1976), Arthur (1981), Hagvar and Hofsvang (1991) and Mackauer et al. (1996).

Host recognition may involve changes in the female's behaviour, and directed responses towards a host. Once a female has encountered a potential host, she examines its quality and suitability, by antennation and ovipositor probing, for offspring development. The aphid is accepted if its perceived value exceeds the female's response threshold. The parasitoid's inability to recognise a suitable host cannot be distinguished from pre-attack rejection (Mackauer et al., 1996). Before oviposition, adult females have to go through various behavioural patterns regulated by both physical and chemical cues. The females of aphidiine parasitoids seem to search for hosts randomly on plants and aphids are usually finally detected by antennal contact. Rehman (1999) also observed and categorized following behavioural patterns of oviposition by aphid parasitoid *Praon volucre* as shown in Figure 1.

1. RS (random searching): The female walks in the arena randomly in search of hosts,
2. DA (detection and approach): The female shows antennal orientation towards a host,
3. AE (antennal examination) the female encounters a host and examines it with antennae,
4. AB (abdomen bending): The female shows orientation for oviposition and bends her abdomen,
5. OV (oviposition): The female actually stabs the host to lay an egg,
6. Preening: The female cleans and grooms her ovipositor and antennae.

The host represents an essential resource for a parasitoid that is characterised by physical, chemical, and behavioural attributes. These attributes determine the host's recognition and acceptance by a parasitoid. Size, shape, and colour of host are the main visual attributes; odour and chemical composition represent the chemical attributes; and movement and host defence tactics represent the behavioural vioural attributes. A parasitoid's preference for different

(a) Random searching of aphid

(b) Aphid detection and approach

(c) Examination of aphid with antennae

(d) Orientation for oviposition (abdomen bending)

(e) Oviposition

Figure 1. Diagrammatic representation of *Praon volucre* behavioural patterns observed on *Sitobion avenae* (Rehman, 1999).

hosts appears to be innate (Chow and Mackauer, 1992), and probably depends on the host's state variables (Mackauer et al., 1996). Female parasitoids seem to search randomly on a leaf, along the veins and leaf edges and aphids are usually detected by antennal contact (Hofsvang and Hagvar, 1986). In Aphidiinae, the influence of host species, size and age, colour morph, aphid defences, chemicals, and already parasitised hosts on host acceptance has been investigated and discussed as below.

Host species

Recognition of the host species has vital importance for parasitoids to ensure a reasonable production of their offspring. In the presence of many host species, both the quality, abundance and distribution pattern of hosts will affect the parasitoid's selection. The evolutionary trend amongst Aphidiinae is apparently towards oligophagy, involving parasitization of several aphid species from a single genus (Starý, 1988). Related aphids are often attacked by related parasitoids, suggesting that these

parasitoids have coevolved with their hosts (Mackauer and Chow, 1986).

Host species selection is important in biological control. Aphid species on wild plants may act as a reservoir for parasitoids attacking other aphid species on nearby crops (Powell, 1986). Aphid species that are not recognised as potential hosts in the field may be sometimes accepted and prove suitable for parasitoid development in the laboratory. Alternative hosts may therefore prove valuable in the mass propagation of parasitoids (Mackauer and Kambhampati, 1988). *Aphidius smithi* from the pea aphid *A. pisum* was successfully reared on *Myzus persicae* on broad beans (Fox et al., 1967).

Preference for certain host species has been demonstrated in laboratory studies where parasitoids more often oviposit in some species than in others, when both the host species are offered separately or simultaneously (Dhiman and Kumar, 1983; Pungerl, 1984; Powell and Wright, 1988; Rehman, 1999; Chau and Mackauer, 2001). Gardner and Dixon (1985) demonstrated that, in the field, different levels of parasitization on various host species may be a result of parasitoid foraging behaviour rather than preference for certain host species. In some cases, parasitoids do not respond to the presence of a particular aphid species and will not oviposit in it. Carver and Woodlock (1985) showed that the pea aphid parasitoids, A. smithi and A. Eadyi did not respond to the presence of Acyrthosiphon kondoi Shinji and did not oviposit. In contrast, A. ervi, Aphidius pisivorus C.F. Smith and P. volucre readily oviposited and successfully developed in A. kondoi. It seems that this aphid has a small parasitoid spectrum, and that A. ervi is the only known efficient parasitoid of this aphid in the field. Dhiman and Kumar (1983) demonstrated the preference of D. rapae for different hosts and found that Lypaphis erysimi was highly preferred over B. brassicae and Myzus persicae.

Rehman (1999) studied effect of host species; the role of host plants in enhancing their abilities to recognise hosts; and the influence of female conditioning at the time of emergence on the host recognition stage of host selection behaviour of Praon myzophagum and P. volucre. Both the parasitoids expressed their preference at the host recognition stage. However, they attacked at higher rates and deposited more eggs in the aphid species from which they have been reared. A significant effect of conditioning was observed on host recognition by P. volucre. Sitobion avenae-reared females of P. volucre which were excised from mummies before emergence showed a significant reduction in attack rate and took longer to attack the first individual of their original host than did females emerged from undissected mummies. It is possible that either genetic selection occurred during rearing and genotype influenced the response of parasitoids to host-derived cues during recognition of individual host species (Powell and Wright, 1988). Or the emerging females could have been conditioned to the cues associated with the host and its food plant through contact with the mummy skin at the time of emergence (van Emden et al., 1996; Rehman, 1999; Gutiérrez et al. 2007), thereby affecting the subsequent host selection.

Both learning, which is a relatively permanent change in behaviour as a result of reinforced practice, and conditioning, where an organism acquires the capacity to respond to a stimulus with a reflex reaction to another stimulus, have been implicated in having an important influence on host selection (Vinson, 1976). Pungerl (1984) found that the parasitoid A. ervi, collected from the pea aphid Acyrthosiphon pisum, also parasitised S. avenae and M. persicae, whereas the same parasitoid collected from S. avenae did not parasitise A. pisum and M. persicae. Powell and Wright (1988) found such conditioning and different responses between parasitoid populations from the field and from laboratory cultures. They found that A. ervi reared on A. pisum produced fewer mummies on M. carnosum than A. ervi cultured on M. carnosum, whereas the latter readily accepted both hosts. Further, they found that A. rhopalosiphi

from laboratory culture produced more mummies on Metopolophium dirhodum than on S. avenae, regardless of their original host. This preference was not shown in field collected parasitoid populations. This may be due to the fact that field populations are more genetically diverse than laboratory populations and may show different behaviour if genotype influences host acceptance and suitability. Powell and Wright (1988) showed that the host species on which their male parent had been reared often changed the female preference, suggesting a genetic influence on host acceptance and perhaps also on host suitability.

Foraging parasitoids may use visual cues to distinguish between hosts and non-hosts. Michaud and Mackauer (1994, 1995) examined the use of visual cues in host recognition by the aphidiine species, A. ervi, A. pisivorus, A. smithi, Ephedrus californicus Baker, Monoctonus paulensis (Ashmead), and Praon pequodorum. These parasitoids showed similar innate preferences with regard to aphid species and colour; they preferred pea aphids, A. pisum, over alfalfa aphids, Macrosiphum creelii, and the green over the pink colour morph of alfalfa aphids. Host recognition and acceptance by A. ervi are regulated by visual as well as chemical cues; the oviposition response can be elicited by appropriate colour stimuli in the absence of chemical cues (Battaglia et al., 1995, 1999; Langley et al., 2006).

Host plants seem to play an important role in the host recognition and acceptance behaviour of parasitoids. The cereal aphid specialist A. rhopalosiphi attacked the host species S. avenae significantly more when the aphids were presented together with a wheat plant (Braimah and van Emden, 1994). The parasitoid also showed greater response to the non-host M. persicae when this was presented with wheat leaves than when it was presented with Brussels sprouts leaves, indicating the role of plant-derived synomones in aphid-parasitoid interaction. In another study Powell and Wright (1992) observed more oviposition stabs by A. rhopalosiphi in A. pisum, a non-host aphid, when wheat leaves were present. Similar trends in host preference behaviour of generalist aphid parasitoid, P. volucre as influenced by host plants were also observed by Rehman (1999).

It has also been demonstrated that oviposition may be a matter of experience and that female parasitoids with a wide range of hosts often prefer the host species from which they have been reared (Eijsackers and van Lenteren, 1970; Rehman, 1999). It would not be surprising to find that a female's preference for a particular host or a host-plant complex, after being determined by the proper stimuli for habitat and host location and recognition, is strongly influenced by prior exposure and success. Preference for a particular host may be influenced by both genetic factors and conditioning (Rehman, 1999; Poppy and Powell, 2004; Poppy et al., 2008). Prior oviposition experience of A. pisivorus on A. pisum affected the attack rate on Macrosiphum creelii but did not change its innate order of host preference (Chow and Mackauer, 1992).

Host stage/size

Host stage selection may also be important in respect of the rearing techniques used in the mass production of parasitoids and for experimental design in parasitoid studies. The parasitoid may show an evolutionary preference for certain instars (Liu et al., 1984). Morphology and behaviour of an aphid, which may differ between various instars, possibly influences its susceptibility to parasitism. Cuticular thickness and aphid defence behaviour, such as kicking, jerking, walking away and dropping from plants, are examples of age-dependent host qualities which could affect a parasitoid's success.

Several studies have demonstrated that aphid size, age and development stage can influence the probability of host acceptance. However, it seems to be very dependent on both host and parasitoid species. Although all aphid instars are parasitized, parasitoids prefer to attack second and third instar aphids. This may be due to more effective defence behaviour, and so increased handling time when attacking larger fourth instars and adult aphids and varying encounter rates due to different host sizes and instar abundances (Shirota et al., 1983; Liu et al., 1984; Kant et al., 2008). Studies have demonstrated that preference for certain aphid instars may be due to aphid defense reactions, which may vary between instars (Takada, 1975; Singh and Sinha, 1982; Hofsvang and Hagvar, 1986; He and Wang, 2006). The ambiguity of choice tests have been demonstrated by several authors. Mackauer (1973) found that A. smithi changed its preference from 1st to 2nd or older instars when females were given a choice. In P. pequodorum, the preference changed from 3rd to equal preference for 2nd, 3rd and 4th instars in choice tests (Sequeira and Mackauer, 1987). Cheng et al. (2010) explored the potential relationship between aphidiine parasitoid development and the primary endosymbiont in aphids by focusing on specific aphid instars and the relative effects on parasitoid oviposition behavior and progeny development. Mackauer (1983) refers to methodological difficulties in choice tests, stressing that preference is not constant but influenced by test duration and by the parasitoids functional response to density. It is also age dependent, but He et al. (2006) are also of the opinion that oviposition strategy of A. ervi is host density dependent.

Chemicals associated with hosts, which the female detects after oviposition probing, may influence parasitoid oviposition behaviour. Pennacchio et al. (1994) observed that Aphidius microlophii probed non-host aphids with the ovipositor but did not release an egg, indicating that receptors on the parasitoid ovipositor can detect internal cues in the host. Parasitoids may have positive or negative responses to the cornicle secretion of their aphid hosts. A.

ervi showed antennal examination and oviposition behaviour towards glass beads contaminated with the cornicle secretion of its host A. pisum (Battaglia et al., 1993, 1995).

Host quality

Host quality for parasitoid growth and development is often assumed to be associated with host size (Waage, 1986).

The parasitised aphid continues to feed, grow and develop. The host represents an open resource system in the future, as opposed to current resources. However, both current and future resources are functions of host age or stage at the time of parasitization (Mackauer and Sequeria, 1993; Mackauer et al., 1997). Parasitoid larvae grew at different rates in different aphids of similar size, which suggests that quality is a specific attribute of each host species (Sequeria and Mackauer, 1993).

The quality of the aphid host plant has a direct bearing on the host quality for parasitoids. Aphids reared on partially resistant plants often were smaller, and showed increased restlessness, compared with counterparts on susceptible plants (van Emden, 1995). Furthermore, host-plant effects can be cumulative, as pea aphids reared for consecutive generations on nutrient-deprived broad beans were smaller than those reared on plants grown on complete nutrients. The aphid's reduced growth potential on low quality plants was reflected in a longer time to adult and in the increased mortality of parasitoids developing in such hosts (Stadler and Mackauer, 1996). Cheng et al. (2010) suggested that age or body size of host aphids may not be the only cue exercised by Lysiphlebus ambiguus to evaluate host quality and that offspring parasitoids may be able to compensate for the nutrition stress associated with disruption of primary endosymbiotc bacteria in aposymbiotic aphids.

Females parasitoids of aphid are generally larger than males, and this may be the result of sex-specific allocation of offspring to higher and lower quality hosts; the sex-specific exploitation of host resources (Mackauer, 1996). When hosts vary in quality, females gain more in fitness from increased size than males, and so the mother may allocate more daughters to large (or high quality) hosts and more sons to smaller (low quality) hosts (Charnov et al., 1981; Godfray, 1994). In the laboratory, Ephedrus californicus was more likely to accept large than small aphids when these were equally available, and were more likely to deposit fertilised eggs (daughters) in higher quality aphids (Cloutier et al., 1991). The sex ratio in field populations of several species of aphidiines tends to be female-biased (Mackauer, 1976; Singh and Sinha 1982; Sequeira and Mackauer, 1993). Sequeira and Mackauer (1992) reported that, in hosts of equal size, females of A. ervi had a higher growth rate than males, growing to a larger size without a corresponding increase in development time, suggesting that larvae may exploit host resources in a sex-specific manner.

Colour forms are generally considered as variants of the same aphid species, but such forms often differ in attributes other than pigmentation, including fecundity, preferred host plant, and behaviour (Miyazaki, 1987). Hence colour polymorphism can affect parasitoid host acceptance behaviour in several ways. Ankersmit et al. (1986) showed that green forms of *S. avenae* are more frequently parasitised than brown forms by *A. rhopalosiphi.* Langley et al. (2006) reported that aphid parasitoid, *A. ervi* altered its preference for pea aphid colour morphs.

Because insect vision affords no depth of field, size and shape evaluation probably occurs at close range during host recognition. A minimum host size is apparently not critical for host acceptance and suitability. The importance of shape perception in host detection by aphidiine wasps has received little attention. Battaglia et al. (1995) found that green colour alone could induce oviposition in naive females of *A. ervi*, but this response was enhanced by pea aphid shape.

Host suitability

The successful development of the parasitoid depends on the selection of a suitable host, and is directly related to host nutrition, intraspecific larval competition, the host's immunity response and the host's endocrine balance. Different host species may differ in their suitability. Some authors distinguish host suitability (Vinson and Iwantsch, 1980b) and host regulation (Vinson and Iwantsch, 1980a) as separate criteria of host selection by a parasitoid. For clarity, Mackauer et al. (1996) distinguish between host suitability, host quality, and host value. They suggested that host suitability and quality are assessed by means of innate responses to the host species and the host individual, respectively.

Host species

Host acceptance and host suitability are usually correlated, but females in several species are known to accept aphids that are unsuitable for immature development (Griffiths, 1960). Thus, acceptance is insufficient evidence of host suitability, and rejection does not indicate that a candidate host is in fact unsuitable. Moreover, some hosts may be suitable and available but not susceptible to parasitism. Sclerotisation of the host's cuticle can interfere with successful oviposition and larval development.
The host species may influence the rate of development and the survival of a parasitoid. A host may be unsuitable due to the lack of some necessary nutritional or hormonal resource (Carver and Sullivan, 1988; Kant et al., 2008). Different host species appear to have different internal defences against the same parasitoid species. *A. rhopalo-siphi* developed more successfully in *S. avenae* than in *M. dirhodum* (Ankersmit, 1983). Aphids may encapsulate the parasitoid egg or larva as a defence mechanism, but this appears to be rare in aphids. Egg encapsulation has only been demonstrated in *M. ascalonicus* Doncaster and *Aulacorthum circumflexum* (Buckton), both aphid species encapsulated eggs and young larvae of *D. rapae,* and in *S. avenae* which encapsulated *A. rhopalosiphi* larvae (Carver and Sullivan, 1988).

Host size

As stated earlier, nutritional deficiency may affect the parasitoid rate of development and survival inside the host. It also can have noticeable effects on size, sex ratio, longevity and fecundity of the parasitoid (Vinson and Iwantsch, 1980). However, in the parasitoid *Aphidius sonchi* Marshall attacking the aphid *Hyperomyzus lactucae* (L.), no noticeable effect of host size on parasitoid development has been found (Liu, 1985). Parasitoids may develop at a slower rate in earlier host instars than in later instars (Hafeez, 1961; Hagvar and Hofsvang, 1986). Mackauer (1986) and Mackauer and Chow (1986) emphasised that the development rate and adult weight of *A. smithi* depended not only on host size at the time of parasitism, but also on the host's capability to grow while parasitised. A significantly lower parasitoid emergence from mummies has been recorded from aphids parasitised as adults than from aphids parasitised as embryos inside their mother (Mackauer and Kambhampati, 1988).

Sex ratio may be unaffected by host size at parasitisation (Liu, 1985; Hagvar and Hofsvang, 1986), but a higher proportion of female offspring emerge from larger hosts (Cloutier et al., 1981). Wellings (1988) attributed a male-biased sex ratio of *A. ervi* emerging from smaller hosts to better male survival in such small hosts, since there was no evidence of facultative control of the primary sex ratio. The sex ratio of the emerging parasitoids may also be influenced by the parental sex ratio.

Generally, smaller hosts give rise to smaller parasitoids with reduced longevity and fecundity. Such relationships between host size, size of emerging parasitoid and parasitoid fecundity have been demonstrated in the aphidiines *A. sonchi* (Liu, 1985); *A. smithi* (Mackauer and Kambhampati, 1988; He and Wang, 2006) and *A. rhopalosiphi* (Haq, 1997). Wellings (1988) found a correlation between aphid size and the size of emerging parasitoids for *A. ervi*. Since aphid weight may depend on plant quality, parasitoid fecundity may also be influenced by plant quality (Haq, 1997). However, the effect of host size on parasitoid fecundity may be less important than its influence on the developmental rate of the parasitoid (Mackauer, 1986).

INFLUENCE OF GENETICS IN HOST SELECTION

The reproductive success of female parasitoids depends on

their ability to find and select suitable hosts in a changing and diversified environment. Parasitoid-host interactions themselves illustrate the complex dynamics that can arise from genetic variability in host and parasitoid species. A continuous evaluation of such interactions is conceivable only if parasitoids' biological traits are determined by genetic variation on which natural selection can act. Mitchell-Olds (1995) reported that genetic variation affects fitness in wild populations adapted to different environments. Various behavioural traits in hymenopterous parasitoids, such as searching rate for the host, handling time, host acceptance, host suitability, fecundity and sex allocation, that affect their establishment or control of pests, have been documented (Hopper et al., 1993). Cronin and Strong (1996) suggested that the traits comprising the foraging strategy of *A. delicatus* should be amenable to selection, predicting that selection for larger wasps will result in large offspring with greater egg loads and higher oviposition rates. Wasps with this combination of attributes are likely to be more efficient natural enemies for use in biological control. However, genotype-environment interactions may play an important role in maintaining genetic variability in body size in natural populations of the aphid parasitoid *A. ervi* (Sequeira and Mackauer, 1992). Genetic variability of abilities for associative learning of odour has been demonstrated by a number of authors (Tully and Hirsch, 1982; Brandes, 1991; Bhagavan et al., 1994). Evidence for the role of genetics and learning in aphid parasitoid foraging behaviour, and the difficulty in differentiating between genetic responses and those conditioned during parasitoid development, has been discussed by Poppy et al. (1997, 2008).

Genetic factors influence the host recognition and attack behaviour of the closely related aphid parasitoids *A. ervi* and *A. microlophii* (Powell and Wright, 1988). Poppy et al. (1997) argued that like many other behavioural traits, parasitoid responses to semiochemicals vary between individuals and this variation could be influenced by genotype, phenotype, the individual's physiology and the environment. Unfortunately, very few studies have investigated the genetics of host-parasitoid interactions. Mackauer et al. (1996) in their review on the host choice by aphidiid parasitoids have mentioned that 'unfortunately no data are available on the genetic variation in host recognition and acceptance of aphid parasitoids in literature'.

Individuals within populations usually vary genetically, and this variation is often expressed both in insect's morphology and in a range of biological attributes such as behaviour (Roush, 1989). Genetic variation may thus have considerable influence on the parasitoids' efficiency (Powell et al., 1996). The mother-daughter correlation studies on the host recognition and host preference behaviour of the generalist aphid parasitoids *Praon myzophagum* and *P. volucre* were conducted by Rehman (1999; Poppy and Powell, 2004). Host recognition regressions between mother-daughter, daughter-granddaughter and mother-granddaughter showed statistically highly significant (P<

0.001) results. It is suggesting that the parasitoids' host recognition and host preference is partially under genetic control and partially influenced by contact with external factors associated with host that could be used in order to produce more efficient parasitoid strains.

CONCLUSION

Parasitoids can be used more effectively by developing strategies to conserve and manipulate their populations in agricultural ecosystems, which include crops and semi-natural habitats (Powell, 1986). He predicts that populations of natural enemies would be greater in diversified habitats due to increased availability of alternative hosts and food sources. The parasitoid's behaviour of attacking alternative aphid hosts may ensure its population stability in the field. *P. volucre* being a generalist parasitoid, which is behaviourally more flexible, may offer better opportunities for enhancement strategies through habitat and behaviour manipulation than highly specialised species, which are genetically more fixed.

There is considerable potential for the use of semiochemicals to manipulate insect behaviour as part of integrated pest control. More recently, it has been shown that parasitoids of the genus *Praon* are attracted to aphid sex pheromones. Particularly, the females of *P. volucre* showed greater response to pheromone baited-traps. This innate response could be utilised to manipulate *Praon* species in the field to improve aphid control strategies (Powell et al., 1993; Hardie et al., 1994; Lilly et al., 1994; Glinwood et al., 1998, 1999) and in *A. ervi* (He et al., 2006). This raises the possibility of treating mass-reared mummies with specific plant-derived semiochemicals to tailor the foraging preferences of the emerging parasitoids for specific target crops. New aphid control strategies are being developed based on the enhancement of naturally occurring parasitoid populations through manipulation of behaviour and their habitats (Powell, 1986; Cloutier and Bauduin, 1990; Powell et al., 1991, 1998; Storeck et al., 2000; Powell and Pickett, 2003; Blande et al., 2008). However, the development of efficient manipulation strategies must be based on a sound understanding of the aphid-parasitoid systems.

Parasitoids show a remarkable phenotypic plasticity due to associative learning and the interaction between innate, conditioned and learnt behavioural responses (Poppy et al., 1997; Poppy and Powell, 2004). The genetic control of learning and the ability to select parasitoids for learning abilities is a very exciting prospect. The importance of genetic variability in influencing the performance of parasitoids released in "classical" biological control program has often been highlighted (Hopper et al., 1993). However, genetic factors also need to be considered in developing and implementing biological control and integrated pest management (IPM) strategies based on augmentative releases and conservation biological control. Identification of genes that determine behavioural responses to specific chemical cues could advance future possibilities for genetic manipulation of parasitoids. The genetic manipulation of parasitoids has the potential to significantly improve biological control. Considering the tritrophic nature of interactions between plant, host and parasitoids, there are two ways to genetically manipulate the parasitoid. One is to directly manipulate the genetics of the parasitoid itself and other method is to exploit the influence of the plant on parasitoid foraging behaviour and genetically manipulate the plant to improve parasitoid efficiency.

The preliminary surveys conducted in different areas of Pakistan indicate that a number of parasitoids, including *Praon* species attack various aphid species on Important crops (Rehman, 1999). Since very less pesticide is being used to control aphids in Pakistan, particularly none on wheat, these parasitoids can play a significant role in maintaining a natural balance in the agro systems. By enhancing their activity through behaviour and habitat manipulations they could form a valuable input into sustainable agricultural systems.

Aphid parasitoids have considerable potential as biological control agents but their efficiency is dependent upon their presence in the right place at the right time and at right host: parasitoid ratio. Understanding parasitoid behaviour, together with identification of physical and chemical cues regulating the behaviour, is providing exciting opportunities for manipulation of parasitoids in the field, either as natural populations or as populations introduced through inundative releases. The mechanisms underlying behavioural plasticity in parasitoids and genetic basis of parasitoid behaviour provide opportunities for mass production of parasitoid strains suitable for use in specific crop/pest situations. The parasitoids having selectively bred to attack specific hosts and then primed to an appropriate plant volatiles as foraging cues before release, could be used in inundative releases.

ACKNOWLEDGEMENT

This paper is a part of PhD research thesis for which funding was granted by Pakistan Agricultural Research Council (PARC) under the World Bank assisted Agricultural Research Project- Phase-II (ARP-II). The thesis research was carried at IACR- Rothamsted, Harpenden, UK. The authors are highly indebted to both organizations for their support.

REFERENCES

Alam MM, Hafiz IA (1963). Some Natural Enemies of Aphids of Pakistan. Tech. Bull. CIBC, Rawalpindi, Pak. pp. 41-44.

van Alphen JJM, Vet LEM (1986). An evolutionary approach to host finding and selection. In: Insect Parasitoids (eds J.K. Waage and D. Greathead), Academic Press, London, pp. 23-61.

Ankersmit GW (1983). Aphidiids as parasites of the cereal aphids, *Sitobion avenae* and *Metopolophium dirhodum*. In: Aphid Antagonists: Proc. Meeting EC Experts' Group (ed R. Cavalloro), pp. 42-49. Portici, Italy 23-24. Nov. 1982.

Ankersmit GW, Bell C, Dijkman H, Mace N, Rietstra S, Schroder J, Visser CDE (1986). Incidence of parasitism by *Aphidius rhopalosiphi* in colour forms of the aphid *Sitobion avenae*. Entomol. Exp. Appl., 40: 223-229.

Arthur AP (1971). Associative learning by *Nemeritis canescens* (Hymenoptera: Ichneumonidae). Can. Entomol., 103: 1137-1141.

Arthur AP (1981). Host acceptance by parasitoids. In: Semiochemicals, Their Role in Pest Control (eds D. A. Nordlund, R.L Jones and W.J. Lewis), John Wiley and Sons, New York, pp. 97-120.

Ayal Y (1987). The foraging strategy of *Diaeretiella rapae*. I. The concept of the elementary unit of foraging. J. Anim. Ecol., 56: 1057-1068.

Battaglia D, Pennacchio F, Marincola G, Tranfaglia A (1993). Cornicle secretion of *Acyrthosiphon pisum* (Homoptera: Aphididae) as a contact kairomone for the parasitoid *Aphidius ervi* (Hymenoptera: Braconidae). Eur. J. Entomol., 90: 423-428.

Battaglia D, Pennacchio F, Romano A, Tranfaglia A (1995). The role of physical cues in the regulation of host recognition and acceptance behaviour of *Aphidius ervi* Haliday (Hymenoptera: Braconidae). J. Insect Behav., 8(6): 739-750.

Battaglia D, Pennacchio F, Poppy G. Powell W, Romano A, Tranfaglia A (1999). Physical and chemical cues influencing the oviposition behaviour of *Aphidius ervi* Haliday (Hymenoptera: Braconidae). Entomol. Exp. Appl., 94: 219-227.

Bhagavan S, Benatar S, Cobey S, Smith BH (1994). Effect of genotype but not of age or caste on olfactory learning performance in the honey bee, *Apis mellifera*. Anim. Behav., 48: 1357-1369.

Blande JD, Pickett JA, Poppy GM (2008). Host foraging for differentially adapted brassica-feeding aphids by the braconid parasitoid *Diaeretiella rapae*. Plant Signal Behav., 3(8): 580–582.

Bouchard Y, Cloutier C (1984). Honeydew as a source of host-searching kairomones for the aphid parasitoid *Aphidius nigripes* (Hymenoptera: Aphidiidae). Can. J. Zool., 62: 1513-1520.

Bouchard Y, Cloutier C (1985). Role of olfaction in host finding by aphid parasitoid *Aphidius nigripes* (Hymenoptera: Aphidiidae). J. Chem. Ecol. 11: 801-808.

Braimah H, van Emden HF (1994). The role of the plant in host acceptance by the parasitoid *Aphidius rhopalosiphi* (Hymenoptera: Braconidae). Bull. Entomol. Res., 84: 303-306.

Brandes C (1991). Genetic differences in learning behavior in honeybees (*Apis mellifera capensis*). Behav. Genet., 21: 271-294.

Carter N, Mclean IFG, Watt AD, Dixon AFG (1980). Cereal aphids: A case study and review. In: Appl. Biol. (ed T.H. Coaker), Acad. Press, London, V: 271-348.

Carver M, Sullivan DJ (1988). Encapsulative defence reactions of aphids (Hemiptera: Aphididae) to insect parasitoids (Hymenoptera: Aphidiidae) and Aphelinidae) (Minireview). In: Ecology and Effectiveness of Aphidophaga (eds E. Niemczyk and A.F.G. Dixon), SPB Academic Publishing, The Hague, pp. 299-303.

Carver M, Woodlock LT (1985). Interactions between *Acyrthosiphon kondoi* (Homoptera: Aphidoidea) and *Aphelinus asychis* (Hymenoptera: Chalcidoidea) and other parasites and hosts. Entomophaga, 30: 193-198.

Charnov EL, los-den Hartogh RL, Jones WT, van den Assem J (1981). Sex ratio evolution in a variable environment. Nature (London), 289: 27-33.

Chau A, Mackauer M (2001). Preference of the aphid parasitoid *Monoctonus paulensis* (Hymenoptera: Braconidae, Aphidiinae) for different aphid species: Female choice and offspring survival. Biol. Control . 20(1): 30-38.

Cheng RX, Ling M, Li BP (2010). Effects of aposymbiotic and symbiotic aphids on parasitoid progeny development and adult oviposition behavior within aphid instars. Environ. Entomol., 39(2): 389-395.

Chow FJ, Mackauer M (1992). The influence of prior ovipositional experience on host selection in four species of aphidiid wasps (Hymenoptera; Aphidiidae). J. Insect Behav., 5: 99-108.

Cloutier C, Bauduin F (1990). Searching behaviour of the aphid parasitoid

Aphidius nigripes (Hymenoptera: Aphidiidae) foraging on potato plants. Environ. Entomol. 19(2): 222-228.

Cloutier C, Lévesque CA, Eaves DM, Mackauer M (1991). Maternal adjustment of sex ratio in response to host size in the aphid parasitoid *Ephedrus californicus*. Can. J. Zool., 69: 1489-1495.

Cloutier C, Mcneil JN, Regniere J (1981). Fecundity, longevity, and sex ratio of *Aphidius nigripes* (Hymenoptera: Aphidiidae) parasitising different stages of host, *Macosiphum euphorbiae* (Homoptera: Aphididae). Can. Entomol., 113: 193-198.

Cronin, JT and Strong, DR (1996). Genetics of oviposition success of a thelytokous fairy fly parasitoid, *Anagrus delicatus*. Heredity, 76: 43-54.

Dedryver CA, Creach V, Rabasse JM, Nenon JP (1991). Spring activity assessment of parasitoids in western France by experimental exposure of cereal aphids on trap plants in a wheat field. IOBC-WPRS Bull., 14: 82-93.

Dhiman SC, Kumar V (1983). Host preference of *Aphidius rapae* (Curtis) (Hymenoptera: Braconidae). In: Proceedings, Symposium on Insect Ecology and Resource Management (ed S.C. Goel), Muzaffarnager, India, pp. 138-141.

Du YJ, Poppy GM, Powell W (1996). Relative importance of semiochemicals from the first and second trophic level in host foraging behaviour of *Aphidius ervi*. J. Chem. Ecol., 22: 1591-1605.

Du YJ, Poppy GM, Powell W, Wadhams LJ (1997). Chemically-mediated associative learning in the host forging behaviour of the aphid parasistoid *Aphidius ervi* (Hymenoptera: Braconidae). J. Insect Behav., 10: 509-522.

Doutt RL (1959). The biology of parasitic Hymenoptera. Ann. Rev. Entomol., 4: 161-182.

Eijsackers, HJP. and van Lenteren, JC (1970). Host choice and host discrimination in *Pseudeucoila bochei*. Neth. J. Zool., 20: 414.

van Emden HF (1978). Insects and secondary plant substances – an alternative viewpoint with special reference to aphids. In: Biochemical Aspects of Plant and Animal Coevolution (ed B. Harborne), Academic Press, London, pp. 309-323.

van Emden HF (1995). Host plant-aphidophaga interactions. Agri. Ecosyst. Environ., 52: 3-11.

van Emden HF, Spongal B, Wagner E, Baker T, Ganguly S, Douloumpaka S (1996). Hopkins 'host selection principle', another nail in its coffin. Physiol. Entomol., 21: 325-328.

Fox PM, Pass BC, Thurston R (1967). Laboratory studies on the rearing of *Aphidius smithi* (Hymenoptera: Braconidae) and its parasitism of *Acyrthosiphon pisum* (Homoptera: Aphididae). Ann. Entomol. Soc. Am., 60: 1083-1087.

Gardner SM, Dixon AFG (1985). Plant structure and the foraging success of *Aphidius rhopalosiphi* (Hymenoptera: Aphidiidae). Ecol. Entomol., 10: 171-179.

Gently GL, Barbosa P (2006). Effects of leaf epicuticular wax on the movement, foraging behavior, and attack efficacy of *Diaeretiella rapae*. Entomol. Exp. Appl., 121: 115-122.

Glinwood RT, Powell W, Tripathi CPM (1998). Increased parasitization of aphids on trap plants alongside vials releasing synthetic aphid sex pheromone and effective range of the pheromone. Biocontrol Sci. Tech., 8: 607-614.

Glinwood, RT, Du YJ, Powell W (1999). Responses to aphid sex pheromones by the pea aphid parasitoids *Aphidius ervi* and *Aphidius eadyi*. Entomol. Exp. Appl., 92: 227-232.

Godfray HCJ (1994). Parasitoids: Behavioural and Evolutionary Ecology. Princeton University Press, Princeton, NJ., p. 473.

Griffiths DC (1960). The behaviour and specificity of *Monoctonus paludum* Marshall (Hymenoptera: Braconidae), a parasite of *Nasonovia ribis-nigri* (Mosley) on lettuce. Bull. Entomol. Res., 51: 303-319.

Guerrieri E, Pennacchio F, Tremblay E (1993). Flight behaviour of the aphid parasitoid *Aphidius ervi* (Hymenoptera: Braconidae) in response to plant and host volatiles. Eur. J. Entomol., 90: 415-421.

Gutiérrez-Ibáñez C, Villagra CA, Niemeyer HM (2007). Pre-pupation behaviour of the aphid parasitoid *Aphidius ervi* (Haliday) and its consequences for pre-imaginal learning. Naturwissenschaften, 94(7): 595-600.

Hafeez M (1961). Seasonal fluctuations of population density of the cabbage aphid *Brevicoryne brassicae* (L.) in the Netherlands and the role of its parasite *Aphidius* (*Diaeretiella*) *rapae* (Curtis). Tijschrift over Planteziekten,

67: 445-548.

Hagvar EB, Hofsvang T (1986). Parasitism by *Ephedrus cerasicola* (Hym.: Aphidiidae) developing in different stages of *Myzus persicae* (Hom.: Aphididae). Entomophaga, 31: 337-346.

Hagvar EB, Hofsvang T (1991). Aphid parasitoids (Hymenoptera, Aphidiidae): biology, host selection and use in biological control. Biocontrol News Inf., 12(1): 13-41.

Hamid S (1983). Natural balance of graminicolous aphids in Pakistan: survey of populations. Agronomie, 3(7): 665-673.

Hardie J, Nottingham SF, Powell W, Wadhams LJ (1991). Synthetic aphid sex pheromone lures female parasitoids. Entomol. Exp. Appl., 61: 97-99.

Hardie J, Hick AJ, H`Ller C, Mann J, Merritt L, Nottingham SF, Powell W, Wadhams LJ, Witthinrich J, Wright AF (1994). The responses of *Praon* spp. parasitoids to aphid sex pheromone components in the field. Entomol. Exp. Appl., 71: 95-99.

Haq E–U (1997). Interaction of cultural control with biological control of rose grain aphid *Metopolophium dirhodum* (Walker) (Aphididae: Hemiptera) on wheat: The potential of greenhouse simulations. PhD Thesis. University of Reading, UK, p. 169.

Hatano E, Kunert G, Michaud JP, Weisser WW (2008). Chemical cues mediating aphid location by natural enemies. Eur. J. Entomol., 105(5): 797-806.

He XZ, Teulonz DAJ, Wang Q (2006). Oviposition strategy of *Aphidius ervi* (Hymenoptera: Aphidiidae) in response to host density. New Zealand Plant Protect., 59: 190-194.

He XZ, Wang Q (2006). Host age preference in *Aphidius ervi* (Hymenoptera: Aphidiidae). New Zealand Plant Protect., 59: 195-201.

Hofsvang T, Hägvar EB (1986). Oviposition behaviour of *Ephedrus cerasicola* (Hym: Aphidiidae) parasitizing different instars of its aphid host. Entomophaga, 31: 261-267.

Hood-Henderson DE, Forbes AR (1988). Behavioural responses of a primary parasitoid *Praon pequodorum* Viereck (Hymenoptera: Aphidiidae) and a secondary parasitoid *Dendrocerus carpenteri* Kieffer (Hymenoptera: Megaspilidae) to aphid honeydew. Proc., XVIII Int. Congress Entomol. Vancouver, 3-9 July, 1988, 373 pp.

Hopper KR, Roush RT, Powell W (1993). Management of genetics of biological control introductions. Annu. Rev. Entomol., 38: 27-51.

kris AGW, Heimpel GE (2007). Response of the soybean aphid parasitoid *Binodoxys communis* to olfactory cues from target and non-target host-plant complexes. Entomol. Exp. Appl., 123(2): 149-158.

Kant R, Sandanayaka WRM, He XZ, Wang Q (2008). Effect of host age on searching and oviposition behaviour of *Diaeretiella rapae* (M'Intosh) (Hymenoptera: Aphidiidae). New Zealand Plant Protect., 61: 355-361.

Langley SA, Tilmon KJ, Cardinale BJ, Ives AR (2006). Learning by the parasitoid wasp, *Aphidius ervi* (Hymenoptera : Braconidae), alters individual fixed preferences for pea aphid colour morphs. Oecologia, 150(1): 172-179.

Lilley R, Hardie J, Wadhams LJ (1994). Field manipulation of *Praon* populations using semiochemicals. Norwegian J. Agric. Sci. Suppl., 16: 221-226.

Liu S-S (1985). Development , adult size and fecundity of *Aphidius sonchi* reared in two instars of its aphid host *Hyperomyzus lactucae*. Entomol. Exp. Appl., 37: 41-48.

Liu S-S, Morton R, Hughes RD (1984). Oviposition preferences of a hymenopterous parasite for certain instars of its aphid host. Entomol. Exp. Appl., 35: 249-254.

Mackauer M (1973). Host selection and host suitability in *Aphidius smithi* (Hymenoptera: Aphidiidae). In: Perspectives in Aphid Biology (ed A.D. Lowe), Caxton Press, Christchurch, New Zealand, pp. 20-29.

Mackauer M (1976). Genetic problems in the production of biological control agents. Annu. Rev. Entomol., 21: 369-385.

Mackauer M (1983). Determination of parasite preference by choice test: the *Aphidius smithi* (Hymenoptera: Aphidiidae)- pea aphid (Homoptera: Aphididae) model. Annal. Entomol. Soc. Am., 76: 256-261.

Mackauer M (1986). Growth and developmental interactions in some aphids and their hymenopterous parasites. J. Insect Physiol., 32: 275-280.

Mackauer M (1996). Sexual size dimorphism in solitary parasitoid wasps: Influence of host quality. Oikos, 76: 265-272.

Mackauer M, Chow FJ (1986). Parasites and parasite impact on aphid populations. In: Plant Virus Epidemics: Monitering, Modeling and Predicting Outbreaks (eds G.D. McLean, R.G. Garret and W.G. Ruesink), Academic Press, Sydney, pp. 95-117.

Mackauer M, Kambhampati S (1988). Parasitism of aphid embryos by *Aphidius smithi*. Some effects of extremely small host size. Entomol. Exp. Appl., 49: 167-173.

Mackauer M, Michaud JP, Völkl W (1996). Host choice by aphidiid parasitoids (Hymenoptera: Aphidiidae): Host recognition, host quality and host value. Can. Entomol., 128(6): 959-980.

Mackauer M, Sequeira R (1993). Patterns of development in insect parasites. In: Parasites and Pathogens of Insects (eds N.E. Beckage, S.N. Thompson and B.A. Federici), Academic Press, Orlando, FL, pp. 1-23.

Mackauer M, Sequeira R, Otto M (1997). Growth and development in parasitoid wasps: Adaptation to variable host resources. In: Vertical Food Web Interactions: Evolutionary Patterns and Driving Forces. Ecological Studies. (eds K. Dettner, G. Bauer and W. Völkl). Springer-Verlag, Heidelberg.

Michaud JP, Mackauer M (1994). The use of visual cues in host evaluation by aphidiid wasps. I. Comparison between three *Aphidius* parasitoids of the pea aphid. Entomol. Exp. Appl., 70: 273-283.

Michaud JP, Mackauer M (1995). The use of visual cues in host evaluation by aphidiid wasps. II. Comparison between *Ephedrus californicus, Monotonus paulensis* and *Praon pequodorum*. Entomol. Exp. Appl., 74: 267-275.

Mitchell-Olds T (1995). The molecular basis of quantitative genetic variation in natural populations. Tree, 10(8): 324-328.

Miyazaki (1987). Forms and morphs of aphids. In: Aphids: Their Biology, Natural Enemies and Control. World Crop Pests (eds A.K. Minks and P. Harrewijn), Elsevier, Amsterdam, 2A: 27-50.

O'Donnell DJ (1989). A morphological and taxonomic study of first instar larvae of Aphidiinae (Hymenoptera: Braconidae). Syst. Entomol., 14: 197-219.

Pennacchio F, Digilio MC, Tremblay E, Tranfaglia A. (1994). Host recognition and acceptance behaviour in two aphid parasitoid species: *Aphidius ervi* and *Aphidius microlophii* (Hymenoptera: Braconidae). Bull. Entomol. Res., 84: 57-64.

Poppy GM, Powell W, Pennacchio F (1997). Aphid parasitoid responses to semiochemicals - genetic, conditioned or learnt? Entomophaga, 42(1/2): 193-199.

Poppy GM, Powell W (2004). Genetic manipulation of natural enemies: can we improve biological control by manipulating the parasitoid and/or the plant? In: Genetics, Evolution and Biological control (eds L.E. Ehler, R. Sforza and T. Mateille), CABI Publishing, UK, pp. 219-233.

Poppy GM, van Emden H, Storeck AP, Douloumpaka S, Eleftherianos I, Powell W (2008). Plant chemistry and aphid parasitoids (Hymenoptera: Braconidae): Imprinting and memory. Eur. J. Entomol., 105: 477-483.

Powell W (1982). The identification of hymenopterous parasitoids attacking cereal aphids in Britain. Syst. Entomol., 8: 179-188.

Powell W (1983). The role of parasitoids in limiting cereal aphid populations. In: Aphid Antagonists Proc. Meeting EC Experts' Group, Portici, Italy, 23-24 Nov. 1982. (ed R. Cavalloro, A.A. Balkema), Rotterdam, pp. 50-56.

Powell W (1986). Enhancing parasitoid activity in crops. In: Insect Parasitoids. 13th Symp. Royal Ent. Soc. London, 18-19 Sept. 1985 (eds J. Waage and D. Greathead), Acad. Press, London, pp. 319-340.

Powell W, Dean GJ, Wilding N (1986). The influence of weeds on aphid-specific natural enemies in winter wheat. Crop Protect., 5: 182-189.

Powell W, Decker UM, Budenberg WJ (1991). The influence of semiochemicals on the behaviour of cereal aphid parasitoids (Hymenoptera). In: Behaviour and Impact of Aphidophaga (eds L. Polgár, R.J. Chambers, A.F.G. Dixon and I. Hodek), SPB Academic Publishing, The Hague, Netherlands, pp. 67-71.

Powell W, Dewar AM, Wilding N, Dean GJ (1983). Manipulation of cereal aphid natural enemies. Int. Congr. Plant Protect., Brighton, UK, 10(2): 780.

Powell W, Hardie J, Hick AJ, Holler C, Mann J, Merritt L, Nottingham SF, Wadhams LJ, Witthinrich J, Wright AP (1993). Responses of the parasitoid *Praon volucre* (Hymenoptera: Braconidae) to aphid sex pheromone lures in cereal fields in autumn: implications for parasitoid manipulation. Eur. J. Entomol. 90: 435-438.

Powell W, Pennacchio F, Poppy GM, Tremblay E (1998). Strategies involved in the location of hosts by the aphid parasitoid *Aphidius ervi* Haliday (Hymenoptera: Braconidae: Aphidiidae). Biol. Control, 11: 104-112.

Powell W, Pickett JA (2003). Manipulation of parasitoids for aphid pest management: progress and prospects. Pest Manage. Sci., 59: 149-155.

Powell W, Walton MP, Jervis MA (1996). Population and communities. In: Insect Natural Enemies: Practical Approach to Their Study and Evolution (eds M. Jervis and N. Kidd), Chapman and Hall, London, pp. 223-292.

Powell W, Wright AF (1988). The abilities of the aphid parasitoids *Aphidius ervi* Haliday and *A. rhopalosiphi* De Stefani Perez (Hymenoptera: Braconidae) to transfer between different known host species and the implications for the use of alternative hosts in pest control strategies. Bull. Entomol. Res., 78: 683-693.

Powell W, Wright AF (1992). The influence of host food plants on host recognition by four aphidiinae parasitoids (Hymenoptera: Braconidae). Bull. Entomol. Res. 81: 449-453.

Powell W, Zhang ZL (1983) The reactions of two cereal aphid parasitoids, *Aphidius uzbeckistanicus* and *A. ervi* to host aphids and their food plants. Physiol. Entomol., 8: 439-443.

Pungerl NB (1984). Host preferences of *Aphidius* (Hymenoptera: Aphidiidae) populations parasitising pea and cereal aphids (Hemiptera: Aphididae). Bull. Entomol. Res., 74: 153-161.

Quicke DLJ (1997). Parasitic Wasps. Chapman and Hall, London, p. 470.

Read DP, Feeny PP, Root RB (1970). Habitat selection by the aphid parasite *Diaeretiella rapae* (Hymenoptera: Braconidae) and hyperparasite *Charips brassicae* (Hymenoptera: Cynipidae). Canad. Enomol., 102: 1567-1578.

Reed HC, Tan SH, Haapanen K, Killmon M, Reed DK, Elliott NC (1995). Olfactory responses of the parasitoid *Diaeretiella rapae* (Hymenoptera: Aphidiidae) to odour of plants, aphids, and plant-aphid complexes. J. Chem. Ecol., 21(4): 407-418.

Rehman A (1999). The host relationships of aphid parasitoids of the genus *Praon* (Hymenoptera: Aphidiidae) in agro-ecosystems. PhD Thesis, University of Reading, Reading UK, p.291.

Reuter OM (1913). Habits and instincts of insects. Friedlander, Berlin.

Roush RT (1989). Genetic variation in natural enemies: Critical issues for colonization in biological control. In: Critical Issues in Biological Control (eds M. Mackauer, L.E. Ehler and J. Rolands), Intercept, Andover, pp. 263-288.

Schmidt JM (1991). The role of physical factors in tritrophic interactions. Redia. 74: 43-93.

Sequeira R, Mackauer M (1987). Host instar preference of the aphid parasite *Praon pequodorum* (Hymenoptera: Aphidiidae). Entomologia Generalis, 12: 259-265.

Sequeira R, Mackauer M (1992). Covariance of adult size and development time in the parasitoid wasp *Aphidius ervi* in relation to the size of its host, *Acyrthosiphon pisum*. Evol. Ecol., 6: 34-44.

Sequeira R, Mackauer M (1993). Seasonal variation in body size and offspring sex ratio in field populations of the parasitoid wasp, *Aphidius ervi* (Hymenoptera: Aphidiidae). Oikos, 68: 340-346.

Sheehan W, Shelton AM (1989). Parasitoid response to concentration of herbivore food plants: Finding and leaving plants. Ecol., 70: 993-998.

Shirota Y, Carter N, Rabbinge R, Ankersmit GW (1983). Biology of *Aphidius rhopalosiphi*, a parasitoid of cereal aphids. Entomol. Exp. Appl., 34: 27-34.

Singh R, Sinha TB (1982). Bionomics of *Trioxys* (*Binodoxys*) *indicus* Subba Rao & Sharma, an aphidiid parasitoid of *Aphis craccivora*. XIII. Host selection by the parasitoid. J. Appl. Entomol., 93: 64-75.

Stadler B, Mackauer M (1996). Influence of plant quality on interactions between the aphid parasitoid *Ephedrus californicus* (Hymenoptera: Aphidiidae) and its host, *Acyrthosiphon pisum* (Homoptera: Aphididae). Can. Entomol., 128: 27-39.

Starý P (1976). Aphid Parasites of the Mediterranean Area. Dr. W. Junk, The Hague, Netherlands, p. 95.

Starý P (1987). Subject bibliography of aphid parasitoids (Hymenoptera: Aphidiidae) of the world 1758-1982. Monogr. Appl. Entomol., Suppl. J. Appl. Entomol., 25: 101.

Starý P (1988). Aphidiidae. In: Aphids. Their Biology, Natural Enemies and Control. World Crop Pests (eds A.K. Minks and P. Harrewijn), Vol. 2B, pp. 171-184. Elsevier, Amsterdam, p. 364.

Storeck A, Poppy GM, van Emden HF, Powell W (2000). The role of plant chemical cues in determining host preference in the generalist aphid parasitoid *Aphidius colemani*. Entomol. Exp. Appl., 97: 41-46.

Takada H (1975). Differential preference for *Myzus persicae* (Sulzer) of two parasites, *Diaeretiella rapae* (M'Intosh) and *Aphidius gifuensis* Ashmead. Jpn. J. Appl. Entomol. Zool., 19: 260-266.

Tully T, Hirsch J (1982). Behaviour-genetic analysis of *Phormia regina*. I. Isolation of lines pure-breeding for a high and low levels of central excitatory state (CES) from an unselected population. Behav. Genet., 12: 395-415.

Turlings TCJ, Wäckers F, Vet LEM, Lewis WJ, Tumlinson JH (1993). Learning of host-finding cues by hymenopterous parasitoids. In: Insect Learning: Ecological and Evolutionary Perspectives (eds D. R. Papaj and A. Lewis), Chapman and Hall, New York, pp. 51-78.

Vet LEM (1995). Parasitoid foraging: the importance of variation in individual behaviour for population dynamics. In: Frontiers and Applications of Population Ecology. (eds R.B. Floyd and A.W. Sheppard). CSIRO, Melbourne.

Vet LEM, Lewis WJ, Papaj DR, van Lenteren JC (1990). A variable-response model for parasitoid foraging behaviour. J. Insect Behav., 3: 471-489.

Vet LEM, Dicke M (1992). Ecology of infochemical use by natural enemies in a tritrophic context. Ann. Rev. Entomol., 37: 141-172.

Vickerman GP (1982). Distribution and abundance of cereal aphid parasitoids (*Aphidius* spp.) on grassland and winter wheat. Ann. Appl. Biol., 101: 185-190.

Vinson SB (1975). Biochemical coevolution between parasitoids and their hosts. In: Evolutionary Strategies of Parasitic Insects and Mites (Ed W.C. Price), Plenum Press, New York, pp. 14-48.

Vinson SB (1976). Host selection by insect parasitoids. Ann. Rev. Entomol., 21: 109-133.

Vinson SB (1981). Habitat location. In: Semiochemicals, Their Role in Pest Control (eds D.A. Nordlund, R.L Jones and W.J. Lewis), John Wiley and Sons, New York, pp. 51-77.

Vinson SB (1984). Parasitoid-host relationship. In: Chemical Ecology of Insects (eds W.J. Bell and R.T. Cardé), Chapman and Hall, London, pp. 205-233.

Vinson SB Iwantsch GF (1980a). Host regulation by insect parasitoids. Q. Rev. Biol., 55: 143-165.

Vinson SB, Iwantsch GF (1980b). Host suitability for insect parasitoids. Ann. Rev. Entomol., 25: 397-419.

Vorley WT, Wratten SD (1987). Migration of parasitoids (Hymenoptera: Braconidae) of cereal aphids (Hemiptera: Aphididae) between grassland, early-sown cereals and late-sown cereals in southern England. Bull. Entomol. Res., 77: 555-568.

Waage JK (1986). Family planning in parasitoids: Adaptive patterns of progeny and sex allocation. In: Insect Parasitoids (eds J. Waage and D. Greathead), Academic Press, London, pp. 63-95.

Wellings PW (1988). Sex-ratio variations in aphid parasitoids: some influences on host-parasitoid systems. In: Ecology and Effectiveness of Aphidophaga (eds E. Niemczyk and A.F.G. Dixon), SPB Academic Publishing, The Hague, pp. 243-248.

Weseloh RM (1981). Host location by parasitoids. In: Semiochemicals: Their Role in Pest Control (eds D.A. Nordlund, R.L. Jones and W.J. Lewis), John Wiley & Sons, New York, pp. 79-95.

Wickremasinghe MGV, van Emden HF (1992). Reactions of adult female parasitoids, particularly *Aphidius rhopalosiphi*, to volatile chemical cues from the host plants of their aphid prey. Physiol. Entomol., 17: 297-304.

14

Sustainable use of plant protection products in Nigeria and challenges

E. D. Oruonye[1]* and E. Okrikata[2]

[1]Department of Geography, Taraba State University, P. M. B. 1167, Jalingo, Taraba State, Nigeria.
[2]Department of Biology Education, School of Science Education, Federal College of Education, Technical, P. M. B. 1013, Potiskum, Yobe State, Nigeria.

Nigeria's drive to boost food security and to fight off insect pests and yield-limiting crop pathogens has led to an unintended consequence: the mass importation and build-up of obsolete and toxic pesticides. Nigerian farmers have been relying heavily on these agrochemicals for the control of various weeds, insect pests and pathogens, leading to the high importation of these products. Although synthetic-chemical pesticides can be used to control some pests economically, rapidly and effectively; most of them cause serious negative impacts to the ecosystem. This paper highlights the need to improve the sustainability of the use of plant protection products. This can be achieved by integrating existing plant protection measures (chemical, mechanical, physical, biological, host-plant resistance, use of pheromones, cultural, etc.) under the framework of what is termed Integrated Pest Management (IPM) while identifying, advocating for and promoting the use of botanical pesticides in the pest management process.

Key words: Plant protection, plant protection products, sustainable use, integrated pest management (IPM), botanical pesticides/botanicals.

INTRODUCTION

Nigeria's drive to boost food security and to fight off insect pests and yield-limiting crop pathogens has led to the build-up of obsolete and toxic pesticides and chemicals (Schwab et al., 1995). Several thousand tons of pesticides and other heavily contaminated materials have accumulated over the years, while inadequate legislation and lax regulations have exacerbated the problem (Lale, 2002). In general, plant insect pests, diseases and weeds impose a serious threat to crop production in Nigeria. Population of weeds, insect pests and diseases have increased over the years especially by the introduction of monoculture farming in the country (Emosairue and Ubana, 1998). Traditionally, Nigerian farmers have been relying heavily on pesticides for the control of various weeds, insect pests and diseases, leading to the high importation of these products and their price have become so high that it is becoming impossible for local farmers to afford (Nwanze, 1991; Schwab et al.,

1995; Van den Berg and Nur, 1998; Okrikata and Anaso, 2008b). These have created the need for alternatives to synthetic pesticides. But inadequate infrastructure for research and extension remains a constraint to the advancement and continuity of such important activity in the country (Bugaje et al., 2007).

Although synthetic-chemical pesticides can be used to control some pests economically, rapidly and effectively; most of them cause serious negative impacts, such as: toxicity and residual effects to humans, target plants, foods and other living things; induction of insect/pathogen resistance resulting to ineffectiveness of pesticides; harmful effects to non-target beneficial organisms and unbalanced ecosystem due to pollution of soil, water and environment (Deedat, 1994; Gupta and Shyam, 1996).

The challenge to sustainable use of plant protection products is a long-term one. It involves helping farmers to properly control/manage pests in such a way that they do not only minimise immediate risks from dangerous pesticides but also reduce the possibility of accumulating future stocks. Synthetic pesticides pose a threat to sustainable development. The adverse impact of persistent organic pollutants, or POPs as many of the pesticides are

*Corresponding author. E-mail: eoruonye@gmail.com.

known, on the environment and health are serious. POPs do not degrade easily, but remain intact in the environment for a long period of time. The pollutants disperse easily across wide geographic areas, retain their toxicity and have a tendency to accumulate in the fatty tissues of organisms (FAO, 2007).

Human health and environmental safety are the two most important issues in the long-term application of pesticides. Therefore, the reduction in the amount of pesticides used in agricultural production has been a major issue to environmentalist. In order to maintain the balance of the ecosystem based on the conservation of natural resources and minimization of harmful effects to the environment; measures other than chemical control for pest management are seriously considered. These measures could partially replace and minimize the risk of using synthetic chemical pesticides to meet the requirment of organic farming in Nigeria. While empirical data on conventional and organic agriculture in Nigeria are scarce, it is worth noting that organic agriculture is at present poorly developed in Nigeria. Though, there are recent moves for the production of organic crops in Nigeria (Olabiyi et al., 2008). Much agricultural production in Nigeria may be described as "organic by default or neglect" as there is a lack of policy or regulation covering organic agriculture in Nigeria. In a bid to advocate for organic farming, Olabiyi (2009) reported that DDT, lindane, endosulfan, toxaphane, chlorpyrifos and many other persistent synthetic chemicals have been found in underground water and deep wells in Nigeria. This, no doubt butress the need for reliable alternatives to synthetic pesticides.

Laboratory and field research in Nigeria have proved that botanicals could be economically feasible, technically effective and environmentally friendly alternatves to synthetic pesticides. Yusuf et al. (1998) observed that powders of leaves of *Azadiracta indica, Melia azaderach, Zingiber officinale, Eucalyptus camaldulensis, Ocimum basilicum, Capsicum frutescens* and wood ash of *Khaya senegalensis* are effective in the control of maize weevil (*Sitophilus zeamais*) in stored maize. Abdul-azeez (2009) applied cashew nut shell extract on cowpea pods infested with aphids and reported that the treatment has insecticidal effects on aphids. Insecticidal activity of some Nigerian plants such as *Eugenia aromatica, Piper umbellatum, Erythrophleum guineense, Aframomum melegueta, Hyptis sauvolens, Allium cepa, Carica papaya, Uvaria afzelli* and *Vernonia amygdalina* against maize weevil was reported by Lajide et al. (1998). Okrikata and Anaso (2008a) observed that different formulations of neem seed kernel powder was effective in controlling *Sesamia calamistis* in sorghum. Lale and Yusuf (2001) reported the efficacy of *Piper guineense* seed oil in controlling *Tribolium castaneum* in stored millet seed. Anaso (1999) reported the efficacy of neem kernel oil in controlling the pest complex of okro.

In 1988, local farmers in Borno state of Nigeria were advised by extension workers to use aqueous extract of neem kernel, sprayed with twigs to ward off grashoppers (Baba Yamta, personal communication). Emosairue and Ukeh (1996) reported the effectiveness of 3% neem oil and 5% neem seed kernel extract in the control of *Podagrica* spp. in south eastern Nigeria. Emosairue and Ubana (1998) also reported the effectiveness of neem seed kernel extract on *Maruca* pod borer. The potential of botanical pesticides cannot be overemphasised as they have also been found to be safe for humans and friendly to the environment. They are also characterized by low mammalian toxicity, lack of mutagenic activity and high rate of biodegradability (Kloss and McCullough, 1987). Therefore, the aim of this paper is to highlight the need to improve the sustainability of the use of plant protection products in Nigeria with special emphasis on the use and integration of botanicals in the pest management system.

PLANT PROTECTION AND PLANT PROTECTION PRODUCT

Plant protection means activities involving employment of biological, chemical, quarantine and other measures that protect plants from disease agents, insect pests and weeds, or decrease their harmful effects and also preserve their ecological equilibrium in nature (Anaso, 1999). Plant protection product on the other hand, means an active substance or a preparation containing one or more active substances, put up in the form in which it is supplied to the user, intended to: protect plants or plant products against all harmful organisms or prevent the action of such organisms; influence the life processes of plants, other than as a nutrient (for example, as a growth regulator); destroy undesired plants; or check or prevent the undesired growth of plants (Dorn, 1998).

In exercise of the powers conferred on the Governing Council of the National Agency for Food and Drug Administration and Control (NAFDAC) of Nigeria by section 8 of the Drugs and Related Products (Registration, etc.) Act of 1996 (As Amended); NAFDAC came up with the Pesticide Registration Regulations of 2005. The regulations lays down that no pesticide is to be manufactured, formulated, imported, exported, advertised, sold, or distributed in Nigeria unless it has been registered in accordance with the provisions of these regulations. Samples of pesticides for registration may, however, be manufactured, formulated, or imported with the approval of the Agency. Other sections deal with, inter alia, the submission of registration applications, the issuance of registration certificates and the withdrawal and cancellation of such certificates (The Pesticide Registration Regulations, 2005). The purpose of developing this national strategy was to act as a driver to enhance environmental protection within the context of sustainable use of plant protection products. The regulatory system is thus designed to ensure that plant protection products do not endanger the health of people (operators, consumers or bystanders) and the environment. However, the Pesticide Registration Regulations of Nigeria has not been able to deliver well because of the following gaps or defects:

(1) It is silent on the need for re-registration of older pesticides under which the pesticides are to be re-examined to make sure that the supporting data for a registered pesticide satisfy current requirements for registration.

(2) The Pesticide Registration Regulations and Related Laws of Nigeria do not capture the need for retailers of pesticides to be trained. Most of them are illiterates and as such sell banned, expired and/or poorly stored pesticides to the end users (farmers) who in most cases know less and yet apply the pesticides themselves as trained applicators/sprayers of pesticides are rare to find and the farmers cannot afford to pay for their services.

(3) The Registration Regulations does not also consider the need for manufacturers or agents to write instructions and warnings on pesticide labels in commonly understood languages of end users (eg. English, Hausa, Igbo, Yoruba and Pidgin English) as most of them are not learned.

(4) Weak regulations as to banning the importation and use of dangerous pesticides and the inactivity or absence of government and non-government agencies of control is also another challenge.

(5) Monitoring and implementation of the Pesticide Regulations as it is, is also weak. For example, some pesticides such as, aldrin, binapacryl, captafol, chlordane, chlordimeform, DDT, dieldrin, dinoseb, ethylene dichloride, heptaclor, lindane, parathion, phosphamidon, monocroptophos, methamidophos, chlorobenzilate, toxaphane, merix endosulphan, delta HCH and ethylene oxide have been banned in Nigeria after having caused deaths in Nigeria (Inalegwu, 2008), yet; some of them are still found in the Nigerian market.

(6) Efforts to review the Pesticide Regulations as well as to work on how to effectively enforce and implement the regulations is yet to be seen.

All these have been combined to allow for the sale, misuse and abuse of dangerous pesticides with its attendant human and environmental (biotic and abiotic) hazards. Given the odds against the continued use of synthetic pesticides, the urgency in the need to find acceptable alternatives is obvious. The Nigerian Government should therefore re-strategise and make laws or policies to govern the processing and use of botanicals which have been found to be safer alternatives

Strategies/policies for sustainable use of plant protection products

Sustainable use, in the context of this paper means, minimising the hazards and risks to both man and his

environment from the use of plant protection products without compromising the necessary crop protection (Lale, 2002). Governments in developed countries and some developing countries have, for many years, operated a policy of 'pesticide minimisation' (Mailu, 1997).

Given the rate of increase in human health related problems such as cancer, genetic disturbances and damage to the immune system, which are caused in part by some of the toxic substances contained in pesticides (Deedat, 1994; Ermel et al., 2002), there is need for the Nigerian government to strategise and consider the effective regulation of these plant protection products. Research findings in Nigeria and other parts of the world prove that integrating botanical pesticides in our pest management system will promote safer agricultural practice. In Nigeria for example, Botanical insecticides have been extracted from various plants including neem (*Azadiracta indica*), Pyrethrum (*Chrysanthemum cinarariaefoliun*), Tobacco (*Nicotiana tabacum*), Derris (*Derris elliptica*), Pawpaw (*Carica papaya*), Tomato (*Lycopersicon esculentum*), Cashew nut (*Anarcardium occidentale*), Garlic (*Allium sativum*), Aligator pepper (*Aframomum melegueta*), Curry leaves (*Hyptis sauvolens*), Onions (*Allium cepa*), Basil (*Ocimum basilicum*), Bitter gourd (*Momordica charatia*), Ginger (*Zinigiber officinale*), Bitter leaf (*Vernonia amygdalina*), Siam weed (*Chromolaena odorata*), and pepper fruit (*Uvaria afzelli*). Their biological properties have been tested and found to include insecticidal and repellent effects against insect pests. Some have also been found to have antifeedant, growth regulatory, oviposition inhibitory, sterility inducing, antifungal and nematicidal properties (Lajide et al., 1998; Anaso, 1999; Abdul-azeez, 2009). Therefore, the following strategies/policies which centers on identifying/discovering, advocating for and promoting the use of botanical pesticides under the framework of integrated pest management will no doubt enhance sustainable use of plant protection products in Nigeria:

(1) The Nigerian government should consider making favorable laws and policies to govern the processing and use of selected botanical pesticides.

(2) There should be a purposeful encouragement of indigenous private sector participation in the formulation, testing and marketing of botanicals.

(3) As a matter of policy, the government should ensure that the period required to register botanicals is far less than that required to register conventional/synthetic pesticides.

(4) Another important policy issue is the possibility of indiscriminate harvesting of botanical materials to meet surge in demand. This could lead to environmental degradation and loss of biodiversity as is already happening in the case of neem in parts of South Asia (Ahmed and Stoll, 1996). To safeguard against this, a species of plant should not be promoted as a source of

biopesticide until adequate arrangements have been made for its increased production (Lale, 2002).

(5) The government should also design a policy to encourage and protect local companies that may be involved in the processing and marketing of botanical pesticides so that the citizens could derive maximum benefit from these locally available resources.

(6) Presently, many dangerous pesticides are being used in Nigeria for crop protection (Lale and Okunade, 1996). These classes of pesticides are either highly restricted or banned by law, but the laws not enforced. The strict enforcement of such law followed by a planned policy to adopt and promote the use of botanicals; which are comparably safer both to the applicator, food consumer as well as to the environment (Schwab et al., 1995; Lale and Mustapha, 2000), will stem the tide of deaths associated with pesticide poisoning (Lale, 2002).

(7) The Nigerian government should also promote aggressive enlightenment campaigns through the efforts of both public and private sectors to create a high level awareness amongst the citizenry about the available alternatives to synthetic pesticides.

Challenges to sustainable use of plant protection products

Intensive synthetic pesticide use, or overuse, in an effort to control pests and disease vectors can reduce the efficacy of pesticides (pesticide resistance). Often when pests are resistant to a certain pesticides, farmers will simply apply more or different pesticides, thereby increasing the residues on food crops and strengthening the pest's resistance even further. In the end, when a pesticide is no longer effective, farmers often face the need to purchase newer, often more expensive products, which can be especially problematic in developing countries. Negative effects on human health can be caused by direct or indirect exposure to pesticides and pesticide exposure can have either acute or chronic effects.

Developing countries and Nigeria in particular face the most challenges in achieving the sound management of pesticides. A large proportion of the population in Nigeria is directly engaged in agricultural work, often on a very small scale. Farmers will purchase pesticide products for individual use, but may not be sufficiently literate to read the instructions or be comfortable in the language the instructions are written in. Particularly in remote areas, the only source of advice may be the pesticide seller, who may also be poorly informed and whose advice may be guided by commercial self-interest. These populations are often not able to afford the newest minimum-risk pesticides, instead using older and often more dangerous products which are usually cheaper because they can be produced as generic products off-patent. Even appro-priate products may be adulterated or have deteriorated

because their shelf life expired while they were in storage or because they were stored improperly. Farmers using such pesticides are at risk of developing pesticide related problems. Most farmers today rely on the use of pesticides to control pests. This sole reliance on pesticides is unsustainable. The lack of awareness and resources can also lead to improper disposal of pesticides and reuse of pesticide containers, thereby posing a threat to humans and the environment.

While evidences abound that botanical pesticides are generally safe and effective (Mallya, 1986; Ahmed and Grainge, 1986, Marandu et al., 1987; Ascher, 1993; Mordue and Blackwell, 1993; Emosairue and Ubana, 1998; Okrikata et al., 2008), their use in Nigeria as in other parts of Africa is still hampered by some challenges which include:

(1) Most data on botanical pesticides are obtained from laboratory trials; field data are rare.
(2) There is still hardly developed any appropriate technology for the application of botanicals, especially the oil and dust formulations (Lale, 2002).
(3) Compared with synthetic insecticides, the effects of botanical insecticides are short-lived. So frequent appilications are required to obtain a reasonable degree of crop protection.
(4) Botanical pesticides formulations are yet to be available in usable forms to farmers in commercial quantities so as to serve as alternatives to synthetic pesticides.
(5) There is the problem of farmers' acceptability of this seemingly new dimension/technology in pest control (Okrikata and Anaso, 2008a).

From the foregoing, the need to advocate for and implement integrated pest management strategies both on field pests, stored product pests, structural pests and domestic pests is indispensable. Therefore, the discovery, advocating for, adoption and promotion of the use of botanical pesticides in an integrated pest management framework is quite relevant.

CONCLUSION AND RECOMMENDATIONS

The key to attaining food sufficiency in Nigeria is ensuring that crops stay healthy and protected from damages by pests and diseases. In order to permanently maintain the productivity, ability to function, regenerative power and the buffering capacity of the open system within which plants are cultivated, plant protection measures must be generally acceptable, economically feasible, technically effective, environmentally friendly and easy to use or apply. National pesticide programmes have various goals, most of which can be met by an effective legal framework. The need to ensure the efficacy of pesticide products for their proposed use, while at the same time protecting pesticide users, consumers, crops, livestock and the environment cannot be overemphasized. In this regard, the importance of recruiting botanicals in plant protection or pest control is imperative.

This study recommends the need to launch a nation-wide effort to clean up these harmful chemicals, prevent future accumulations, promote safe-handling techniques by working directly with farmers and strengthen the country's institutional capacities to tackle the issue. Farmers should be enlightened on how to identify the actual pests and diseases, determine the level of infestation and the symptoms to describe the pathological conditions of crops, to be able to establish the economic injury level values for pests and diseases in order to derive a farm-level relevant methods In the context of effective plant protection. Farmers should also be advised to include the critical use of right pesticides and other protective measures in order to satisfy the criteria of effective plant protection. Plant protectionists should also include in their research, the development of crop resistant varieties to pests and diseases. Finally, the importance of food sufficiency, agricultural policy makers in Nigeria should advocate for a revision of the laws to facilitate the effective supervision of pesticide quality and monitoring of residues due to the use of agricultural chemicals. Adequate funding of researches in crop protection which amidst others is geared toward discovering botanical pesticides is also indispensable.

REFERENCES

Abdul-azeez A (2009). The Effect of Cashew Nut Shell Extracts on Pest (Aphid) of Cowpea. Biol. Environ. Sci. J. 6(3), 25-28.

Ahmed S, Grainge M (1986). Potential of the neem tree (*Azadirachta indica*) for pest control and rural development. Econ. Bot. 40: 201-209.

Ahmed S, Stoll G (1996). Biopesticides. In: *Building on Farmers' Knowledge* (Bunders J, Haverkort B, Hiemstra W, eds.), Macmillan Education Ltd., London and Basingstoke. pp. 52-79.

Anaso CE (1999). Evaluation of neem extracts for control of major insect pests of Okro (*Abelmoschus esculentus*). Ph.D Thesis. University of Maiduguri p. 131.

Ascher KRS (1993). Non conventional insecticidal effects of pesticides available from neem tree. Arch. Insect Biochem. Physiol. 22: 433-449.

Bugaje SM, Kuta DD, Magashi AI, Ubale AS (2007). Food security in Nigeria. Paper presented at the conference of young agricultural specialists, organized by SELHOZ (A Nigerian Non Governmental Oreganization Based in Russia). Held on 22nd July – 5th August at The People's Friendsip University, Russia, Moscow.

Deedat YD (1994). Problems associated with the use of pesticides: An overview. Insect Sci. Applic. 15: 247-251.

Dorn S (1998). Integrated stored product protection as a puzzle of mutually compatible elements. IOBC Bulletin 21: 9-12.

Emosairue SO, Ukeh DA (1996). Field trial of neem product for control of Okra flea beetles (*Podagrica spp.*) in south eastern Nigeria. Afri. J. Plant Protect 6: 22-26.

Emosairue SO, Ubana UB (1998). Field evaluation of neem for the control of some cowpea pests in South Eastern Nigeria. Global J. Pure Appl. Sci. 4(3): 237-241.

Ermel K, Schmutterer H, Kleeberg H (2002). Commercial product standardisation and problems of quality control. In: Schmutterer H(ed.). "The Neem Tree (*Azadirachta indica* A. Juss) and Other

Meliaceous Plants. Sources of Unique Natural Products for Integrated Pest Management, Medicine, Industry and Other Purposes", 2nd ed. Neem Foundation, Mumbai p. 893.

FAO (2007). Designing National Pesticide Legislation. Food and Agriculture Organization of the United Nations, Rome.

Registration Regulations (The year). The Federal Military Government of Nigeria. (Supplement to Official. Gazette Extraordinary, No.27, Vol 83, 18th June, 1996, Part B, pp. B303-B307).

GUpta SK, Shyam KR (1996). Antisporulant activity of some fungicides against *Pseudoperonospora cubensis* on cucumber. Indian J. Mycol. Plant Pathol. 26(3): 293-295.

http://www.businessonline.com/index.php?option=com_content&view=article8id=4501:agribusiness-is-organic.

Inalegwu S (2008). 30 Agrochemical Products Banned in Nigeria After Deaths. *Vanguard*. Retrieved July 10, 2010, from http:www.organicconsumers.org/articles/article_12416.cfm.

Kloss H, McCullough FS (1987). Plants with recognized moluscidal activity. In: Plant Molluscicides (Mott, K,E., ed.), UNDP/World Bank/WHO Special Programme fo Research and Training in Tropical Diseases. pp. 45-108.

Lajide L, Adedire CO, Muse WA, Agele SO (1998). Insecticidal activity of powders of some Nigerian plants against the maize weevil (*Sitophilus zeamais* Motsch). ESN Publications. 31: 227-235.

Lale NES, Okunade SO (1996). A survey of some aspects of fish processing in the Lake Chad district of Nigeria. Ann. Borno. 13: 362-368.

Lale NES, Mustapha A (2000). Potential of combing neem (*Azadirachta indica* A.Juss) seed oil with varietal reistance for the management of cowpea bruchid, *Callosobruchus maculatus* (F.) (Coleoptera:Bruchidae). J. Plant Dis. Protec. 107: 399-405.

Lale NES, Yusuf BA (2001). Potential of varietal resistance and *Piper guineense* seed oil to control infestation of stored millet seeds and processed products by *Tribolium castaneum* (Herbst). J. Stored Prod. Res. 37: 63-75.

Lale NES (2002). Stored Product Entomology and Acarology in Tropical Africa.Mole Publications, Maiduguri, Nigeria p. 204.

Mailu AM (1997). Review of Kenyan Agricultural Research. Pest Plants. 29: 3-11.

Marandu WYF, Temu AEM, Kabango D (1987). Plant Protection. In: Maize Improvement Programme Annual Progress Report 1987. UAC, Mbeya, Tanzania, p. 3.

Mallya GA (1986). Maize entomology. In: Maize Improvement Progress Report 1985-1986. UAC, Mbeya, tanzania, pp. 27-31.

Mordue AJ, Blackwell A (1993). Azadirachtin: an update. J. Insect Phys. 39: 903-924.

Nwanze KF (1991). Components for the management of two insect pests of pearl millet in Sahelian West Africa. Insect Sci. Applic. 12: 673-678.

Okrikata E, Bukar SM, Anaso CE (2008). Evaluation of Used Engine Oil and Other Nature Based Materials on the Emergence of Harvester Ants (*Messor galla* F.) in Maiduguri, Borno State. Academic Publications and Research Association of Nigeria (APRAN). A J. Manage. Sci. Technol. Edu. 4(1): 88-94.

Okrikata E, Anaso CE (2008a). Influence of some inert diluents of neem kernel powder on protection of sorghum against pink stalk borer (*Sesamia* calamistis, Homps) in Nigerian Sudan savanna. J. Plant. Protect. Res. 48(2): 161-168.

Okrikata E, Anaso CE (2008b). Bioefficacy of Various Neem Dust Formulation for the Control of Sorghum Stemborers II: Effect on Stalk and Peduncle in the Semi-arid zone of Nigeria. Yobe J. Environ. Dev. 1(1): 29-38.

Olabiyi TI, Okusanya AO, Harris PJ (2008). Accessing the World Market for Organic Food and Beverages from Nigeria. In: 16th IFOAM Organic World Congress, Modena, Italy. June 16-20, 2008. Retrieved July 11, 2010 from http://orgprints.org/11713/1/olabiyi 11713 ed. doc.

Olabiyi TI (2009). Agribusiness: Is organic farming the best option? Retrieved July 11 2010, http://www.businessdayonline.com/index.php.option=com_content&view=article8id=4501:agribusiness-is-organic.

Schwab AI, Jager I, Stoll G, Gorgen R, Prexterschwab S, Attenburger R (1995). Pesticide in tropical agriculture: hazards and alternatives.

PAN ACTA Trop. Agroecol. No. 131.

The Pesticide Registration Regulations (2005). Retrieved July 10, 2010 from

http//www.nafdac.gov.ng/index.php?option=com_docman&task...12...

Van den BJ, Nur AF (1998). Chemical control. In: Andrew P (ed.) African Cereal Stemborers: Economic Importance, Taxonomy, Natural Enemies and Control. CAB International in association with the ACP-EU Technical Centre for Agricultural and Rural Co-operation (CTA). pp. 319-332.

Yusuf SR, Ahmed BI, Chaudhary JP, Yusuf AU (1998). Laboratory evaluation of some plant products for the control of maize weevil (*Sitophilus zeamais* Mots.) in stored maize. ESN Occassional publications. 31: 203-213.

15

Use of a liquid inoculum of the arbuscular mycorrhizal fungi *Glomus hoi* in rice plants cultivated in a saline Gleysol: A new alternative to inoculate

F. Fernández[1,2], J. M. Dell'Amico[2], M. V. Angoa[3]* and I. E. de la Providencia[4]

[1]SYMBORG S.L. Campus de Espinardo No 7. Edificio CEEIM. Murcia. España. CP 30100.
[2]Instituto Nacional de Ciencias Agrícolas. (INCA), Habana, Cuba.
[3]Centro Interdisciplinario de Investigación para el desarrollo Integral Regional. (CIIDIR) IPN, Justo Sierra No. 28, CP 59519, Jiquilpán Michoacán, México.
[4]Université catholique de Louvain, Unité de microbiologie, 3 Place Croix du Sud, 1348 Louvain la Neuve, Belgium.

The functioning and efficacy of a liquid inoculum containing spores of the arbuscular mycorrhizal fungi (AMF) *Glomus hoi* previously isolated from saline soil, were evaluated over growth and production of *Oryza sativa* plants cultivated under flooding conditions. Growth dynamics, grains weight, panicles number, as well as panicles weight were determined in the rice plants. Other mycorrhizal parameters such as biomass of ectophyte (EcB), arbuscular endophyte biomass (AEB), the ratio EcB:AEB, spore density and visual fungal density were also determined. The treatments evaluated were Mycorrhizal plants and Non-mycorrhizal plants (control). Results showed that rice plants inoculated with AMF grew higher than control plants. This was related to the AEB and EcB development. For the yield parameters the AMF treatment presented the highest values compared to control. These results showed the capability of a liquid inoculum to promote growth, development and improve the production of rice plants under saline conditions. This study showed the potential of using a liquid inoculum in field given its ability to store more spores without affecting their viability. In addition, the high applicability and easier manipulation of a liquid inoculum compared to the traditional solid inocula would reduce costs and maximize benefits in field.

Key words: Growth, increase production, *Oryza sativa*, liquid biofertilizer, saline soil.

INTRODUCTION

Oryza sativa is the second most important cereal in the world after wheat, and the principal crop in Asia, serving as food for about 50% of the world's population (Ladha et al., 1997; Sass and Cicerone, 2002). It is predicted that a 50 to 60% increase in rice production will be required to meet demand from population growth by 2025 (Zhang and Wang, 2005). World rice consumption increased 40 % in the last 30 years, from 61.5 kg per capita to about 85.9 kg per capita (milled rice) (UNCTAD- FAO, 2008). Protected horticultural crops as well as those planted in

open fields such as rice, are coping with increasing salinization of irrigation water (Al-Karaki, 2006). This water contributes to generate more salinization of agriculture fields. Salinization is a process of soil degradation that is increasing in importance throughout the world (Keren, 2000; Liang et al., 2005).

Saline soils occupy more than 7% of the earth land surface and represent a major limiting factor in crop production (Feng et al., 2002; Nourbakhsh and Sheikh-Hosseini, 2006). Excessive amounts of salt have a range of adverse effects on the physical and chemical properties of soil, microbiological processes and plant growth (Yuang et al., 2007; Zhu, 2002). Salinization is often associated to poor fertility level of land (González-Nuñez et al., 2004). Plants photosynthesis, its efficiency

*Corresponding author. E-mail: valeangoa@hotmail.com.

and the rate of protein accumulation could be inhibited by low water and low fertilizer availability (Hasegawa et al., 2000). Nevertheless, the exploitation of soil microbes used in salt stressed lands or irrigated with saline water could be an alternative for plants development under these extreme conditions (Giri et al., 2003). It has been demonstrated that mutual associations of plants with soil microorganisms may improve plant tolerance to stressful conditions (Carvalho et al., 2003).

Several studies have shown that inoculation with arbuscular mycorrhizal fungi (AMF) can diminish the stress caused by salinity (Al-Karaki et al., 2001; Azcón-Aguilar and Barea, 1997; Dixon et al., 1993; Hartmond et al., 1987; Juniper and Abbott, 1993; Rao, 1998; Ruiz-Lozano et al., 1996; Singh et al., 1997). AMF improvement of salt resistance has been usually associated with AM-induced increases in phosphorous acquisition and plant growth (Rosendahl and Rosendahl, 1991), as well as in the resistance of the crop (Al-Karaki, 2000; Cho et al., 2006). Ruiz-Lozano et al. (1996) and Azcón and El-Atrash (1997), suggested that plants could be more effectively protected against salinity stress by AMF symbiosis rather than by phosphorous supplementation. However, responses to deficiency of oxygen in soil, such as a decline in photosynthetic capacity, stomatal conductance and nutrient uptake (Kozlowski and Pallardy, 1984), may induce a lower allocation of carbohydrates from the host plant to the fungi (Carvalho et al., 2003).

It is known that wetland plants can increase the flow of oxygen to their rhizospheres through aerenchyma development (Chabbi et al., 2000). It is likely that the aerobic AMF in wetlands obtain oxygen via this route, which could also depend partially on air spaces in the cortex (Carvalho et al., 2003; Ipsilantis and Sylvia, 2007; Smith and Smith, 1997). In addition, there is a limited knowledge of AMF interactions with wetland plants although its presence is well known (Khan and Belik, 1995; Thormann et al., 1999). Apparently, there is no relationship between the percentage of root length colonized by AMF and the plant hydrological category (Aziz et al., 1995; Turner et al., 2000).

Flooding conditions may suppress the mycorrhizal association (Miller, 2000) but do not affect root colonization (Ipsilantis and Sylvia, 2006; Stevens and Peterson, 1996). An oxygen restriction could either delay the extension of external mycelium into flooded soil (Beck-Nielsen and Madsen, 2001) or promote the adherence of high amounts of hyphae to the roots (Hildebrandt et al., 2001). However, knowledge on the extent to which salinity and flooding affect the growth and efficacy of the indigenous AMF is scarce (Carvalho et al., 2003). Feng et al. (2002) showed that selected AM species increased the resistance to osmotic stress due to a significant increase in soluble sugars and electrolyte concentration in maize roots. Ruiz-Lozano and Azcón (2000) reported an improvement in the salinity tolerance

of lettuce plants inoculated with AMF and a specific strain of Glomus isolated from saline conditions. This inoculum protected the plants against the detrimental salt effects through an increase on the radical system.

Although there are some reports about inoculation of AMF in rice crops under field conditions (Fernández et al., 1997; Secilia and Bagyaraj, 1992), its use is really scarce and it always involves a solid substrate to pelletize the seed (Rivera and Fernández, 2005). The solid substrate technique is useful; nevertheless, it is complex due to the great volume of seeds needed to sowing (150 and 200 kg ·ha^{-1}). According to the variety of plant, soil and fertilization program, great machines would be necessary to make the inoculation process in a homogeneous way; therefore, the use of a liquid inoculum could represent a more convenient option to ease inoculation in field.

Glomus hoi has been characterized by a high production of external mycelium, spores, frequent arbuscules even at the end of crop (Fernández et al., 2006a); it has been developed in different host plants as Sorghum vulgare and Brachiaria decumbens. But its potential for being used to counteract stress caused by salinity is a derivative of the fact that this fungus has always been used in saline soils with a high electric conductivity of 8315 μS cm^{-1} and sodium contents of 6.62 cmol kg^{-1} soil (Dell´Amico et al., 2007).

The objective of this work was to evaluate whether the inoculation of rice plants with AMF species isolated from a heavily saline soil could improve their development and performance under saline edaphic conditions; additionally we tested the efficacy of a liquid inoculum of AMF, G. hoi rather than a solid traditional inoculum to investigate a potentially more efficient alternative.

MATERIALS AND METHODS

Description of the study area

This work was conducted in the central Greenhouse of the National Institute of Agricultural Sciences (INCA), in San Jose de las Lajas, Havana, Cuba which is located at the geographic coordinates: 23° 0.12 .06 "N Latitude and 82° 8'31 .46" W Longitude.

Soil

Soil used in the experiments was a Gleysol containing: 2.3% organic matter, 0.5 g kg^{-1} total nitrogen, 13.2 cmol kg^{-1} total P, 10.2 cmol kg^{-1} of calcium, 5.6 cmol kg^{-1} of magnesium, 0.9 cmol kg^{-1} of potassium, 6.9 cmol kg^{-1} of sodium, an electric conductivity of 8789 ụS cm^{-1} and a pH of 8.2 in H_2O.

Mycorrhizal fungi inoculums

G. hoi (Berch and Trappe) previously isolated from pasture under saline soils (Fernández et al., 2010) at Departamento del Chaco, Bolivia, was used in this study. It was selected as an efficient isolate for improving plant growth under salinity stress conditions

Table 1. Plant height (cm), root system length (cm) and mycorrhizal colonization (%) quantified during the 90 days after germination (DAG) in AMF-treated and control plants.

DAG	1	5	10	20	25	35	40	60	90
Plant height (cm)									
AMF	10.5 b	10.5 b	10.9 b	15.1 a	19.8	23.8 a	32.8 a	63.5 a	74.1 a
Control	11.1 a	11.2 a	12.2 a	17.6 b	18.9	20.7 b	25.7 b	58.6 b	68.1 b
Es x	0.2***	0.12***	0.1***	0.2***	0.9 ns	0.1***	0.2***	0.6***	0.2***
Root length (cm)									
AMF	0.6	0.6	1.4 b	6.18b	6.82b	7.9 a	17.08a	23.9 a	26.9 a
Control	0.6	0.6	2.1 a	6.29a	6.4a	6.4 b	7.23b	10.4 b	13.1 b
Es x	0.0 ns	0.3 ns	0.2***	0.3 ns	0.4 ns	0.2***	0.3 ns	0.1***	0.2***
Mycorrhizal colonization (%)									
AMF	2 a	6 a	13 a	17 a	18 a	24 a	35 a	38 a	44 a
Control	0 b	3 b	3 b	12 b	13 b	15 b	20 b	19 b	22 b
Es x	0.3***	0.2**	0.13***	0.1***	0.2***	0.1***	0.2***	0.2***	0.4***

Values within a column followed by the same letter do not differ significantly (*P*<0.05) using a Tukey HSD test n=10.

(Fernández et al., 2006b); inoculum was propagated using *S. vulgare* for four months. Spores were isolated from substrate by a sedimentation process. First, 250 g of substrate and propagules were mixed with water in a tank by mechanical stirring for 15 min. This mix was decanted in a filtration cylinder to collect the supernatant.

Supernatant was centrifuged at 2000 g for 5 min to obtain the spores along with the mycelium fragments and it was stored in a protective osmotic solution (number of international publication of patent WO 2006/060968 AI) (Fernández et al., 2006c).

Evaluation of the mycorrhizal inoculum efficacy on growth and production of plants

The experiment was conducted in a greenhouse at a temperature range of 20 to 35°C. Soil (1000 g per pot) was placed into 20 × 15 ×15 cm plastic pots. 15 seeds of *O. sativa* L (cv.LP5) were sown into each container. Each pot was considered a single sample, and there were ten sample replicates per treatment. Treatments were divided into:

1) Non-mycorrhizal control; and
2) Inoculation with a liquid mycorrhizal application.

The mycorrhizal treatment was inoculated with a 100 ml of the protective osmotic solution that contained 3000 spores and mycelium fragments whereas the non-mycorrhizal treatment was rinsed with 100 ml of the same protective osmotic solution free of mycorrhizal spores. An amount of 100 ml of a nutrient solution containing 100 mg L^{-1} of nitrogen, 50 mg L^{-1} of phosphorous and 50 mg L^{-1} of potassium was added to the soil every week (Jeon et al., 2004; Minagri, 1999). Tap water was supplied daily during 20 days after sowing, and a water film was implanted during the entire assay to maintain the flooding conditions.

Plant height, root length and mycorrhizal activity (root colonization, visual fungal density, spore density and fungal biomass) of five plants per pot per treatment were measured at day 1, 5, 10, 15, 20, 25, 40, 60 and 90. Note that mycorrhizal activity could be measured at day 1 due to the presence of residual mycelium and roots remaining in the suspension inoculated.

These propagules established communication with the seed and started the colonization through the issuance of both extramatric and internal mycelium at the following days. It has been observed that the biochemical activity between inoculum and host plant could be developed 24 h after inoculation (Pérez-Ortega, 2010). After 90 days, the plants were harvested and the yield components (panicle length, spikelet number per panicle, and grain yield) were quantified. The experiment was based on a completely random design. A one-way analysis of variance (ANOVA) was performed in order to determine if there were significant differences between treatments. Post-hoc comparisons were then analyzed using a Duncan's multiple-range test (p<0.05).

Analyses of mycorrhizal colonization and visual fungal density

Mycorrhizal colonization and the visual fungal density were tested according to the grid line-intersect method described by Giovannetti and Mosse (1980) and Trouvelot et al. (1986). Fine roots were collected and washed with water. The washed roots were cut into 1 cm segments and thoroughly mixed. A sub-sample of 0.5 g was bleached with 10% (w/v) KOH at 90°C for 2 h and stained with trypan blue (Phillips and Hayman, 1970).

Quantification of ectophyte/arbuscular endophyte biomasses and total spore density

Ectophyte biomass (EcB) and the total spore densitiy in soil were analyzed based on the method by Herrera-Peraza et al. (2004). The ratio EcB: arbuscular endophyte biomass (AEB) was calculated as an indicator of mycorrhizal function. Each experiment was repeated three times.

RESULTS

Efficacy of the mycorrhizal inoculum on growth and production of plants

Plant growing

The application of AMF liquid inoculum had a positive effect on the growth dynamics of plants and produced a remarkable effect on the variables analyzed (Table 1).

Table 2. Yield and some of its components: Panicles.plant^{-1} (P), panicles weight (PW) and 100 grains weight (GW) determined in AMF-treated and control plants.

Treatments	P	PW (g)	100 GW (g)	Yield (g plant^{-1})
AMF	8.33 a	2.70 a	3.69 a	21.66 a
Control	5.40 b	1.87 b	2.70 b	15.40 b
Es x	0.12 ***	0.05***	0.001***	1.34***
C.V (%)	11.2	9.2	6.5	14.6

Values within a column followed by the same letter do not differ significantly ($P<0.05$) using a Tukey HSD test, n=10.

Nevertheless, up to 20 days, the highest growth values corresponded to the control plants; whereas at the end of their life cycle, mycorrhizal-inoculated plants performed significantly better than the control group ($p< 0.05$). Differently, the root growth was very slow and little compared to the high of the stem. At an early stage, both groups showed a similar rate of growth; at day 10 the control root was longer than the mycorrhizal-inoculated treatment and this effect was maintained for 25 days. Roots of inoculated plants had a significant two fold-increase lengthwise compared to the controls ($p< 0.05$).

The main yield components for both treatments of rice plants are shown in Table 2. An increase in yield and its components in the AMF treatment was noteworthy, as a clear sign of the efficiency of this type of fungal inoculum on the plant productivity even under saline conditions.

Root colonization, visual fungal density and spore density

The analysis of AMF colonization was a progressive process and a high value was obtained (Table 1). In inoculum-free plants there was a poor colonization because of the unsterilized soil used in this experiment (data not shown). Since root colonization only indicates the presence of the mycorrhizal fungi hyphae in the root but not the fungal intensity in the inner root, a fungal visual density test was conducted according to the method reported by Trouvelot et al. (1986) This test expresses more adequately the efficacy of the mycorrhizal fungi to promote growth and production in the plants. The fungal occupancy had a typical microbial behavior where the symbiont slowly colonized the inner root from 0 up to 20 days (Figure 1). An exponential growth of mycelium with a latent phase very well defined was observed after 40 days, when the fungal activity reached a stationary phase to the end of crop. This behavior was related with the spore dynamics (Figure 1).

During the first day, a low number of spores appeared in the soil samples, as a cumulative result of the pool of spores from the initial inoculation. This number increased along the next days, due to the germination process in adequate conditions of humidity and temperature. Accumulation of the new spores production began at 30

days, as a direct consequence of the increase in the production of mycelium and the symbiosis development. The increment of the spore population continued to grow up until reaching 12 spores g soil^{-1} (Figure 1).

EcB and AEB

EcB and AEB (Figure 2) are some of the main indicators of the fungal symbiont presence in annual crops, due to the strong relationships between their functioning and the plant symbiosis process. The biggest EcB was presented during the first development stages of symbiosis. In contrast, the arbuscular endophyte developed low biomass content over the first 25 days, after which, high values were obtained. This stage could be considered as a transition phase. After 30 days, the EcB decreased but then achieved a stabilization process. AEB had a different performance: a gradual increment associated with plant growth and symbiosis development. The EcB:AEB ratio (Figure 3), indicates the actual mycorrhizal functioning (Fernández, 2003). The performance was initially characterized by a strongly decreasing curve, which means a high EcB and a low endophyte presence. This was the result of hyphal spread in soil from host root.

After 20 days, a slight transitional stage takes place and a mutualistic phase was immediately implanted up to the end of the crop. Once the inner fungal components were in equilibrium with the ectophyte, the biomasses values fell down to below one, or even lower. From our point of view, this kind of mycorrhizal performance under this saline soil conditions have two marked phases:

First, an initial parasitic-transitional phase (from 0 to 25 days), with a great amount of ectophyte presence; second, a phase characterized by an intense endophyte presence and significant plant growth (starting from 25 days).

DISCUSSION

Plant growth

The plant growth in response to AMF colonization varied

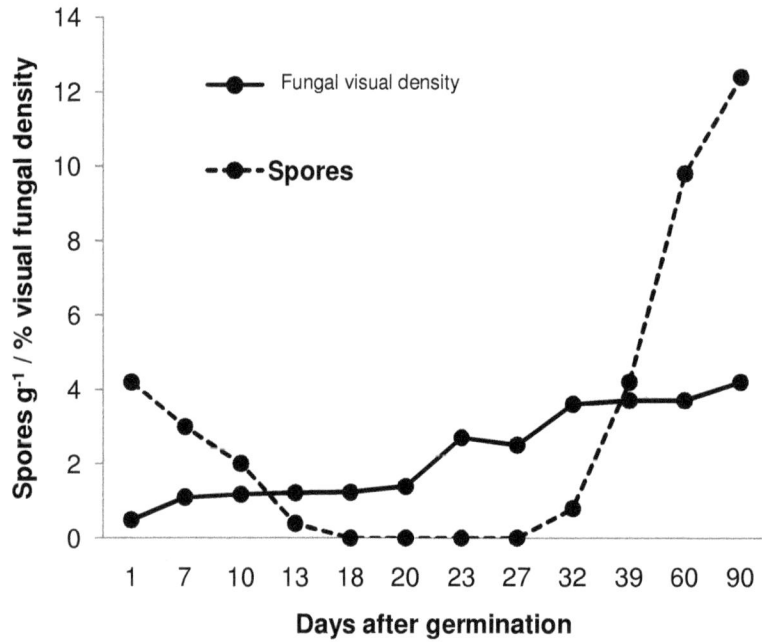

Figure 1. Spore dynamics and fungal visual density (%), in rice plants inoculated with liquid AMF under saline soil.

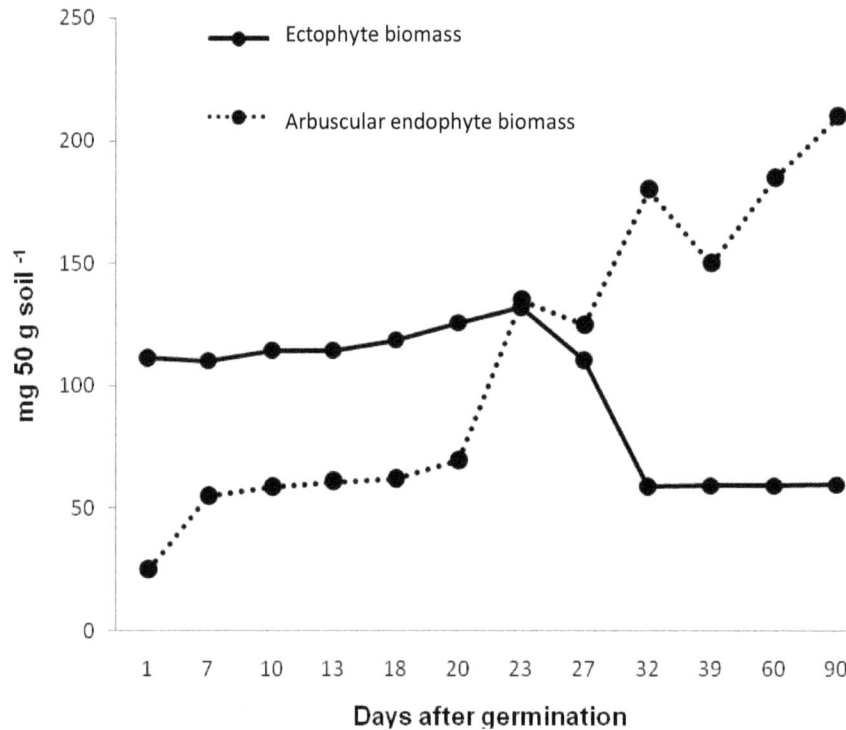

Figure 2. Dynamics of the Endophyte biomasses (EcB) and the Arbuscular ectophyte biomasses (AEB) for 100 days after germination in rice plants under saline soil, inoculated with a liquid AMF suspension under saline soil.

widely in its timing (Table 1). In this work this response was observed after 20 days and it was sustained until 90 days, in contrast to the reported by Bethlenfalvay et al. (1982), who found in soybean that the response was 42

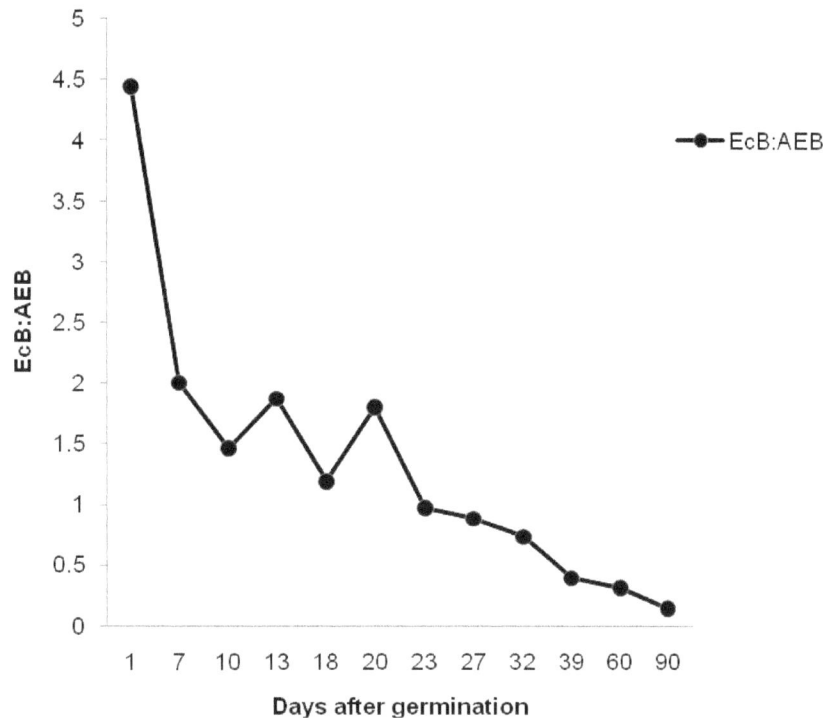

Figure 3. Ectophyte biomass (EcB): Arbuscular endophyte biomass (AEB) ratio found in rice plants inoculated with a liquid AMF suspension under saline soil.

days after planting and the growth was delayed 70 days. The promoter effect of AMF on the height and development of aerial part and the root system of the plants have been reported by several authors (Bethenfalvay and Linderman, 1992; Fernández et al., 1997; Fernández, 1999; Llonin and Medina, 2002; Terry et al., 1998). Indeed, salts could affect the capacity of the plant to supply carbohydrates, required for hyphal growth, by decreasing photosynthesis and plant growth under salinity (Mc Millen et al., 1998; Munns et al., 1995). However, beneficial effects of AMF on plant growth were observed after the establishment of the symbiont. This response could be close related to the mycorrhizal effectiveness, as the symbiosis process involves fluxes of photosynthates to the root system. Therefore, an adequate mycorrhizal development due to the use of these substances implicates a strong root system (Bonfante and Perotto, 1992; Bowen, 1987).

Results reported by Debouba et al. (2006) showed a greater shoot and root dry matter in tomato plants pre-inoculated with AMF irrigated with both saline and non saline water. Similar shoot and root weight increases were reported in banana (Yano-Melo et al., 2003), cotton (Tian et al., 2004), zucchini (Colla et al., 2007), *Lotus glaber* (Sannazzaro et al., 2006), soybean (Sharifi et al., 2007) and *Acacia auriculiformis* (Giri et al., 2003). However, our results differed from those studies in showing that the shoot was higher than the root (Table 1). Debouba et al. (2006) found that the increase of the

root/shoot ratio in tomato was due to the higher sensitivity to salt of the leaves compared to the roots. The increase in this root/shoot ratio supports the hypothesis that the tomato assigns more dry weight to roots in order to maximize its capabilities for nutrient and water absorption. This is consistent with the results of Zhang (1995) who found that an increase in the root/shoot ratio reduced the shoot growth and yield, as well as the plant efficiencies in the use of water and nutrients (Debouba et al., 2006). This may also indicate that salt conditions induced nutritional deficiencies (Shangguan et al., 2004).

Studies in rice plants showed a bigger increment in the height of the shoot compared to the root (Maggio et al., 2004). This could be explained as a result of an inhibition of the root growth induced by salinity (Maggio et al., 2004). However, it was found in tomato plants that the maximum root density was reached at 60 days after transplanting and it gradually decreased until harvesting (Al-Karaki 2006). Our results show that the increment in rice growth started at 20 days after transplant, and it continued until 90 days. It is clear that salinity did not increase plant mortality, but it caused a reduction of root development (Table 1). In this sense, the response of inoculated rice to salinity was higher than the control. In the most commonly cultivated rice areas, young seedlings are very sensitive to salinity (Flowers and Yeo, 1981; Heenan et al., 1988). However, our data show a high production of rice grains (21.66 g plant^{-1}) compared to the control (15.40 g plant^{-1}) (p<0.05), as can be observed in all yield components related to final grain (Table 2).

Similar results were found in zucchini fruits where mycorrhizal colonization enhanced the fruit dry matter with the highest values recorded on inoculated plants (average 5.9%) compared to non-treated plants (average 5.6%). It is well established that crop growth and yield decrease with high salinity (Colla et al., 2007). AMF alleviate the detrimental effect of salinity on growth and productivity. Improved nutritional and leaf water status might have assisted the plants to translocate minerals and to assimilate them as well as to alleviate the impacts of salinity on fruit production (Colla et al., 2007). Contrary to the studies reported by other researchers who found that yield was severely affected by salinity, in this work was observed that some parameters such as panicle length, spikelet number per panicle, and grain yield were significantly improved in salt treatments (Table 2) (Cui et al., 1995; Heenan et al., 1988; Khatun et al., 1995; Zeng and Shannon, 2000). Salinity also delayed the panicle emergence and flowering (Khatun et al., 1995). These results show the positive effect of the mycorrhizal inoculation to improve growth under these unfavourable conditions. This is consistent with the observed by Al-Karaki (2006) in pre-inoculated tomato plants with AMF irrigated with saline water, where yield (5.26 kg m^2) and fruit fresh yield (23 g) were greater than in non-inoculated plants (3.28 kg m^2 and 15 g respectively). This improvement in fruit fresh yield due to AMF inoculation was 29% under non saline and 60% under saline water conditions (Al- Karaki, 2006).

Some authors have found crops with low specificity for certain AMF strains in a specific soil condition (Fernández, 1999; Fernández-Martín et al., 2004; Rivera and Fernández, 2005; Ruiz, 2001; Sánchez, 2001). It points out the need of a competent mycorrhizal strain to establish an efficient colonization with any mycorrhizal-dependent crop. Indeed, crops show different quantitative effects to the mycorrhizal inoculation, but the competent strains could be used in the majority of crops. In this regard, several authors had found a strong mycorrhizal dependency in maize and cotton plants using the fungal strain *Glomus mosseae* under saline conditions (Feng et al., 2002). Cotton plants inoculated with different isolates of *G. mosseae* increased significantly their shoot dry weight by 68% under 1 to 3g NaCl kg^{-1}, and by 27% under 2g NaCl kg^{-1}.

Nevertheless, their shoot dry weight improved significantly by 31% when the NaCl level was 3g kg^{-1}, while root growth was not significantly affected (Tian et al., 2004).

Root colonization, fungal visual density and spore density

Structures characteristic of AMF were observed in the roots after inoculation at all levels of salinity (Figure 1). The percentage of root colonization (Figure 1), was higher in mycorrhizal than non-mycorrhizal plants, similar to the observed by Giri et al. (2003). This plant dependence seems to have a significant ecological importance, related to the adaptation of AMF to the saline stressful conditions. The fungal strain used in this assays, was isolated from a saline soil. It seems that the performance of the particular AM species under saline conditions is closely related to its adaptability to the ecological niches, the places, the original conditions of isolation and the substrates where it has been preserved up to this use. The AMF could follow a strategy to keep the functional memories in the symbiotic capacity to be used under similar conditions to the ones where it was originally found.

Since AMF require oxygen to succeed, stressful environments regularly flooded with saline water may be detrimental for their survival and infectivity (Saint-Etienne et al. 2006). Nevertheless, some AMF are able to persist in flooded soils and to colonize wetland plants (Khan, 1993; Landwehr et al., 2002; Miller and Bever, 1999; Turner et al., 2000). More than 50% of the plant's populations have been colonized by AMF in some wetlands conditions (Ragupathy et al., 1990). Brown and Bledsoe (1996), observed AMF in the aerenchymatous tissue of saline marsh plants, suggesting that AMF are adapted to life in oxygen-deficient soils.

Results observed in rice related to an increase in the ratio of root growth of AMF treatment with the increase in water level could be a result of development of an aerenchyma, such as the one reported by Carvalho et al. (2003) for *Aster tripholium* plants. These particular plants have the capability to induce a well-developed aerenchyma, suggesting that AMF could thus overcome the lack of oxygen in soil by colonizing these more oxygenated portions of the root. Therefore the effect of flooding on fungal growth was greater in EcB than in AEB (Figure 3). If we take into account that the assay was performed under flooding conditions starting from 15 days, it could be supported that mycorrhizal colonization was very effective. At the end of the life cycle of the crop a root colonization percentage of 44 could be observed. We considered it elevated compared to other experiments in rice with inocula in solid base such as MicoFert and EcoMic, where the maximum mycorrhizal colonization values were around 25% (Fernández et al., 1997).

Similar results were reported by Sannazzaro et al. (2006) who found a reduction in mycorrhizal fungus colonization of *Lotus glaber* plants grown under saline conditions. This could be explained as it has been shown that AMF may be affected by salinity during spore germination (Juniper and Abbott, 1993), hyphal growth (McMillen et al., 1998) and arbuscule formation (Pfeiffer and Bloss, 1988). However, the increment in root colonization in rice in presence of the liquid inoculum has been already observed by others authors in tomato, lettuce and maize, as well as in different type of soils (Sánchez –Blanco et al., 2004) and (Fernández et al., 2006a). Native spore population was low (less than 1g soil^{-1}), very typical of agricultural saline soil with high sodium content and other salinity conditions in which the AMF presence and diversity always decreases because of the intensive till and overexploitation, among others (Rao, 1998).

The AMF spore production has been well documented in the literature, especially in *Glomus* species, which are great spore producers. Studies are facilitated because of the fungal development in the arbuscular-mycorrhizal associations and the fungal life cycle. In our study, large amounts of spores could be obtained under conditions of inoculum production. Once the vegetable tissue is dead, the spore contents can be multiplied up to twice due to the fast translocations of nutrients from roots to the hyphal network in order to form new ones. Values obtained for *G. hoi*, the species used in this assay, were nearest to 300 spores g substrate^{-1} (Fernández, 2003).

EcB and AEB

The increment in the spore production could be related to the biomasses of the ectophyte and endophyte (Figure 2). Initially, there was a rather more external mycelium than inner one but this condition was reversed after day 20. At the final harvest, endophyte biomass was almost three times bigger than the external component. These results were similar to the data reported by Bethlenfalvay et al. (1982). The biggest EcB presented during the first development stages of symbiosis (Figure 2), was caused by the parasitic fungal abilities to live at expenses of the plant, as previously reported by other authors as a parasitic phase (Bethlenfalvay et al., 1989). This phase is considered a derivative of the exuberant fungal mycelium growth at early stages of plant growth with a low photosynthetic rate and a high metabolic cost.

AEB (Figure 2) had a different performance: a gradual increment associated with plant growth and symbiosis development

Some authors have previously described that process using fungal strains in solid substrate (Bethenfalvay and Linderman, 1992; Fernández, 2003). Therefore, this work is the first report where AMF conserved in a liquid suspension has been used. The EcB: AEB ratio (Figure 3) was characterized by a strong declining curve, which means a high EcB and low endophyte presence. This was provoked by hyphal spread in soil from host root to explore the surrounding environment looking for nutrients for itself to maintain the symbiosis with plant and to increase the reciprocal interchanges with the host tissue. This particular behavior has been recently described under *in vitro* conditions as a common event (Bago and Cano, 2005). A diminished plant growth was also observed during early stages up to 18 to 20 days, as a clear sign of the efforts that plants need to do in order to maintain the symbiosis (Table 1). An evident increase in the AEB was presented derived from the high nutrient demand and the arbuscular increase in the inner cell host (Figure 2).

Previous studies had reported that salinity reduced the mycorrhizal colonization, caused an inhibition of germinative power in spore of genus *Glomus*, decreased the ectophyte growth in soil (Hirrel, 1981; Rao, 1998), reduced the normal spread of hyphal network after the colonization (McMillen et al., 1998) and decrease the arbuscular number in host plant (Pfeiffer and Bloss, 1988). This work showed that the inoculation with a liquid suspension of AMF abolished the major limitations blocking the fungus development and the subsequent nutritional benefits conferred to the rice plants to which it was associated. We can point out that the use of this type of liquid inoculum as well as the AMF strain were very effective under soil condition where high salinity and deficient nutrients are conjugated to act in a negative way on the plant growth. Moreover, the use of a liquid inoculum along with an efficient AMF that shows an intensive response under adverse conditions is a feasible option and opens new strategies to be used as tools for improving the development of crops under stress conditions.

ACKNOWLEDGEMENTS

We would like to thank "Instituto de Ciencia y Tecnologia de Cuba" for funding this work and Instituto Nacional de Ciencias Agrícolas (INCA) de la Habana, Cuba for lending us their facilities to conduct this research. We thank Professor David Thomas for critically and exhaustively reviewing this manuscript.

REFERENCES

Al-Karaki GN (2000). Growth of mycorrhizal tomato and mineral acquisition under salt stress. Mycorrhiza, 10: 51–4.

Al-Karaki GN, Hammad R, Rusan M (2001). Response of two tomato cultivars differing in salt tolerance to inoculation with mycorrhizal fungi under salt stress. Mycorrhiza, 11: 41–47.

Al-Karaki GN (2006). Nursery inoculation of tomato with arbuscular mycorrhizal fungi and subsequent performance under irrigation with saline water. Sci. Hortic., 109: 1–7.

Azcón-Aguilar C, Barea JM (1997). Applying mycorrhiza biotechnology to horticulture: significance and potentials. Sci. Hortic., 68: 1–24.

Azcón R, El-Atrash F (1997). Influence of arbuscular mycorrhiza and phosphorus fertilization on growth, nodulation and N$_2$ (N-15) in *Medicago sativa* at four salinity levels. Biol. Fert. Soils, 24: 81–6.

Aziz T, Sylvia DM, Doren RF (1995). Activity and species composition of arbuscular mycorrhizal fungi following soil removal. Ecol. Appl., 5: 776–784.

Bago B, Cano C (2005). Breaking myths on Arbuscular Mycorrhizas in vitro biology. In: Declerk S, Strullu DG and Fortin JA (Eds). *In vitro* culture of Mycorrhiza. Soil Biol. Springer-Verlag, Berlin Heidelberg. Pp. 111-138.

Beck-Nielsen D, Madsen TV (2001). Occurrence of vesicular–arbuscular mycorrhiza in aquatic macrophytes from lakes and streams. Aquat. Bot., 71: 141–148.

Bethlenfalvay GJ, Pacovsky RS, Brown MS, Fuller G (1982).

Mycotrophic growth and mutualistic development of host plant and fungal endophyte in an endomycorrhizal symbiosis. Plant Soil, 68: 43-54.

Bethlenfalvay GJ, Brown MS, Franson RL, Mihara KL (1989). The glycine-*Glomus-Bradyrhizobium* symbiosis - IX. Nutritional, morphological and physiological response of nodulated soybean to geographic isolates of the mycorrhizal fungus of *Glomus mosseae*. Physiol. Plant, 76: 226–232.

Bethenfalvay GJ, Linderman JA (1992). Mycorrhizae and crop productivity. Horticultural Crops Research Laboratory. USDA-ARS.

Bonfante-Fassolo P, Perotto S (1992). Plants and endomycorrhizal fungi: The cellular and molecular basis of their interaction. In: Verma DP (Eds). Molecular signals in plant-microbe communications. CRC press, Boca Raton, pp: 445–470.

Bowen GD (1987). The Biology and Physiology of infection and its development. In: Safir GR (Ed), Ecophysiology of VA Mycorrhizal Plants. CRC Press, Boca Raton, pp. 27-57.

Brown AM, Bledsoe C (1996). Spatial and temporal dynamics of mycorrhizas in *Jaumea carnosa*, a tidal saltmarsh halophyte. J. Ecol., 84: 703–715.

Carvalho LM, Correia PM, Caçador I, Martins-Louçao MA (2003). Effects of salinity and flooding on the infectivity of salt marsh arbuscular mycorrhizal fungi in *Aster tripolium* L. Biol. Fert. Soils, 38: 137–143.

Chabbi A, Mckee KL, Mendelssohn IA (2000). Fate of oxygen losses from *Typha domingensis* (Typhaceae) and *Cladium jamaicense* (Cyperaceae) and consequences for root metabolism. Am. J. Bot., 87: 1081–1090.

Cho K, Toler H, Lee J, Ownley B, Stutz JC, Moore JL, Augé RM (2006). Mycorrhizal symbiosis and response of sorghum plants to combined drought and salinity stresses. J. Plant Physiol., 163: 517-528.

Colla G, Rouphael Y, Cardarelli MT, Tullio M, Rivera CM, Rea E (2007). Alleviation of salt stress by arbuscular mycorrhizal in zucchini plants grown at low and high phosphorus concentration. Biol. Fert. Soils, pp. 1-9.

Cui H, Takeoka Y, Wada T (1995). Effect of sodium chloride on the panicle and spikelet morphogenesis in rice. Jpn. J. Crop Sci., 64: 593–600.

Debouba M, Gouia H, Suzuki A, Ghorbel MH (2006). NaCl stress effects on enzymes involved in nitrogen assimilation pathway in tomato "*Lycopersicon esculentum*" seedlings. J. Plant Physiol., 163: 1247-1258.

Dell'Amico JM, Fernández F, Nicolás E, López LF, Sánchez-Blanco MJ (2007). Physiological response of tomato to the application of twoinoculants *Glomus sp1* (INCAM 4) by two different routes of inoculation. Trop. Crops, 28: 33-38.

Dixon RK, Garg VK, Rao MV (1993). Inoculation of *Leucaena* and *Prosopis* seedlings with *Glomus* and *Rhizobium* species in saline soil: Rhizosphere relations and seedlings growth. Arid Soil Res. Rehabil., 7: 133–144.

Feng G, Zhang FS, Li XL, Tian CY, Tang C, Rengel Z (2002). Improved tolerance of maize plants to salt stress by arbuscular mycorrhiza is related to higher accumulation of soluble sugars in roots. Mycorrhiza, 12: 185–190.

Fernández F, Ortiz R, Martínez MA, Costales A, Llonín D (1997). The effect of commercial arbuscular mycorrhizal fungi (AMF) inoculants on rice (*Oryza sativa*) in different types of soils. Cult, Trop., 18(1): 5-9.

Fernández F (1999). Managing arbuscular mycorrhizal associations on the production of coffee seedlings (*C. arabica* L. var. Catuaí) in some soil types, [Thesis Phd.], INCA, p. 128.

Fernández F (2003). Advances in the production of arbuscular mycorrhizal inoculants. In: Rivera R and Fernández K (Eds), Effective management of themycorrhizal symbiosis, a path to sustainable agriculture. Caribbean Case Study. Foreign Ministry in Havana, Chapter IV, pp. 144-166.

Fernández-Martín F, Rivera Espinosa RA, Providencia I, Fernández K, Rodríguez Y (2004). Effectiveness of mycorrhizal inoculation by seed dressing. Fungal functioning and agrobiological effect. In: Frías-Hernández JT, Olalde-Portugal V and Ferrera-Cerrato R (Eds), Advances in the understanding of the biology of mycorrhizae. University of Guanajuato, ISBN: 968-864-333-5, pp. 252-267.

Fernández-Martín F, Dell'Amico RJM, Rodríguez P (2006a). Effectiveness of some types of mycorrhizal inoculants in tomato crop. Trop. Crops, 27: 25-30.

Fernández-Martín F, Gómez Alvarez R, Venegas LLF, Martínez SMA, de la Noval PBM, Rivera ERA (2006b). Mycorrhizal inoculant product. Patent CU 22641 A1. Oficina Cubana de la Propiedad Industrial (OCPI), Available at: http://www.ocpi.cu/doc/2002/t7696.pdf, viewed 5 April.

Fernández F, Dell'Amico JM, Pérez Y (2006c). Liquid Mycorrhizal Inoculant. http://wipo/int/pctdb. Patente WO/2006/060968.

Fernández F, Dell Amico JM, Alarcón JJ, Nicolás E, Pedrero F (2010). Improving rice yield through the application of mycorrhizal isolated in saline conditions. Agriculture, 935: 916-919.

Flowers TJ, Yeo AR (1981). Variability in the resistance of sodium chloride salinity within rice (*Oryza sativa* L.) varieties. New Phytol., 88: 363–373.

Giri B, Kapoor R, Mukerji KG (2003). Influence of arbuscular mycorrhizal fungi and salinity on growth, biomass, and mineral nutrition of *Acacia auriculiformis*. Biol. Fert. Soils, 38: 170–175.

Giovannetti M, Mosse B (1980). An evaluation of techniques to measure vesicular-arbuscular infection in roots. New Phytol., 84: 489-500.

González-Núñez LM, Tóth T, García D (2004). Integrated management for the sustainable use of salt-affected soils in Cuba. Integrated management for sustainable use of salt-affected soils in Cuba. Sci. Univ., 20(40): 85-102.

Hartmond U, Schaesberg NV, Graham JH, Syverten JP (1987). Salinity and flooding stress effects on mycorrhizal and non mycorrhizal citrus rootstock seedlings. Plant Soil, 104: 37–43.

Hasegawa PM, Bressan RA, Zhu JK, Bohnert HJ (2000). Plant cellular and molecular responses to high salinity. Annual Review of Plant Physiology. Plant Mol. Biol., 51: 463–499.

Heenan DP, Lewin LG, McCaffery DW (1988). Salinity tolerance in rice varieties at different growth stages. Aust. J. Exp. Agric., 28: 343–349.

Herrera-Peraza RA, Furrazola E, Ferrer RL, Fernández R, Torres Y (2004). Functional strategies of root hairs and arbuscular mycorrhiza in an evergreen tropical forest, Sierra del Rosario, Cuba. Revista CNIC. Biol. Sci., 2: 113-123.

Hildebrandt U, Janetta K, Ouziad F, Renne B, Nawrath K, Bothe H (2001). Arbuscular mycorrhizal colonization of halophytes in Central European salt marshes. Mycorrhiza, 10: 175–183.

Hirrel MC (1981). The effect of sodium and chloride salts on the germination of *Gigaspora margarita*. Mycol., 43: 610–617.

Ipsilantis I, Sylvia DM (2006). Abundance of fungi and bacteria in a nutrient-impacted Florida wetland. Appl. Soil Ecol., 35: 272–280.

Ipsilantis I, Sylvia DM (2007). Interactions of assemblages of mycorrhizal fungi with two Florida wetland plants. Appl. Soil Ecol., 35: 261–271.

Jeon WT, Park CY, Cho YS, Park KD, Yun ES, Kang U G, Park ST, Choe ZR (2004). Root growth characteristics of rice grown under long-term fertilization of chemical fertilizer and compost in paddy. Korea J. Crop Sci., 48: 484-489.

Juniper S, Abbott L (1993). Vesicular–Arbuscular mycorrhizas and soil salinity. Mycorrhiza, 4: 45–57.

Khan AG (1993). Occurrence and importance of mycorrhizae in aquatic trees of New South Wales, Australia. Mycorrhiza, 3: 31–38.

Khan AG, Belik M (1995). Occurrence and ecological significance of mycorrhizal symbiosis in aquatic plants. In: Varma A and Hock B (Eds.), Mycorrhiza: Structure, Function, Molecular Biology and Biotechnology. Springer-Verlag, Berlin, pp. 627–666.

Keren R (2000). Salinity. In: Sumner, M.E. (Ed.), Handbook of Soil Science. CRC Press, Boca Raton, pp. G3–G25.

Khatun S, Rizzo CA, Flowers TJ (1995). Genotypic variation in the effect of salinity on fertility in rice. Plant Soil, 173: 239–250.

Kozlowski TT, Pallardy SG (1984). Effect of flooding on water, carbohydrate and mineral relations. In: Kozlowski TT (Ed) Flooding and plant growth. Academic Press, London, pp: 165–193.

Ladha JK, Bruijin FJ, Malik KA (1997). Introduction: Assessing opportunities for nitrogen fixation in rice a frontier project. Plant Soil, 194: 1–10.

Landwehr M, Hildebrandt U, Wilde P, Nawrath K, Toth T, Biro B, Bothe H (2002). The arbuscular mycorrhizal fungus *Glomus geosporum* in European saline, sodic and gypsum soils. Mycorrhiza. 12: 199–211.

Liang Y, Nikolic M, Peng Y, Chen W, Jiang Y (2005). Organic manure stimulates biological activity and barley growth in soil subject to secondary salinization. Soil Biol. Biochem., 37: 1185–1195.

Llonin D, Medina N (2002). Mineral nutrition with N, P and K in the symbiotic mycorrhizal fungi, tomato (*Lycopersicon esculentum* Mill) in Ferralsol. Trop. Crops, 23: 83-88.

Maggio A, De Pascale S, Angelino G, Ruggiero C, Barbieri G (2004). Physiological response of tomato to saline irrigation in long-term salinized soils. Eur. J. Agron., 21: 149–159.

Mc Millen BG, Juniper S, Abbott LK (1998). Inhibition of hyphal growth of a vesicular–arbuscular mycorrhizal fungus in soil containing sodium chloride limits the spread of infection from spores. Soil Biol. Biochem., 30: 1639–1646.

Miller SP, Bever JD (1999). Distribution of arbuscular mycorrhizal fungi in stands of the wetland grass *Panicum hemitomon* along a wide hydrologic gradient. Oecologia, 119: 586–592.

Miller S (2000). Arbuscular mycorrhizal colonization of semi-aquatic grasses along a wide hydrologic gradient. New. Physiol., 145: 145-155.

Minagri (1999). Rice Research Institute. Technical Instructions Rice. Havana, Cuba, p. 119.

Munns R, Schachtman DP, Condon AG (1995). The significance of a two-phase growth response to salinity in wheat and barley. Aust. J. Plant Physiol., 22: 561–569.

Nourbakhsh F, Sheikh-Hosseini AR (2006). A kinetic approach to evaluate salinity effects on carbon mineralization in a plant residue-amended soil. J. Zhejiang Univ. Sci. B., 7(10): 788-793.

Pfeiffer CM, Bloss HE (1988). Growth and nutrition of guayule (*Parthenium argentatum*) in a saline soil as influenced by vesicular–arbuscular mycorrhiza and phosphorus fertilization. New Phytol., 108: 315–321.

Phillips DM, Hayman DS (1970). Improved procedures for clearing roots and staining parasitic and vesicular arbuscular mycorrhizal fungi for rapid assessment of infection. Trans. Brit.. Mycol. Soc., 55: 158-161.

Pérez-Ortega EJ (2010). Arbuscular mycorrhizal fungi (AMF) for the biosecurity of pathogensin tomato crop (*Solanum lycopersicum* L.). PhD. Thesis. Havana University, Faculty of Biology, National Institute of Agricultural Sciences. Havana, Cuba, p. 138.

Rao DLN (1998). Biological amelioration of salt-affected soils. In: Microbial Interactions in Agriculture and Forestry. Science Publishers, Enfield, USA, 1: 21–238.

Ragupathy S, Mohankumar V, Mahadevan A (1990). Occurrence of vesicular-arbuscular mycorrhizae in tropical hydrophytes. Aquat. Bot., 36: 287-291.

Rivera RA, Fernández F (2005). Inoculation and management of mycorrhizal fungi within tropical agroecosystems. In: Uphof N (Ed), Biological approaches to sustainable soil systems. Cornell University, New York, pp. 479-488.

Rosendahl CN, Rosendahl S (1991). Influence of vesicular arbuscular mycorrhizal fungi (*Glomus* spp.) on the response of cucumber (*Cucumis sativis* L.) to salt stress. Environ. Exp. Bot., 31: 313–318.

Ruiz-Lozano JM, Azcón R (2000). Symbiotic efficiency and infectivity of an autochthonous arbuscular mycorrhizal *Glomus* sp. from saline soils and *Glomus deserticola* under salinity. Mycorrhiza, 10: 137–143.

Ruiz-Lozano JM, Azcón R, Gómez M (1996). Alleviation of salt stress by arbuscular-mycorrhizal *Glomus* species in *Lactuca sativa* plants. Physiol. Plantarum, 98: 767–772.

Ruiz L (2001). Effectiveness of mycorrhizal associations on plant roots and tubersin brown soils with carbonates and Reds Ferrallitic central Cuba. Thesis in Science Degree option to Doctor in Agricultural Sciences. INCA, p. 117.

Sánchez C (2001). Use and management of mycorrhizal fungi and green manures in the production of coffee seedlings in some soils of Guamuhayasolid."Thesis in Science Degree option to D octor in Agricultural Sciences. INCA, p. 105.

Sánchez-Blanco MJ, Conejero W, Navarro A, Ortuño MF, Alarcón JJ, Torrecillas A, Morte A, García-Mina JM, López LF (2004). Implementation of a new mycorrhizal inoculant fluid through drip irrigation on lettuce. Agric. Murcia, 3: 14-15.

Sannazzaro AI, Ruiz AO, Albertó EO, Menéndez AB (2006). Alleviation of salt stress in *Lotus glaber* by *Glomus intraradices*. Plant Soil, 285: 279–287.

Sass RL, Cicerone RJ (2002). Photosynthate allocations in rice plants: Food production or atmospheric methane? PNAS, 99: 11993–11995.

Secilia J, Bagyaraj DJ (1992). Selection of efficient vesicular-arbuscular mycorrhizal fungi for wetland rice. Biol. Fert. Soils, 13: 108-111.

Shangguan ZP, Shao MA, Ren SJ, Zhang LM, Xue Q (2004). Effect of nitrogen on root and shoot relations and gas exchange in winter wheat. Bot. Bull. Acad. Sin., 45: 49–54.

Sharifi M, Ghorbanli M, Ebrahimzadeh H (2007). Improved growth of salinity-stressed soybean after inoculation with salt pre-treated mycorrhizal fungi. J. Plant Physiol., 164: 1144-1151.

Singh RP, Choudhary A, Gulati A, Dahiya HC, Jaiwal PK, Sengar RS (1997). Response of plants to salinity in interaction with other abiotic and factors. In: Jaiwal PK, Singh RP and Gulati A (Eds.), Strategies for Improving Salt Tolerance in Higher Plants. Science Publishers, Enfield, USA, pp. 25–39.

Smith FA, Smith SE (1997). Structural diversity in (vesicular)–arbuscular mycorrhizal symbioses. New Phytol., 137: 380–388.

Stevens KJ, Peterson RL (1996). The effect of a water gradient on the vesicular-arbuscular mycorrhizal status *of Lythrum salicaria* L. (purple loosestrife). Mycorrhiza, 6: 99–104.

Terry E, Pino MA and Medina N (1998). Agronomic effectiveness of Azofert and Comic in tomato crop (*Lycopersicon esculentum* Mill). Cult. Trop., 19: 33-37.

Tian C, Feng G, Li XL, Zhang FS (2004). Different effects of arbuscular mycorrhizal fungal isolates from, saline or non-saline soil on salinity tolerance of plants. Appl. Soil Ecol., 26: 143–148.

Thormann MN, Currah RS, Bayley SE (1999). The mycorrhizal status of the dominant vegetation along a peatland gradient in southern boreal Alberta, Canada. Wetlands, 19: 438–450.

Trouvelot A, Kough J, Gianinazzi-Pearson V (1986). Measuring the rate of VA mycorrhization of root systems. Research Methods for Estimating havinga functional significance. In: Gianinazzi-Pearson V and Gianinazzi S (Eds), Proceedings of the 1st European Symposium on Mycorrhizae: Physiological and Genetical Aspects of Mycorrhizae, Dijón, INRA, Paris, pp. 217-222.

Turner SD, Amon JP, Schneble RM, Friese CF (2000). Mycorrhizal fungi associated with plants in ground-water fed wetlands. Wetlands, 20: 200–204.

UNCTAD Secretariat from the Food and Agriculture Organization of the United Nations (FAO, 2008).

Yano-Melo AM, Saggin OJ, Costa MJr. L (2003). Tolerance of mycorrhized banana (*Musa sp. cv. Pacovan*) plantlets to saline stress. Agric. Ecosyst. Environ., 95: 343–348.

Yuang B, Li Z, Liu H, Gao M, Zhang Y (2007). Microbial biomass and activity in salt affected soils under arid conditions. Appl. Soil Ecol., 35: 319–328.

Zeng L, Shannon MC (2000). Salinity Effects on Seedling Growth and Yield Components of Rice. Crop Sci., 40: 996–1003.

Zhang DY (1995). Analysis of growth redundancy of crop root system in semi-arid area. Act Bot, Boreali-Occidentalia Sin., 15: 110–114.

Zhang Q, Wang G (2005). Studies on nutrient uptake of rice and characteristics of soil microorganisms in a long-term fertilization experiments for irrigated rice. J. Zhejiang Univ. Sci., 68: 147-154.

Zhu JK (2002). Salt and drought stress signal transduction in plants. Ann. Rev. Plant Biol., 53: 247–273.

16

Natural incidence and infectivity level of three nepoviruses in ornamental crops in Iran

T. Ghotbi and N. Shahraeen[*]

Department of Plant Virus Research, Iranian Research Institute for Plant Protection (IRIPP), Agricultural Research and Education Organization, Tehran, Iran.

Damage to ornamental crops by nepoviruses has occurred sporadically in Iran in the past. However, since 2006, outbreaks of nepoviruses have been recorded every year. The most affected ornamental crops were surveyed in two main cultivation areas in provinces of Markazi (Mahallat) and Tehran in 2006 - 2007. In all, 420 samples (with or without any conspicuous virus symptoms) were collected and analyzed by double- antibody sandwich enzyme-linked immunosorbent assay (DAS-ELISA) with polyclonal antibody to Tomato ring spot virus (ToRSV), Tobacco ring spot virus (TRSV) and Arabis mosaic virus (ArMV) . These viruses frequently were detected in samples of many different ornamentals (33 species) and often in mixed infections. Where as 8 samples found to be infected by one virus, 3 samples double infection and 6 samples were mix infected by three viruses. ArMV, ToRSV and TRSV were mechanically transmitted to _Vigna unguiculata, Nicotiana tabacum, Chenopodium amaranticolor, C. quinoa, Petunia hybrida, Datura stramonium_ and _D. metel_ indicator host plants and virus recovery was rechecked by ELISA. Of the total of 420 samples tested, 106 samples were reacted with the above virus antiserum. In Tehran, ToRSV was identified in 19 samples (4.52%), TRSV in 7 samples (1.66%), and ArMV in 5 samples (1.19%). In Markazi province, ToRSV was identified in 20 samples (4.76%), TRSV in 20 samples (4.76%), and ArMV in 23 samples (5.47%). In all, TRSV was shown to be prevalent nepovirus infecting ornamentals in these regions.

Key words: Tomato ring spot virus, tobacco ring spot virus, arabis mosaic virus, DAS-ELISA.

INTRODUCTION

The economic importance of ornamental horticulture is shown in a number of ways, in terms of the absolute size of the industry and world-wide sales. Many plant virus diseases cause significant losses in the production and quality of ornamental crops are difficult to control. New diseases occur as different crops are introduced or grown in new areas. Many crops are susceptible to multiple viruses, each of which may cause serious economic losses, and infected plant material may not be acceptable for export (Rakhshandehroo et al., 2006; Loebenstein et al., 1995). However, when occurring in complexes, they possibly exacerbate the symptoms induced by other in the other hand the antagonistic reactions also is not studied. At least 125 different viruses have been identified that in-

fect and cause disease in ornamental plants. Identification of the virus and vector is important for development of practical disease prevention or control methods, and to minimize pesticide usage (Hsu and Maroon, 2003). At the final 'production' stage of growing and distributing ornamental plants, losses due to viral infections can range from 10 to 100%, depending upon the virus-host combination (Brunt et al., 1996; Loebenstein et al., 1995). Viruses of serious consequence recently identified by the floral and nursery industry in key ornamental crops include, but are not limited to: tospoviruses, nepoviruses, potyviruses, fabaviruses, closteroviruses, potexviruses, carlaviruses, cucumoviruses, Caulimoviruses. There is therefore a need for research on these new and emerging virus and virus-like problems. Among these viruses, nepoviruses (Comoviridae) are one the main industrial group (Brunt et al., 1996; Loebenstein et al.,1995).The main nepoviruses reported to infect ornamentals are: Arabis mosaic virus, Tomato

*Corresponding author. E-mail: shahraeen@yahoo.com

ring spot virus and Tobacco ring spot virus (Loebenstein et al., 1995; Card et al., 2007). ArMV is reported from *Arabis hirsute* and *Crocus* spp. in England (Smith et al., 1944). This virus is transmitted by a nematode vector (*Xiphinema and Dorylamidae spp.*), mechanical inoculation; grafting and seed (Murant, 1983). In Iran ArMV and ToRSV from *Gladiolus* spp. were reported from ornamentals for the first time (Ghotbi et al., 2005; Ghotbi and Shahraeen, 2005; Rakhshandehroo et al., 2006; Kamran et al., 1981). ToRSV reported from America on *Nicotiana tabacum* (Price, 1936) and *Gladiolus* spp. (Leobenstein et al., 1995). ToRSV and ArMV transmitted by the vector *Xiphinema* and Dorylamidae spp. ToRSV and ArMV are transmitted by mechanical inoculation, grafting, seeds and pollen (Brunt et al., 1996; Card et al., 2007). TRSV was first reported in *Nicotiana tabacum* (Fromme et al., 1927) *Anemone* spp. In England (Hollings and Stone, 1963; Leobenstein, 1995) on *Gladiolus* spp. TRSV transmitted also, non-specifically by insects and mites, aphids and thrips. All nepoviruses produce necrotic or chlorotic local lesions following mosaic, ring spots or mottle symptoms on infected host plants but TRSV symptoms disappear soon after infection. Serological assay ,electron microscopy and RT-PCR techniques is reported to be a common tests to characterize nepoviruses including ToRSV and TRSV (Anonymous,2009) The aim of this study was to determine the prevalence and percent infection of important nepoviruses occurring on main cultivated ornamental crops in two region of Iran using routine biological and serological techniques.

MATERIALS AND METHODS

A total of 420 samples were collected in 2006 - 2007 from fields and glasshouses of mainly Tehran and Markazi Provinces. These were from 24 different family and 36 plant species. Samples, comprised young and fresh leave and stems of each of ornamentals and flowering weeds with various symptom types including leaves and stem deformation, stunting , necrosis of stem and leaves, veinal discoloration, general yellowing of leaves, systemic chlorotic and necrotic spots and general mosaic mottle on leaves or without any conspicuous symptom. Samples for each plant species were selected in random and on the basis of general plant appearance at the time of sampling. In this study, nepoviruses detection in suspected plant species was carried regardless of symptoms relation analysis. The number and location of the sample species are listed in Table 1.

Standard double antibody sandwich enzyme-linked immunosorbent assay (DAS-ELISA) (Clark and Adams, 1977) was performed with polyclonal antiserum for ToRSV, TRSV and ArMV nepovirus species. All serological reagents against nepopoviruses used were from the Bioreba Plant Virus Antiserum collection Company, (Switzerland) including the respective positive controls for each ELISA. Absorbance at 405 nm was measured with Labsystem multiskan ELISA microplate reader (Denmark). Healthy *N. tabacum* triturated in general extraction buffer was used as negative control. A reaction was considered positive only if the absorbance was more than three times the background mean of negative control. The serological reagents used in ELISA did not reveal any considerable cross reactivity with other virus speciesof the genus, hence permitted an accurate species identification.

Mechanical transmissions to test plants were done for selected ELISA positives. Samples were prepared by grinding 1 g of leaf triturated in ice cold 0.1 M potassium phosphate buffer, pH 7.0 containing 0.15% of 2-mercaptoethanol. Samples were inoculated to *Nicotiana tabacum, Datura stramonium, D. metel, Petunia hybridda, Chenopodium amaranticiolor, Ch. Quinoa* and *Vigna unguiculata* (cv. local Mashad) (Table 2). The test plants were kept in an insect proof greenhouse at a constant temperature of 23 - 25°C. Symptoms on indicator hosts were recorded every two days for 14 days following inoculation and twice a week for the following 30 days. The indicator plants then tested to confirm the presence of a particular virus also to test for any symptom less infection.

RESULTS

Three nepoviruses ArMV, ToRSV and TRSV were detected infecting ornamentals in two main ornamental growing regions of Iran (Table 1, Figure 1). Of the 420 samples assayed 94 were reacted positively with ELISA. ArMV was identified in 6.66% of samples, ToRSV in 9.28% of samples and TRSV in 6.42% of samples. Besides ELISA results, symptoms descriptions of indicator test plants for these viruses are given in Table 2. ToRSV induced short flower and stunting in gladiolus crops. ArMV inoculated to *C amaranticolor* produced numerous pin point chlorotic local lesions followed by severe systemic mosaic and leaf deformation. *P. hybida* test plant reacted with mosaic and systemic brown necrotic spots on trifoliate leaves when inoculated by TRSV. Mechanically TRSV inoculated *N. glutinosa* produced systemic veinal necrosis followed by tissue wilting and plant stunting. Field collected *Gerbera sp.* with symptoms of general yellowing and leaf discoloration were shown to be TRSV infected. ToRSV infected Dahlia sp. was with symptoms of leaf discoloration mosaic and stunting (Figure 2). *C. amaranticolor* and N. glutinosa were found to be good diagnostic hosts for the nepoviruses and symptoms expressed were informative for a preliminary differentiation of the 3 nepoviruses. *Datura stramonium* and *D. metel* did not produce any symptom upon inoculation by ArMV, ToRSV and TRSV. Virus disease may cause reduction in the market value; new introduction may carry seed borne of vegetative-borne viruses. If this is combined with the economic importance of ornamental and flower production, it will explain the important of epidemiology of the viruses in ornamentals (Loebenstein et al., 1995). Seed transmission of viruses in ornamental crops is of minor economic importance, as many of these crops are vegetative propagated. Data on percent losses or estimation of reduction in true yield of ornamental crops are limited. Recent researches indicated the presence and spread of tospoviruses on ornamentals and of other agricultural crops in Iran (Farzadfar et al., 2003; Ghotbi et al., 2005). Presence of ArMV from ornamental crops and its ornamental host range from Tehran (Varamin) and Markazi (Mahallat) regions is reported for the first time from Iran (Ghotbi and Shahraeen, 2005).

DISCUSSION

On the basis of the present study ToRSV, ArMV and TRSV

Table 1. Natural distribution of nepoviruses in Markazi and Tehran provinces on important cultivated ornamental crops.

Plant species	ArMV	TRSV	ToRSV	*Total tested		Mixed infections
				Markazi	Tehran	
Amaranthacea *Amaranthus cruentus*	**1(M)	1(M)	2(M)	8	4	1M(ArMV + TRSV + ToRSV)
Amaryllidaceae *Polianthes* spp.	-	-	-	11	7	-
Araceae *Sheflera* spp.	-	1(M)	-	6	5	-
Arum oriental	2(M)	2(M)	2(M)	7	3	-
Balsaminaceae *Impatiens* spp.	-	-	12(M)	15	4	-
Caryophyllaceae *Dianthus* spp.	-	-	-	9	3	-
Centianaceae *Lesianthus* spp.	3(M)	-	-	5	7	-
Compositae *Calandula* spp.	-	-	1(T)	5	8	-
Chrysanthemum morifulium	10(M)	7(M)	11(T)	21	11	2M(ArMV + TRSV + ToRSV)
Chrysanthemum prutescene	-	1(M)	-	15	10	-
Dahlia spp.	1(T)	-	1(T)	4	8	-
Gazania spp.	-	-	-	5	4	-
Gerbera spp.	-	1(M)	-	4	9	-
Helianthus annus	3(T)	5(T)	3(T)	-	12	1T(ArMV + TRSV + ToRSV)
Rudbekia spp.	-	-	-	8	-	-
Tanaketum spp.	-	-	-	9	-	-
Zinia elegans	5(M)	-	-	7	5	-
Crucifreae *Cheiranthus cheiri*	-	-	-	-	14	-
Ericaceae *Erica carnea*	-	-	-	9	-	-
Gesneraceae *Saintpaulia ioantha*	-	-	-	-	10	-
Graniaceae *Pelargonium X hortorum*	-	-	-	5	7	-
Pelargonium odoratissimum	-	-	-	9	6	-
Iridaceae *Gladiolus* spp.	1(M)	4(M)	4(M)	9	5	1M(ArMV + TRSV + ToRSV) 1M(TRSV + ToRSV)
Malvaceae *Althea* spp.	-	2(M)	-	3	8	-
Moraceae *Ficus benjamine*	-	-	-	5	8	-
Nyctaginaceae *Bougainvillea* spp.	-	-	-	-	9	-
Primulacea *Primula* spp.	1(T)	1(T)	1(T)	-	8	1T(ArMV + TRSV + ToRSV)
Rosaceae *Rosa damascena*	-	-	-	6	10	-
Scrophulariaceae *Anthirrhinum* spp.	-	-	-	-	9	-
Solanaceae *Petunia hybridagrandiflora*	-	1(T)	2(T)	5	6	1T(TRSV + ToRSV)
Tropaeolaceae *Tropaeo majus*	-	-	-	8	-	-
Umbeliferae *Tagetis patula*	1(M)	1(M)	-	8	5	1M(ArMV + ToRSV)
Violaceae *Viola* spp.	-	-	-	-	9	-
TOTAL	28 (23M,5T)	27 (20M,7T)	39 (20M,19T)	420 (206 M,214 T)		

*Total samples tested was not calculated on the basis of mixed infections
** Number of infected plant
 M: Markazi provine, T: Tehran province

nepoviruses were the most prevalent viruses on ornamental crops reporting from Iran. ArMV is reported to be transmitted at low rates through seeds of *Viola tricolor* and *Petunia hybrida*. Seed transmission in this case could be of economic importance. Seed transmission of ToRSV and TRSV has been reported to occur in *Pelargonium*. TRSV and ToRSV in pelargonium species are reported as two common nepoviruses infecting green house cultivated pelargonium (Anonymous, 2009). TRSV is also transmitted in seed of *P. hybrida* and *Zinnia elegans*, at 20% and 5% respectively (Loebenstein et al., 1995; Brunt et al., 1996). ToRSV and TRSV occurred mostly on a wide range of perennial crops. TRSV and ToRSV were probably disseminated from North America to other countries in infected planting material; they were reported from Japan in imported corms (Fakomoto et al.,

Table 2. Host range studies of different nepovirus species

Host	ToRSV		TRSV		ArMV	
	L	S	L	S	L	S
Nicotiana rustica	CL	-	NL	LD	-	LD
Nicotiana glutinosa	MNL	-	-	LD, SNL	-	-
Vigna anguiculata	CL	-	NL	LD	NL	-
Datura metel	-	-	-	-	-	-
D. stramonium	-	-	-	-	-	-
Petunia hybrida	-	-	-	SNL. Mo	-	CL
Chenopodium amaranticolor	CL	SNL	CLNL	-	NL	Mo, LDMNL
Cucumis sativus	CL	LD	CL	RS LD	NL	CL

CL: chlorotic lesion, LD: leaf deformation, MNL: mild necrotic lesion, Mo: mosaic, NL: necrotic lesion, RS: ring spot, SNL: sever necrotic lesion

Nepoviruses in Tehran province

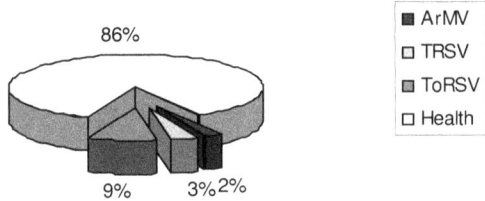

Nepoviruses in Markazi province

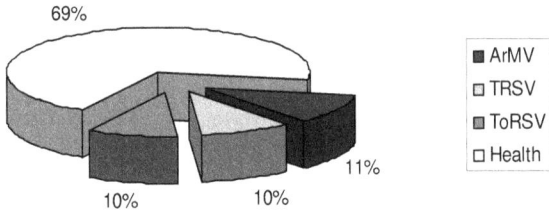

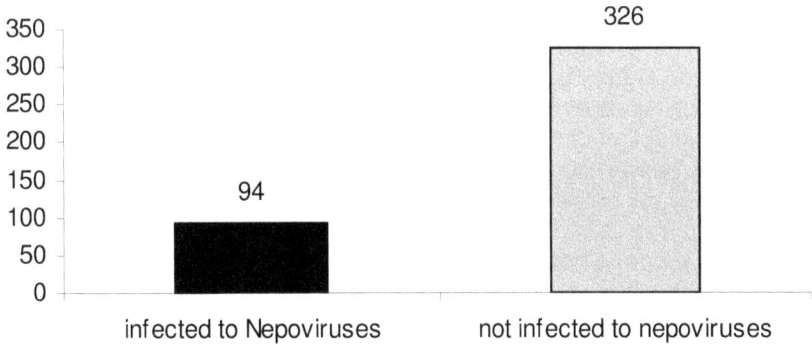

Figure 1. Percent infection of nepoviruses on ornamentals in Tehran and Markazi provinces.

Figure 2. a: Local chlorotic pin point spot on *C. amaranticolor* infected by ArMV, b: Virus inoculated *P. hybrida* showing systemic mosaic and necrotic lesion, c: Naturally TRSV infected *Gerbera* sp. with yellowing, discoloration and growth reduction symptoms, d: ToRSV infected *Dahlia* sp. Showing mosaic, leaf discoloration and stunting, and e: ToRSV inoculated N.glutinosa indicator host plant showing mosaic, systemic veinal necrosis and stunting

1982). ToRSV was also reported from Australia (Randles and Francki, 1965) and Iran (Kamran et al., 1982). Nepoviruses is an example of seed transmitted embryoinfecting viruses, where creating loci of infection can, however cause severe damages in the field if an efficient vector is present. However as these viruses are present and reported on other cultivated plants in Iran including cucurbits (Cucurbitaceae),soybean (Fabaceae),potato (Solan--aceae) and Vitis vinifera (Vitaceae) (Farzadfar et al., 2003).The use of virus free seeds and vegetative materials for propagation and integrated pest and disease control management and breeding for virus disease resistant is recommended.

ACKNOWLEDGEMENT

Thanks to Mr. Engineer Bayat (Agricultural Research Centre Markazi, province) for collecting samples. Mr. M. Afzali for his assistance with greenhouse plants.

REFERENCE

Anonymous (2009). Enhancing greenhouse soil less culture production with improved disease and pest managements .Dept. of Agril. Research Project-3607-21000-015-13. www.ars.usda.gov/research/ project .htm/

Brunt AA, Crabtree K, Dallwitz MJ, Gibbs AJ, Watson L, Zurcher EJ (1996). Plant viruses online: Descriptions and lists from the VIDE Database. Version: 16th, Jan 1997; URL.

Card SD, Pearson MN, Clover GRG (2007). Plant pathogens transmitted by pollen. Aust, Plant Path. 36:455-461.

Clark MF, Adams AN (1977). Chracterization of microplate method of enzyme-linked immunosorbent assay for the detection of plant viruses. J. Gen. Vir. 34: 475-483.

Fakumoto F, Ito Y, Tochihara H (1982). Viruses isolated from Gladiolus in Japan. Ann. Phytopathol. Soc. Japan 48: 68-71.

Farzadfar Sh, Golnaraghi AR, Pourrahim R (2003). Plant viruses of Iran. Saman Company Publication, Tehran, pp. 203

Fromme FD, Wingard SA, Priod CN (1927). Ring spots of tobacco; an infectious disease of unknown cause. Phytopathology.17: 321.

Ghotbi T, Shahraeen N (2005). First report on incidence of *Arabis mosaic virus* (ArMV) on ornamental plants in Iran. Iran. J. Plant Path. 41(2): 304.

Ghotbi T, Shahraeen N, Winter S (2005). Occurrence of tospoviruses in ornamental and weed species in Markazi and Tehran provinces in Iran. Plant Dis. 89(4): 425-429.

Hollings M, Stone OM (1963). Anemone necrosis virus (ANV). Rep. Glasshouse crops Res. Inst. 1962 p.89.

Hsu HT and Maroon-Lango C (2003). Management of viral diseases in floral and nursery crops in: Advances in Plant Disease Management (eds). Res. Signpost. pp. 413-429.

Kamran R, Izadpanah K (1981). Isolation and Identification of BYMV and ToRSV from *Gladiolus* in Shiraz. Iran. J. Plant Path. 17 (1-4): 1-7

Loebenstein G, Lawson RH, Brunt AA (1995). Virus and virus-like disease of bulb and flower crops. Wiley and Sons Publ. pp. 543.

Murant AF (1983). Seed and pollen transmission of nematode-borne viruses. Seed Sci. Technol. 11: 973-979

Price WC (1936). *Tomato ring spot virus*. Phytopathology. 26: 665.

Rakhshandehroo F, Zamani Zadeh H, Modarresi A (2006). Occurrence of Prunus necrotic ring spot virus and Arabis mosaic virus in Rose in Iran. Plant Dis. 86:1982

Randles JW Franki RIB (1965). Some properties of tobacco ring spot virus isolated from South Australia. Aust. Bio. Sci. 18: 979-986.

Smith KM, Markham R (1994). The new viruses affecting tobacco and other plants. Phytopathology 34: 32.

17

Management of inherent soil fertility of newly opened wetland rice field for sustainable rice farming in Indonesia

Sukristiyonubowo[1]*, Ibrahim Adamy Sipahutar[1], Tagus Vadari[1] and Agus Sofyan[2]

[1]Indonesian Soil Research Institute, Jalan ir. Juanda 98 Bogor 16123, Indonesia.
[2]Direktorat Jendral Prasarana dan Sarana Pertanian, Jalan M.T. Harjono, Jakarta Selatan.

Study on management of inherent soil fertility of newly opened wetland rice for sustainable rice farming in Indonesia was conducted in Bulungan District, from 2009 to 2010. The aims were to know the soil fertility status and properly manage its fertility status to improve rice yield and sustain rice farming. Six treatments were imposed including T0: farmers practices, T1: farmer practices + compost + dolomite, T2: NPK recommended rate, N and K were split two times, T3: NPK recommended rate, N and K were split three times, T4: NPK recommended rate, N and K were split three times + compost + dolomite and T5: NPK recommended rate, N and K were split two times + compost + dolomite. The residual effect of dolomite and compost applied in 2009 was continually assessed in 2010. The results indicated that inherent soil fertility was categorised poor with high level of iron and manganese. Application of 250 kg urea, 100 kg super phosphate-36 and 100 kg potassium chloride ha^{-1}, in which N and K fertilisers were split three times plus 2 tons dolomite and 2 tons compost ha^{-1} also the residual effect of dolomite and compost improved soil fertility, rice growth and biomass production.

Key words: Soil fertility, newly wetland rice, rice farming, dolomite, compost.

INTRODUCTION

Soil quality assessment including chemical, physical and biological properties has become a model to determine soil function. According to Doran and Parkin (1994) soil quality is defined as the capacity of the soil to function within ecosystem boundaries to sustain biological productivity, maintain environmental productivity and promote plant and animal health. So far, Sharma et al. (2008) reported that in the past soil quality is understood as inherent soil capacity to supply essential plant nutrients.

Combining use of mineral and organic fertilisers (recycling rice straw, crop residues, compost and manure) are recommended to improve soil function (Fenning et al., 2005; Hasegawa et al., 2005; Khai et al., 2007; Yang et al., 2007; Sukristiyonubowo and Tuherkih,

2009).

In Indonesia over two third of total population depend on agricultural sector within which wetland rice play an important role in sustaining food security and building rural livelihood like providing job and income. However, agricultural practices particularly in rice farming, application of fertilizers rate and crop residue management differ among farmers within sub district, resulting variability in production and soil fertility properties.

Furthermore, the shrinking of agricultural land and harvest areas in Indonesia as well as in the countries producing rice due to; (a) increasing agricultural land conversion to non-agricultural purposes, (b) increasing water competition among agricultural sector and industrial as well as domestic purposes and (c) water pollution reducing total harvest areas leading to rice production (Anonymous, 2002; Bhagat et al., 1996; Bouman and Tuong, 2001; Sukristiyonubowo, 2007).

---

*Corresponding author. E-mail: sukristiyonubowo@yahoo.com.

Table 1. The detail treatment of the effect of NPK fertilization, dolomite and compost made of straw on soil chemical properties and rice yield of newly opened rice fields.

Code	Treatment	Urea (kg ha^{-1})	SP-36 (kg ha^{-1})	KCl (kg ha^{-1})	Dolomite (kg ha^{-1})	Compost (kg ha^{-1})
T0	Farmer Practices (as control)	100	100	-	-	-
T1	Farmer Practices + Compost + Dolomite	100	100	-	2000	2000
T2	NPK with recommendation rate	250	100	100	-	-
T3	NPK with recommendation rate (N and K were split 3 x)	250	100	100	-	2000
T4	NPK with recommendation rate + Compost + Dolomite (N and K were split 3x)	250	100	100	2000	2000
T5	NPK with recommendation rate + Compost + Dolomite	250	100	100	2000	2000

Hence, the Indonesian agricultural challenge ahead especially for rice is producing more rice with limited soil and water.

Highly weathered Indonesian soils, especially ultisols and oxisols are mainly granted for extending newly opened wetland rice field to meet rice growing demand in Indonesia, besides potential acid sulphate soils. Furthermore, these soils are acidic with low natural level of major plant nutrients, but having toxic levels of Al, Mn and Fe (Sudjadi, 1984). Theoretically, soils fertlity status can be effectively addressed with addition of mineral fertilisers.

However, for the smallholder farmers, for instance farmers living in transmigration areas the costs to purchase the fertilisers are problem. The chemical fertiliser in sufficient quantity is beyond the financial reach of smallholder farmers. Practically, to sustain crop production, proper management practices using more organic matter plus liming and application of appropriate inorganic fertiliser is often proposed (Fageria and Baligar. 2001; Yan et al., 2007; Sukristiyonubowo et al., 1993; Sukristiyonubowo and Tuherkih, 2009). In addition, Whitbread et al. (2003), Sukristiyonubowo 2007, 2010 also reported that sustainable farming system requires stable soil fertility through balance fertilization and crop residue management.

The aims of the study were to know the inherent soil fertility status and its management to improve productivity and sustain rice under newly opened wetland rice field.

MATERIALS AND METHODS

To determine soil fertility status of newly opened wetland rice, soils samples were taken from the newly opened wetland rice of Bulungan District established in 2007. Composite samples of topsoil, 0 to 20 cm layers, were taken in March 2009, before land preparation. Eight composite samples were analysed. Every composite sample was randomly collected from ten sampling points mixed thoroughly to 1 kg as a composite sample. These samples were submitted to the Analytical Laboratory of the Indonesian Soil Research Institute at Laladon Bogor to determine chemical properties and texture. Soil chemical analyses included the measurement of pH (H_2O and KCl), organic matter (organic carbon and total nitrogen), phosphorus, potassium, base saturation and cation exchange capacity (CEC) as well as iron (Fe) and manganese (Mn) contents.

Organic matter was determined using the Walkley and Black method, pH (H_2O and KCl) was measured in a 1:5 soil-water suspension using a glass electrode method, total P and available P were measured colorimetrically using HCl 25% and Olsen methods, respectively. The total potassium (K) was extracted using Chloride Acid 25% (HCl 25%) and subsequently determined by flame-spectrometry. The CEC was determined using an Ammonium Acetate 1 M, (pH 7.0) extraction and expressed in cmol$^+$ kg^{-1} soil. Base saturation was computed based on the sum of Ca^{++}, Mg^{++}, K$^+$, and Na$^+$ relative to CEC. Available iron (Fe) was measured using DTPA method (Soil Research Institute, 2009).

Field experiment was carried out in Panca Agung Village, Bulungan District in 2009 and 2010. Six treatments were tested including T0: farmers practices (as control), T1: farmer practices + straw compost + dolomite, T2: NPK with recommendation rate, in which N and K were split two times, T3: NPK with recommendation rate, in which N and K was split three times, T4: NPK with recommendation rate in which N and K were split three times + straw compost + dolomite and T5: NPK with recommendation rate in which N and K were split two times + straw compost + dolomite. They were arranged into randomized complete block design (RCBD) and replicated three times. The plot sizes were 5m x 5m with the distance among plot was 50 cm and between replication was 100 cm. NPK fertiliser used originated from single fertiliser namely urea, super phosphate-36 (SP-36) and potassium chloride (KCl). Based on the direct measurement with Soil Test Kits, the recommendation rate was determined about 250 kg urea, 100 kg SP-36 and 100 kg KCl ha^{-1}, while the farmer practices rate was 100 kg urea and 100 kg SP-36 ha^{-1}. For the treatment T2 and T5, urea and KCl were applied two times, 50% at planting time and 50% at 21 days after transplanting (DAT). For the treatment T3 and T4, urea and KCl were split three times namely; 50% at planting time, 25% at 21 DAT and the last 25 % was given at 35 DAT. Dolomite as much as two tons ha^{-1} and rice straw compost of about two tons ha^{-1} were broadcasted a week before planting. The detail treatment is presented in Table 1.

In 2010, the residual effect of dolomite and compost was continually assessed, but the similar rates of mineral fertiliser in all treatments were still added. Like the previous experiment, for the treatment R2 and R5, urea and KCl were applied two times, namely 50% at planting time and 50% at 21 DAT, while for treatment R3 and R4 urea and KCl were split three times, 50% at planting time, 25% at 21 DAT and the last 25 % was added at 35 DAT.

Ciliwung rice variety was cultivated as plant indicator. Transplanting was conducted in the end of June 2009 and harvest in the beginning of October 2009. While for the residual experiment transplanting was conducted in the end of March 2010 and harvest in the beginning of July 2010. Twenty-one-day old seedlings were transplanted at about 25 x 25 cm plant and row spacing with about three seedlings per hill. Rice biomass productions including grains, straw and residues were observed. On a hectare basis, biomass productions were extrapolated from sampling areas of 1 x 1 m. These sampling units were randomly selected at every plot. Rice plants were cut about 10 to 15 cm above the ground surface. The samples were manually separated into rice grains, rice straw, and rice residues. Rice residues included the roots and the part of the stem (stubble) left after cutting. Fresh weights of rice grain, rice straw and rice residue were immediately weighed at each sampling unit. Meanwhile, for rice growth parameter namely plant height and tiller number were monitored at 30 DAT, 60 DAT and at harvest. The common crop rotation is rice to rice fallow.

All data were statistically examined by analysis of variance (ANOVA) and computed using software SPSS program. Means were compared to Duncan Multiple Range Test with a 5% degree of confidence.

RESULTS AND DISCUSSION

Soil fertility of newly opened wetland rice fields

Originally, the land of about 30 ha was used to grow upland rice and cash crop, but since the last decade it was left out as the land productivity was considered low due to lack of water and lack of evidence of economic profitability. It was converted to wetland rice area in 2007, when the water from upper site (forest) was flew through pipes and collected to the agriculture water reservoir.

Chemical and physical soil properties of newly wetland rice fields of Panca Agung opened in 2007 are presented in Table 2. The texture of the soil varied from silty, clay, loam to clay, and classified into medium to fine textures. The pH of soils was acidic, varying between 4.62 and 4.70. The cation exchange capacities (CEC) values ranged between 5.81and 9.51 cmol$^+$ kg^{-1} suggesting uniformity in clay mineralogy and low in organic matter contents. The CEC may be categorised as low. The levels of soil organic carbon (OC) and total nitrogen were very low, ranging from 0.71 to 1.29% and from 0.03 to 0.05 %, respectively. This may be due to the fact that in the past, all rice straw or crop residue was removed from the field to be used as cattle feed and/or was burnt. Sommerfeldt et al. (1988) and Clark et al. (1998) also observed higher soil organic matter (OM) levels in soils managed with animal manure and cover crops than in soils without such inputs.

Total P ranged from 31 to 58 mg P$_2$O$_5$ kg^{-1}. These values, observed in all newly opened rice field in Panca Agung, classified as very low. Furthermore, available P is also considered low, ranging between 1.09 and 2.69 mg

P$_2$O$_5$ kg^{-1}. It can be suggested that application of recommended fertilisers rate as much as 100 kg SP-36 ha^{-1} season^{-1} increase the availability of P. Total K extracted with HCL 25% is considered low, varying from 55 to 138 mg K$_2$O kg^{-1}. This also suggests that application of recommended fertilisers rate as much as 100 kg KCl ha^{-1} season $^{-1}$ will increase the total K in the soil. Clark et al. (1998), Rasmussen and Parton (1994) and Wander et al. (1994) also reported similar findings.

Base saturation is relatively low, varying between 16 and 39%. This is mainly due to the low concentrations of exchangeable Ca (1.04 to 1.83 cmol$^+$ kg^{-1}) and K (0.05 to 0.11 cmol$^+$ kg^{-1}). So far, exchangeable Mg concentrations are relatively low (0.21 to 0.27 cmol$^+$ kg^{-1}). Looking to the ratio of exchangeable calcium, magnesium and potassium percentage, the data also indicated an imbalanced ratio. In normal conditions, the ratio ranges from 60 to 65% of calcium, 10 to 15% of magnesium and 5 to 7% of potassium (Sukristiyonubowo, 2007). Looking at the iron (Fe) and manganese (Mn) concentrations, both elements are in toxic condition meaning that to minimize that problem application of organic matter as well as lime is must to enhance rice growth and yield. Therefore, it may be concluded that in general the chemical soil fertility is considered low due to low pH, organic matter content, available P and exchangeable potassium concentrations and very low to low P and K total. In addition, the soil properties variability was small as indicated by range of parameters values. Recommendation of agronomic practices to enhance rice yield of newly opened wetland rice include application of soil ameliorant like addition of manure, compost, lime, besides addition of mineral fertilisers.

Effect of lime, compost and mineral fertiliser on soil chemical properties

The influence of dolomite, compost made from rice straw, and NPK fertiliser on soil chemical properties is presented in Table 3. The results showed that in general addition of dolomite, compost and NPK fertiliser improved the total nitrogen (total N), available phosphorous (available P) measured with Bray I method, potential phosphorous (potential P) and potassium (potential K) extracted with Chloride Acid 25% (HCl 25%).

Furthermore, they also reduced Fe and Mn contents when they compared with control (farmer practices) and the original soil (taken in 2009 before land ploughing). So far, the increase in total N, potential P and K extracted with HCl 25%, available P measured with Bray I method in the treatments T3 and T4, where urea and KCl were split three times, was higher than other treatments. In addition, reduction in Fe and Mn concentration was also more than others. This shows that availability of plant nutrient especially nitrogen, phosphorous and potassium for rice growth and development when urea and KCl were split three times, was higher than other treatments.

Table 2. Chemical and physical soil properties of newly opened wetland rice in Tanjung Palas Utara Sub district, Bulungan District established in 2007 (Soils were sampled in March 2009).

Soil parameters	Unit	Value	Criteria
pH		4.62 to 4.70	Very acid
Organic matter			
C-organic	%	0.71 to 1.29	Very low
N Total	%	0.03 to 0.05	Very low
C/N ratio		20 to 26	
P Total (HCl 25%)	ppm	31 to 58	Very low
K Total (HCl 25%)	ppm	55 to 138	Low
P Bray I	ppm	1.09 to 2.69	Very low
CEC	cmol (+) kg^{-1}	5.81 to 9.53	Low
Base Saturation	%	16 to 39	Low
K	cmol (+) kg^{-1}	0.05 to 0.11	Very low
Ca	cmol (+) kg^{-1}	1.04 to 1.83	Very low
Mg	cmol (+) kg^{-1}	0.21 to 0.27	Low
Na	cmol (+) kg^{-1}	0.05 to 0.19	Low
Fe	ppm	170 to 210	High
Mn	ppm	50.40	High
Texture 1:			
Sand	%	6.1	
Silt	%	64.8	Silty Clay Loam
Clay	%	29.1	
Texture 2:			
Sand	%	1.3	
Silt	%	18.3	Clay
Clay	%	80.4	

Table 3. Effect of dolomite, compost and NPK fertiliser on soil chemical properties of newly opened wetland rice in Panca Agung Village, Bulungan District, East Kalimantan Province (soils were sampled at harvest time in October 2009).

Treatments	Chemical soil properties					
	N Total (%)	P HCl 25% (ppm P$_2$0$_5$)	K HCl 25% (ppm K$_2$0)	P Bray I (ppm P$_2$0$_5$)	Fe (ppm)	Mn (ppm)
Before experiment	0.05	58	31	1.09	170	50
T0	0.10	62	37	9.14	185	16.09
T1	0.12	171	29	7.40	183	17.46
T2	0.10	172	28	7.06	190	19.58
T3	0.14	149	39	9.63	157	16.80
T4	0.16	195	38	10.62	167	13.52
T5	0.14	154	31	9.25	171	18.31

T0: Farmer Practices (as control). T1: farmer practices + compost + dolomite. T2: NPK with recommendation rate. T3: NPK with recommendation rate (N and K were split 3 x). T4: NPK with recommendation rate + compost + dolomite   (N and K were split 3x). T5: NPK with recommendation rate + compost + dolomite.

Besides, application of dolomite and compost to poor soil fertility was important. Combination of chemical fertilisers with compost and dolomite is important not only to improve nutrient supply, but also to minimise the

Table 4. Effect of dolomite, compost, and NPK fertiliser addition on rice plant growth of ciliwung variety at 30 DAT, 60 DAT and harvest cultivated at newly opened rice of Panca Agung site, Bulungan District.

Treatments	Plant height (cm)			Tiller number	
	30 DAT	60 DAT	Harvest	30 DAT	60 DAT
T0	39.50 ± 2.0 [c]	65.20 ± 4.2 [c]	87.67 ± 2.7 [b]	15.75 ± 0.3 [bc]	18.90 ± 2.2 [b]
T1	42.17 ± 3.2 [bc]	69.88 ± 4.5 [bc]	83.63 ± 2.5 [b]	13.72 ± 1.0 [c]	21.30 ± 2.3 [ab]
T2	45.67 ± 2.7 [ab]	71.10 ± 2.0 [abc]	90.20 ± 3.8 [a]	16.93 ± 0.1 [ab]	20.98 ± 1.0 [ab]
T3	47.50 ± 2.5 [ab]	74.73 ± 3.0 [ab]	90.47 ± 1.5 [a]	15.47 ± 0,1 [bc]	24.28 ± 3.0 [a]
T4	49.67 ± 4.2 [a]	76.38 ± 2.4 [a]	94.17 ± 1.8 [a]	18.57 ± 1.0 [a]	24.37 ± 2.0 [a]
T5	46.67 ± 2.7 [ab]	76.27 ± 3.1 [a]	89.72 ± 2.0 [a]	17.50 ± 2.5 [ab]	21.00 ± 2.0 [ab]
CV (%)	9.48	7.88	5.83	11.56	12.37

Note: The mean values in the same column followed by the same letter are not statistically different. T0: farmer practices (as control). T1: farmer practices + compost + dolomite, T2: NPK with recommendation rate. T3: NPK with recommendation rate (N and K were split 3 x). T4: NPK with recommendation rate + compost + dolomite (N and K were split 3x). T5: NPK with recommendation rate + compost + dolomite.

environmental damage.

Effect of dolomite, compost and NPK fertiliser on rice growth

The effect of dolomite, compost and NPK fertiliser addition on rice growth is presented in Table 4. At 30 DAT all treatments (T3 to T5) significantly increased the rice plant height. The T4 indicated the best treatment and significantly reached the highest rice plant height of about 49.67 ± 4.2 cm. The similar finding was observed at 60 day after transplanting. At 60 DAT, all treatments also improved rice plant height compared to control (Farmer Practices). So far, the treatment T4 also indicated the highest rice plant, meaning at this treatment nutrients supply, especially nitrogen and potassium for rice growth and development was more available. The highest rice plant was about 76.38 ± 2.4 cm (Table 4).

Furthermore, the similar results were also found at harvest time. Compared to farmer practices, all treatments, especially treatments T2 to T5 significantly increased the rice plant height. The highest rice plant was about 94.17 ± 1.8 cm, reached by T4. These data demonstrated that improvement of soil function were not only due to application of compost, dolomite and NPK fertiliser, but also the split application of N and K into three times providing more available nitrogen and potassium when they required by rice plant. Therefore, it can be said that application of NPK at recommendation rate in which N and K were split three times plus two tons ha^{-1} dolomite and two tons ha^{-1} not only improved soil function, but also rice plant height.

The similar phenomenon was also observed for the rice tiller number. At 30 DAT all treatments significantly enhanced the rice tiller number (Table 4). Interestingly, the best treatment was also showed by T4, reaching of rice tiller number as much as 18.57 ± 1.0. This data revealed that the time of fertilizer application is urgent, especially when the soil has low pH and high Fe, Al and Mn concentrations to supply more nutrients for rice growth and development.

Furthermore, at 60 DAT the similar finding was observed. The treatment T4 also indicated the highest rice tiller number of about 24.37 ± 2.0. This was significantly different with farmer practices. These data demonstrated that application of soil ameliorant and inorganic fertilisers are prerequisite for newly opened wetland rice fields to improve soil productivity. Therefore, it can be concluded that application of lime, compost and mineral fertiliser are must to improve soil function leading to enhance rice growth and yield.

Effect of dolomite, compost and NPK fertiliser on rice production

The effect of dolomite, compost and NPK fertiliser addition on rice biomass production is given in Table 5. As demonstrated in rice growth and development, compared to farmer practices, all treatments significantly enlarged the dry rice straw production. In addition to this, the treatment T4 showed the highest dry rice straw production, about 5.20 ± 0.4 t ha^{-1}. The similar results were observed for rice grain yield. All treatments significantly increased the rice grain yield, ranged between 3.09 and 4.29 t ha^{-1}. According to the farmers until one year after conversion to wetland rice areas, the rice yield was considered good, reaching 2.0 to 2.5 t ha^{-1}, afterward reduced to 1.0 to 1.5 t ha^{-1}.

It is interesting to note, the highest rice grain yield was about 4.29 ± 0.4 t ha^{-1} reached by T4, elevating yield about 1.78 t ha^{-1} or 71% compared to farmer practices (Table 5). The constant improvement of T4 to the soil function, rice growth and biomass production proved that is not only due to addition of compost, dolomite and mineral fertilisers, but also the way and time to apply the organic and especially inorganic fertilisers. The similar finding was reported by previous scientists (Clark et al., 1998; Fageria and Baligar, 2001; Yan et al., 2007;

Table 5. Effect of dolomite, compost, and NPK fertiliser addition on dry rice biomass production of Ciliwung variety cultivated at newly opened wetland rice of Panca Agung site, Bulungan District.

Treatments	Biomass production (t ha^{-1})			Increasing rice grain yield	
	Rice residue	Rice straw	Rice grain	t ha^{-1}	%
T0	2.76 ± 0.4 c	3.83 ± 0.4 b	2.51 ± 0.6 d	-	
T1	3.05 ± 0.4 bc	3.94 ± 0.7 b	2.97 ± 0.03 cd	0.46	18
T2	3.61 ± 0.2 ab	5.02 ± 0.5 a	3.09 ± 0.3 bc	0.58	23
T3	3.90 ± 0.2 a	4.64 ± 0.6 ab	3.68 ± 0.4 abc	1.17	47
T4	4.01 ± 0.1 a	5.20 ± 0.4 a	4.29 ± 0.4 a	1.78	71
T5	4.18 ± 0.3 a	5.24 ± 0.2 a	3.80 ± 0.3 ab	1.29	51
CV (%)	16.55	15.47	20.41		

The water content of dry rice grain was 12 to 14%. The mean values in the same column followed by the same letter are not statistically different.

Table 6. Residual effect of dolomite and compost on rice plant height of ciliwung variety at 30 DAT, 60 DAT and harvest cultivated at newly opened wetland rice of Panca Agung site, Bulungan District.

Treatments	Plant height (cm)			Tiller number		
	30 DAT	60 DAT	Harvest	30 DAT	60 DAT	Harvest
R0	39.50 ± 3.0 ^{c}	46.63 ± 7.1^c	72.30 ± 5.0 ^{c}	11.33 ± 1.3 ^{c}	14.35 ± 3.0 ^{b}	11.70 ± 0.9 ^{b}
R1	43.26 ± 2.9 ^{bc}	50.03 ± 4.6 ^{b}	74.70 ± 2.3 ^{bc}	11.15 ± 0.6 ^{c}	15.67 ± 0.4 ^{ab}	13.47 ± 0.5 ^{ab}
R2	47.33 ± 2.6 ^{ab}	68.77 ± 5.5 ^{a}	85.07 ± 4.1 ^{a}	13.60 ± 0.5 ^{b}	15.33 ± 1.0 ^{ab}	12.67 ± 0.7 ^{ab}
R3	49.83 ± 2.8 ^{ab}	66.37 ± 2.4 ^{a}	84.63 ± 3.4 ^{a}	13.80 ± 1.5 ^{b}	14.73 ± 1.9 ^{b}	13.57 ± 2.2 ^{ab}
R4	51.67 ± 6.9 ^{a}	70.33 ± 6.0 ^{a}	87.00 ± 7.7 ^{a}	17.83 ± 1.0 ^{a}	19.10 ± 3.8 ^{a}	14.70 ± 0.4 ^{a}
R5	47.12 ± 2.9 ^{ab}	64.80 ± 2.9 ^{ab}	83.70 ± 6.2 ^{ab}	14.50 ± 0.5 ^{b}	16.20 ± 0.6 ^{b}	14.27 ± 0.5 ^{a}
CV (%)	11.11	15.33	9.82	15.52	15.27	10.10

Note: The mean values in the same column followed by the same letter are not statistically different. R0: farmer practices (as control). R1: farmer practices + residue compost + residue dolomite. R2: NPK with recommendation rate. R3: NPK with recommendation rate (N and K were split 3 x). R4: NPK with recommendation rate + residue compost + residue dolomite (N and K were split 3x). R5: NPK with recommendation rate + residue compost + residue dolomite.

Rasmussen and Parton, 1994; Sukristiyonubowo et al., 1993; Sukristiyonubowo and Tuherkih, 2009; Wander et al., 1994).

Residual effect of dolomite and compost on rice growth

The residual effect of dolomite and compost on rice growth (rice plant height and rice tiller number) is given in Table 6. The results indicated that at 30 DAT, 60 DAT and at harvest time, the residue of dolomite and compost significantly improved the rice plant height. Compared to control, it increased about 9 to 31%, 7 to 51% and 3 to 21%, respectively (Table 6).

Interestingly, the highest rice plant in every observation (30 DAT, 60 DAT and at harvest) was also achieved by R4. This meant addition of 2 tons ha^{-1} dolomite and 2 tons ha^{-1} compost made from rice straw gave an effect up to two years maybe more in improving of soil fertility as well as rice growth. It can be concluded that dolomite and compost as much as two ton ha^{-1}, therefore can be applied at least every two years.

Furthermore, this was also found for rice tiller number. At 30 DAT, 60 DAT and at harvest time, the tiller numbers at treatments R1, R4 and R5 were significantly enhanced. Compared to farmer practices (R0), tiller numbers increased about 28 to 57%, 9 to 34 and 15 to 26%, respectively. The highest rice tiller number in every observation (30 DAT, 60 DAT and at harvest) was also achieved by R4.

Residual effect of dolomite and compost on rice biomass production

Similar findings were also observed for rice biomass production. The residue of dolomite and compost significantly enhanced the rice biomass production including fresh rice residues, fresh rice straw and rice grain yields (Table 7). Compared to the farmer practices, no addition of dolomite and compost (R0), the residual effect of dolomite and compost at R1, R4 and R5 significantly improved the fresh rice residue, fresh rice straw and rice grain yields.

The best rice residue, rice straw and rice grain yield of

Table 7. Residual effect of dolomite and compost on rice biomass production of ciliwung variety at 30 DAT, 60 DAT and harvest cultivated at newly opened wetland rice of Panca Agung site, Bulungan District.

Treatments	Biomass production (t ha^{-1})		
	Fresh rice residue	Fresh rice straw	Dry rice grain
R0	3.54 ± 0.5 b	4.72 ± 0.7 c	1.90 ± 0.2 c
R1	4.46 ± 0.4 ab	6.38 ± 0.4 b	2.03 ± 0.3 c
R2	4.86 ± 0.6 ab	7.34 ± 0.8 ab	3.32 ± 0.2 ab
R3	5.04 ± 0.6 ab	6.96 ± 0.2 b	3.25 ± 0.1b
R4	5.84 ± 0.5 a	7.82 ± 0.6 a	3.74 ± 0.3 a
R5	5..79 ± 1.1 a	7.99 ± 1.1 a	3.53 ± 0.3 ab
CV (%)	19.97	18.45	25.96

Note: The water content of dry rice grain was about 12 to 14%. The mean values in the same column followed by the same letter are not statistically   different. R0: farmer practices (as control). R1: farmer practices + residue compost + residue dolomite. R2: NPK with recommendation rate. R3: NPK with recommendation rate (N and K were split 3 x). R4: NPK with recommendation rate + residue compost + residue dolomite (N and K were split 3x). R5: NPK with recommendation rate + residue compost + residue dolomite.

about 5.84 ± 0.5, 7.82 ± 0.6, 3.74 ± 0.3 t ha^{-1} season^{-1} were achieved by R4. These data proved that application of NPK recommended rate (250 kg urea, 100 kg SP-36 and 100 kg KCl ha^{-1}, in which N and K were split 3x) at the residue of 2 tons ha^{-1} dolomite and 2 tons compost ha^{-1} are essential to continually improve poor inherent soil fertility, rice growth and rice biomass production of newly opened wetland rice filed.

Sustainable rice farming in indonesia

Based on the facts finding including on improvement of soil fertility, rice growth and rice biomass production, application of NPK at recommendation rate plus two tons dolomite and two tons compost is important to be socialized as newly opened wetland rice are done in many parts of Indonesia, especially in outer of Java and Bali islands. Application of this technology is not only to improve rice yield, but also to sustain rice farming and increase farmer's income.

Thus, to meet the rice growing demand, the Indonesian government not only open the wetland rice fields, but also enhance their productivity. Otherwise, the newly wetland will be left out and rice import will be happened. Increasing of rice yield will motivate the farmer to work hard to improve the rice yield. Therefore, scaling up of this technology or demonstration plot and training should also be conducted to open their farmer eyes and to change their cultural practices, as the farmers in transmigration came from Java and Bali islands. With these ways, sustainable rice farming under newly opened wetland rice fields leading poverty alleviation will be achieved.

Conclusion

Inherent soil fertility of newly opened wetland rice was

categorised poor, with constrains low in pH, organic matter (soil organic carbon and total nitrogen), potential and available phosphorus, and potential potassium. In addition, it has high level of iron (Fe) and manganese (Mn).

Application of recommended fertiliser rate namely 250 kg urea, 100 kg SP-36 and 100 kg KCl ha^{-1}, in which N and K were split three times combined with 2 tons dolomite ha^{-1} and 2 tons compost made of rice straw ha^{-1} enhanced the soil fertility, rice growth and rice biomass production. Residue of 2 tons dolomite ha^{-1} and 2 tons compost ha^{-1} effectively increased soil function, rice growth and rice biomass production. Training and demonstration plot should be conducted to socialize this technology to sustain rice farming in Indonesia.

ACKNOWLEGEMENTS

This experiment was funded by the Minister of Research and Technology and Directorate of Areal Development and Water Management and also, the Minister of Agriculture. Thanks to Anda Suhanda and Rahmat Hidayat for their work in the field.

REFERRENCES

Anonymous (2002). Statistic of Indonesia. Biro Pusat Statistik. Jakarta. (In Indonesia).

Bhagat RM, Bhuiyan SI, Moody K (1996). Water, tillage and weed interactions in lowland tropical rice: a review. Agric. Water Manage., 31: 165-184.

Bouman BAM, Tuong TP (2001). Field water management to save water and increase its productivity in irrigated lowland rice. Agric. Water Manage., 49: 11-30.

Clark MS, Horwath WR, Shennan C, Scow KM (1998). Changes in soil chemical properties resulting from organic and low-input farming practices. Agron. J., 90: 662-671.

Doran JW, Parkin TB (1994). Defining and assessing soil quality. In: Defining soil quality for a sustainable environment. Eds. J.W. Doran, D.C. Coleman, D.F. Bezdicek, B.A. Stewart, pp. 3-21.

Fageria NK, Balligar CV (2001). Improving nutrient use efficiency of annual crops in Brazilian acid soils for sustainable crop production. Communication Soil Science Plan Analysis. 32(7 and 8): 1301-1319.

Fenning JO, Adjei-Gyapong T, Yeboah E, Ampontuah EO, Quansah G, Danso SKA (2005). Soil Fertility status and potential organic inputs for improving smallholder crop production in the interior savannah zone of Ghana. J. Sustain. Agric., 25(4): 69-92.

Hasegawa H, Furukawa Y, Kimura SD (2005). On farm assessment of organic amandments effect on nutrient status and nutrient use efficiency of organic rice fields in Northeastern Japan. Agric. Ecosys. Environ. J., 108: 350-362.

Khai NM, Quang HP, Oborn I (2007). Nutrient flows in small scale peri urban vegetables farming system in Southeast Asia – a case study in Hanoi. J. Agric. Ecosys. Environ., 122: 192-202.

Rasmussen PE, Parton WJ (1994). Long-term effects of residue management in wheat-fallow: I. Inputs, yields, and soil organic matter. Soil Sci. Soc. Am. J., 58: 523-530.

Sharma KL, Kusuma GJ, Mandal UK, Gajbhiye PN, Srinivas K, Korwar GR, Bindu VH, Ramesh V, Ramachandran K, Yadav SK (2008). Evaluation of long-term soil management practices using key indicators and soil quality indices in a semi-arid tropical Alfisol. Aust. J. Soil Res., 46: 368-377.

Soil Research Institute (2009). Penuntun analisa kimia tanah, tanaman, air dan pupuk (Procedure to measure soil chemical, plant, water and fertiliser). Soil Research Institute, Bogor (in Indonesia), 234 p.

Sommerfeldt TG, Chang C, Entz T (1988). Long-term annual manure applications increase soil organic matter and nitrogen, and decrease carbon to nitrogen ratio. Soil Sci. Soc. Am. J., 52: 1668-1672.

Sudjadi M (1984). Red podzolic soil fertility problems and possible solutions Yellow. In Proceedings of Farming Patterns Supporting Research Transmigration. bodies of Agricultural Research, Jakarta (in Indonesia), pp. 3-10.

Sukristiyonubowo S, Mulyadi, Wigena P, Kasno A (1993). Effect of organic matter, lime and NPK fertilizer added on soil properties and yield og peanut. J. Indonesian Soil Fertilizer (in Indonesia), 11: 1-7.

Sukristiyonubowo S (2007). Nutrient balances in terraced paddy fields under traditional irrigation in Indonesia. PhD thesis. Faculty of Bioscience Engineering, Ghent University, Ghent, Belgium, 184 p.

Sukristiyonubowo S, Tuherkih E (2009). Rice production in terraced paddy field systems. J. Penelitian Pertanian Tanaman Pangan, 28(3): 139-147.

Sukristiyonubowo S, Du Laing G, Verloo MG (2010). Nutrient balances of wetland rice for the Semarang District. J. Sustain. Agric., 34(8): 850-861.

Wander MM, Traina SJ, Stinner BR, Peters SE (1994). Organic and conventional management effects on biologically active organic matter pools. Soil Sci. Soc. Am. J., 58: 1130-1139.

Whitbread A, Blair G, Konboon Y, Lefroy R, Naklang K (2003). Managing crop residue, fertiliser and leaf litters to improve soil C, nutrient balance and grain yield of rice and wheat cropping system in Thailand and Australia. J. Agric. Ecosystems Environ., 100: 251-263.

Yan D, Wang D, Yang L (2007). Long term effect chemical fertiliser, straw and manure on labile organic matter in a paddy soil. Biol. Fertil. Soil J., 44: 93-101.

Yang SM, Malhi SS, Song JR, Xiong YC, Yue WY, Lu LL, Wang JG, Guo TW (2006). Crop yield, nitrogen uptake, and nitrate-nitrogen accumulation in soil as affected by 23 annual applications of fertiliser and manure in the rainfed region of North-western China. Nutr. Cycl. Agroecosys. J., 76: 81-94.

18

In vitro explants regeneration of the grape 'Wink' (*Vitis vinifera* L. 'Wink')

Ping Zhang, Zhi-Ying Yu, Zong-Ming Cheng, Zhen Zhang and Jian-Min Tao*

College of Horticulture, Nanjing Agricultural University, Nanjing, 210095, China.

The effects of different hormones and their concentration combinations, different explant types, dark periods, on the regeneration of the grape 'Wink' was investigated using *in vitro* leaves, petioles, internodes and radicles. The results showed that of all the media prepared, the explants on following media produced the highest regeneration rate: 78.74±1.60% of the leave explants on the medium prepared with MS+18.20 µM, TDZ+0.49 µM IBA regenerated with a mean of 6.75±0.75 shoots per explant; 39.33±1.47% of the petioles on the medium prepared with MS+9.10 µM TDZ+0.49 µM IBA regenerated with a mean of 3.55±0.50 shoots per explant and 41.37±1.13% of internodes on the same medium regenerated with a mean of 4.74±0.64. However, radicles did not generate on any of the media. TDZ enhanced regeneration rate of leaves better than BA. From 0 week to 4 weeks in dark, leave explants displayed a higher rate of regeneration for 2 or 3 weeks. Adventitious shoots were rooted on 3/4 MS+1.73 µM IBA medium, and the rooted plantlets survived after acclimatization and they were transplanted to the greenhouse.

Key words: Grape, explants, regeneration, hormones, adventitious shoots.

INTRODUCTION

Vitis vinifera L. 'Wink', one of the Eurasian species, originated in Japan and is a popular late-maturing variety widely cultivated in China. However, its vine grows too prosperously and requires too much pruning. Therefore, growth inhibition of the plant is necessary. To improving plant characteristics, such as plant growth, genetic engineering techniques are often applied, in which achieving a high rate of regeneration is key step.

Reports have been published on grape regeneration using organ regeneration (Stamp et al., 1990; Tao et al., 2005; Li et al., 2007) or somatic embryogenesis regeneration (Li et al., 2000; Pinto-Sintra 2007; Yuan et al., 2007; Araya et al., 2008; Zhi et al., 2010). Of the two popular methods, somatic embryogenesis regeneration is too tedious. It requires different mediums, takes more time (over half a year) to induce somatic embryo and uses immature pollens which are limited by phenophase

(Tsvetkov et al., 2000; Mulwa et al., 2007; López-Pérez et al., 2008; Dhekney et al., 2008; Vidal et al., 2009). Organ regeneration is also limited because plants regenerated from the explants transferred by exogenous gene may alter heredity, but the variation is very low. Plants regenerated from leaf explants display even no variation (Yang et al., 2006; Jin et al., 2009). Therefore, organ regeneration is relatively simpler and more feasible.

In this study, we used explants from the grape 'Wink' to develop a protocol of regeneration with a purpose of laying a foundation for the application of genetic engineering techniques to research on the growth of *V. vinifera* L. 'Wink'.

MATERIALS AND METHODS

Plant material

Micropropagation of *V. vinifera* L. 'Wink' was established *in vitro* from nodal sections (1 to 2 cm in length) of mature plants. A ¾ MS medium was prepared and supplemented with 1.73 µM IBA, 3% (w/v) sucrose and 0.55% agar (PH 5.8). The medium samples of 30 to 40 ml each were put in jars and autoclaved at 121°C for 20 min. One nodal section with one *in vitro* micro-shoot was proliferated in each jar. Then, the *in vitro* micro-shoots on the medium were

*Corresponding author. E-mail: taojianmin@njau.edu.cn.

Abbreviations: MS, Murashige and Skoog medium (1962)**; TDZ,** thidiazuron (N-phenyl N'1, 2, 3-thidiazol-5-ylurea)**; BA,** 6-benzyladenine**; IBA,** indole-3-butyric acid.

incubated at 25±2°C under a 16-h photoperiod provided by cool-white fluorescent tubes at an intensity of 32 to 40 μmol m^{-2}s^{-1}. The shoots grew into plantlets 4 weeks later and were cut to obtain new nodal sections which were then transferred to fresh media of the same preparation as described above and were multiplied there.

Effects of different hormones

In vitro expanded and immature leaves along with their petioles (0.1 cm in length) were transversely cut along the midrib (0.5 cm in length and 0.5 cm in width). These leaf explants were placed with adaxial or abaxial side in contact with the media and were cultured in MS media supplemented with TDZ in different concentrations of 4.55, 9.10, 13.65 or 18.20 μM and IBA in the concentration of 0.49 μM, or in MS media supplemented with BA in the concentrations of 4.44, 8.88, 13.32 or 17.76 μM and IBA in the concentration of 0.49 μM. Each of the foregoing prepared media was added with 3% (w/v) sucrose and 0.55% agar. All the explants on the media were kept in dark for two weeks to induce callus and/or adventitious shoots. The other culturing conditions were the same as described earlier.

Effects of different explants

Leave explants obtained in the same way as described earlier, were placed with the adaxial sides in contact with MS media supplemented with TDZ in the concentrations of 13.65, 18.20, 22.75 or 27.30 μM and IBA in the concentrations of 0.05, 0.49, 0.99, 1.48 or 2.47 μM. A 4×5 complete factorial design of the four concentrations of TDZ and five concentrations of IBA produced 20 treatments.

Young petioles, young internodes and radicles from 3 to 4 week-old plants were obtained, cut and used as explants. Each of the explants was cultured in MS media supplemented with TDZ in the concentrations of 4.55, 9.10, 13.65 or 18.20 μM and IBA in the concentrations of 0.05, 0.49, 0.99, 1.48 or 2.47 μM. A 4×5 complete factorial design of the four concentrations of TDZ and five concentrations of IBA produced 20 treatments.

The other preparation and culturing conditions for the foregoing explants were the same as described earlier.

Effects of dark periods

Leaf explants were placed with the adaxial side in contact with the MS medium supplemented with 18.20 μM TDZ, 0.49 μM IBA, 3% (w/v) sucrose and 0.55% agar. These explants were dark treated for 0, 1, 2, 3 or 4 weeks, respectively. They were then cultivated under a 16-h photoperiod (32 to 40 μmol m^{-2}s^{-1}) at 25±2°C for 7 weeks (including dark periods). The other culturing conditions were the same as described in the section of "Effects of different hormones".

The experiment was repeated three times. The percentage of callus was recorded after 1 or 2 weeks. The percentage of callus was categorized on a scale from 0 to 4: 0 = 0%, 1 = 25%, 2 = 50%, 3 = 75%, and 4 = 100%. The regeneration percentage and the number of adventitious shoots of each explant were recorded 7 weeks after the beginning of the experiment. The adventitious shoots were cut and transferred to 3/4MS media supplemented with 1.73 μM IBA. These shoots would be rooted in this medium.

Transplant of plantlets

The resulting plantlets were acclimatized in natural light in room for one week. Then the lip of the jar was removed to let the plantlets to adapt to the natural environment for 3 to 4 days. Thirdly, the plantlets were taken out of the jar and cleaned of the medium. Finally, they were transplanted to the medium in pots in the greenhouse. The medium was made of peat, vermiculite and perlite (3:1:1). Proper temperature (25±1°C) and relative humidity (60% RH-80% RH) were provided for them to grow.

Data analysis

Regeneration rate was defined as the average percentage of each treatment with the number of the explants that developed shoots divided by the number of total explants. The mean of shoots was presented as the total number of shoots divided by the number of the explants which regenerated the shoots. The regeneration percentage and the mean of shoots were presented as the mean ± standard error. Data was analyzed by the Duncan's test using DPS data program (version 3.01). The related formulas were as follows:

Regeneration percentage = the number of developed shoots / the number of total explants × 100%

Mean of shoots = the total number of developed shoots / the number of explants regenerated shoots × 100%

RESULTS

Effects of different hormones

The way the leaves were placed on the medium had a great influence on regeneration rate (Table 1). The regeneration rate was higher, and the mean of shoots was greater when the adaxial side of the leaf explant was in contact with the MS medium (Figure 1 A to C) than when the abaxial side was in contact with the same medium. This was true of the cases with media of different preparations (Table 1).

Under the condition of the adaxial side of the leaf explant in contact with the MS medium, when the medium was supplemented with 8.88 μM BA and 0.49 μM IBA, the regeneration rate and the mean of shoots were the highest (Table 2). However, when the concentration of BA was 13.32 μM or more, both the regeneration rate and the mean of shoots began to fall, even to zero, with callus becoming loose and flocculent.

Comparing the results shown in Tables 1 and 2, we can see that the regeneration rate and the mean of shoots were higher when the medium was supplemented with TDZ than when the medium was supplemented with BA, while the concentration of IBA was fixed.

Effects of different explants

Table 3 showed that, after one week of culturing in dark, 13.65 μM TDZ induced 75 to 100% of callus, but the percentages of callus decreased when the concentrations of TDZ increased with the same concentration of IBA. When the duration of culturing in dark lasted for two weeks, all the callus reached 100% with the MS medium

Table 1. Adventitious shoot regeneration response of the grape 'Wink' via *in vitro* leaf explants with different placements as affected by thidiazuron (TDZ) and indole-3-butyric acid (IBA).

TDZ (μM)	IBA (μM)	Adaxial surface in contact with medium		Abaxial surface in contact with medium	
		Regeneration (%) (±SE)	Mean no. shoots (±SE)	Regeneration (%) (±SE)	Mean no. shoots (±SE)
4.55	0.49	16.5±1.38[d]	2.49±0.30[c]	15.48±0.86[d]	1.3±0.16[d]
9.10	0.49	28.65±2.19[c]	3.36±0.18[b]	21.67±2.08[c]	1.8±0.14[c]
13.65	0.49	43.47±1.69[b]	4.16±0.21[b]	31.36±0.88[b]	2.6±0.15[b]
18.20	0.49	78.41±1.97[a]	6.79±0.83[a]	68.55±0.90[a]	3.3±0.15[a]

Different letters in the same column denote significant differences at P<0.05 by Duncan's test.

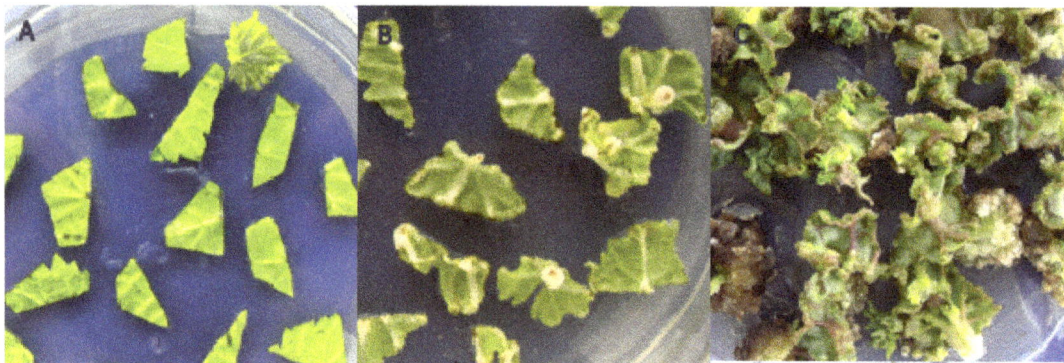

Figure 1 A to C. Adventitious shoot regeneration from *in vitro* explants of the grape 'Wink'. A to C: The process of regeneration of the leaf explant with the adaxial side in contact with the MS medium supplemented with 18.20 μM TDZ and 0.49 μM IBA, and in dark for two weeks.

Table 2. Adventitious shoot regeneration response of the grape 'Wink' via *in vitro* leaf explants as affected by 6-benzyladenine (BA) and indole-3-butyric acid (IBA).

BA (μM)	IBA (μM)	Adaxial surface in contact with medium	
		Regeneration (%) (±SE)	Mean no. shoots (±SE)
4.44	0.49	8.57±1.24[b]	1.38±0.16[b]
8.88	0.49	12.56±2.14[a]	2.39±0.13[a]
13.32	0.49	6.07±0.42[c]	1.22±0.18[b]
17.76	0.49	0[d]	0[c]

Different letters in the same column denote significant differences at P<0.05 by Duncan's test.

supplemented with all the selected concentration combinations of TDZ and IBA.

More importantly, 18.20 μM TDZ induced higher regeneration rate and mean shoots than the other concentrations of TDZ with the same concentration of IBA. At 18.20 μM TDZ, 0.49 μM IBA induced the highest regeneration rate (78.74±1.60) and mean shoots (6.75±0.75).

After one week of culturing in dark, 4.55 μM TDZ TDZ induced 100% of callus in all concentrations of IBA, but the percentage of callus was a slightly decreased when the concentrations of TDZ increased (Table 4). Our data showed that the callus reached 100% with the MS

medium supplemented with all the selected concentration combinations of TDZ and IBA when the culturing period in dark lasted for two weeks.

9.10 μM TDZ induced higher regeneration rate and mean shoots than the other concentrations of TDZ no matter in which concentration of IBA. At 9.10 μM TDZ, 0.49 μM IBA induced the highest regeneration rate (39.33±1.47) and mean shoots (3.55±0.50) (Figure 1 D to F).

In Table 5, the percentage of callus and the rate of regeneration of internode explant were of the same trend as that of the petiole explant (Figure 1 G to I). However, the highest mean of shoots was induced at a different

Table 3. Callus induction and adventitious shoot regeneration response of the grape 'Wink' via *in vitro* leaf explants as affected by thidiazuron (TDZ) and indole-3-butyric acid (IBA).

TDZ (µM)	IBA (µM)	Callus coverage (0-4)	Adaxial surface in contact with medium	
			Regeneration (%) (±SE)	Mean no. shoots (±SE)
13.65	0.05	3-4	37.26±2.72[f]	3.38±0.44[e]
13.65	0.49	3-4	43.51±1.42[e]	4.22±0.28[cd]
13.65	0.99	3-4	42.79±1.37[e]	3.78±0.55[de]
13.65	1.48	2-3	35.35±1.53[f]	3.45±0.30[e]
13.65	2.47	2-3	33.06±1.58[g]	2.58±0.67[f]
18.20	0.05	3-4	71.53±2.46[c]	4.76±0.51[c]
18.20	0.49	3-4	78.74±1.60[a]	6.75±0.75[a]
18.20	0.99	2-3	76.32±2.02[b]	5.76±0.52[b]
18.20	1.48	2-3	69.62±1.89[c]	4.73±0.49[c]
18.20	2.47	1-2	61.68±2.07[d]	3.25±0.38[e]
22.75	0.05	2-3	20.06±1.05[i]	1.22±0.20[h]
22.75	0.49	2-3	24.78±0.01[h]	2.35±0.41[fg]
22.75	0.99	1-2	19.56±0.53[i]	1.87±0.49[g]
22.75	1.48	1-2	0[j]	0[i]
22.75	2.47	1-2	0[j]	0[i]
27.30	0.05	1-2	0[j]	0[i]
27.30	0.49	1-2	0[j]	0[i]
27.30	0.99	1-2	0[j]	0[i]
27.30	1.48	1-2	0[j]	0[i]
27.30	2.47	0-1	0[j]	0[i]

Classification of callus coverage on leaf explants: 0 = 0%, 1 = 25%, 2 = 50%, 3 = 75%, and 4 = 100%. Callus coverage observed on leaf explants for each plant growth regulator combination of concentrations. Different letters in the same column denote significant differences at $P<0.05$ by Duncan's test.

preparation with the MS medium supplemented with 9.10 µM TDZ and 0.99 µM IBA (Table 5).

Only a few radicle explants developed callus on the MS medium with different concentrations of TDZ (13.65, 18.20, 22.75 or 27.30 µM) and IBA (0.05, 0.49, 0.99, 1.48 or 2.47 µM). And none of the callus could induce adventitious shoots.

Effects of dark periods on shoot organogenesis

Table 6 showed that when the best medium preparation was selected and fixed, two or three weeks of culturing in dark induced higher regeneration rate and mean of shoots than the other durations of the dark period. If the leaf explant was cultured directly in light, no adventitious shoots were developed. Four weeks in dark also induced quite high regeneration rate, but, among the adventitious shoots thus developed, only a few survived the long darkness.

DISCUSSION

In the present study, the effects of different hormones and their concentration combinations of the type of the plant explants and of dark periods on the regeneration of the grape 'Wink' were investigated for the purpose of establishing a protocol of regeneration.

Firstly, different placement of leaf explants on the MS medium was studied, and the result was that, when the adaxial side contacted the medium, the explants developed a higher rate of regeneration (Table 1). This was consistent with some previous studies (Li et al., 2002, 2007). However, the rate of regeneration in our study (78.74%±1.60) was obviously higher than those of previous studies, where 33% was observed in Li et al. (2002) and 13.48% in Li et al. (2007). This improvement was attributed to our optimization of hormone combination in the research design.

Secondly, the effect of two cytokinins TDZ and BA on regeneration was compared by experiment. It was observed that TDZ induced higher regeneration rate than BA (Tables 1 and 2), which generally agreed with the reports by Yuan et al. (2007) and Fang et al. (2007).

Thirdly, different types of explants have significant effect on the rate of regeneration (Thomas et al., 2000). The regeneration of leaf, petiole, internode and radicle explants was investigated in the present research. It was found that the leaf explant had the highest regeneration rate but the petiole and internode explants produced stronger adventitious shoots (Tables 3 to 5). The radicle

Table 4. Adventitious shoot regeneration response of the grape 'Wink' via *in vitro* petiole explants as affected by thidiazuron (TDZ) and indole-3-butyric acid (IBA).

TDZ (µM)	IBA (µM)	Callus coverage(0-4)	Regeneration (%) (±SE)	Mean no. shoots (±SE)
4.55	0.05	4	8.03 ± 0.46^{gh}	1.23 ± 0.17^{h}
4.55	0.49	4	9.44 ± 1.30^{g}	1.35 ± 0.14^{h}
4.55	0.99	4	7.27 ± 0.14^{h}	1.20 ± 0.18^{h}
4.55	1.48	4	0^{k}	0^{i}
4.55	2.47	4	0^{k}	0^{i}
9.10	0.05	4	29.55 ± 1.00^{b}	2.85 ± 0.10^{bc}
9.10	0.49	3-4	39.33 ± 1.47^{a}	3.55 ± 0.50^{a}
9.10	0.99	3-4	26.77 ± 1.60^{c}	3.10 ± 0.15^{b}
9.10	1.48	3-4	21.99 ± 0.49^{d}	2.49 ± 0.02^{de}
9.10	2.47	3-4	16.79 ± 2.19^{e}	2.12 ± 0.70^{fg}
13.65	0.05	3-4	6.96 ± 0.45^{hi}	2.01 ± 0.31^{fg}
13.65	0.49	3-4	11.82 ± 0.87^{f}	2.79 ± 0.14^{bcd}
13.65	0.99	3-4	7.87 ± 0.16^{gh}	2.55 ± 0.04^{cde}
13.65	1.48	3-4	5.31 ± 0.39^{ij}	1.32 ± 0.06^{h}
13.65	2.47	2-3	0^{k}	0^{i}
18.20	0.05	3-4	5.41 ± 0.40^{ij}	1.36 ± 0.16^{h}
18.20	0.49	3-4	9.54 ± 0.86^{g}	2.27 ± 0.21^{ef}
18.20	0.99	3-4	4.20 ± 0.78^{j}	1.80 ± 0.11^{g}
18.20	1.48	2-3	0^{k}	0^{i}
18.20	2.47	2-3	0^{k}	0^{i}

Classification of callus coverage on petiole explants: 0 = 0%, 1 = 25%, 2 = 50%, 3 = 75%, and 4 = 100%. Callus coverage observed on leaf explants for each plant growth regulator combination. Different letters in the same column denote significant differences at P<0.05 by Duncan's test.

Figure 1 D to F. Adventitious shoot regeneration from *in vitro* explants of the grape 'Wink'. D to F: The process of regeneration of the petoile explant with the adaxial side in contact with the MS medium supplemented with 9.10 µM TDZ and 0.49 µM IBA and in dark for two weeks.

explant showed no regeneration rate (leaves 78.74%±1.60, petiolets 39.33%±1.47, internodes 41.37%±1.13, radicles 0%). These results were similar to those of the study by Quan and Chang (2005) (leaves 17.9%, petiolets 2.3%, internodes 2.6%, radicles 0%).

Finally, the necessary dark periods of culturing were compared, and two weeks were found to be enough to achieve the optimal regeneration rate (Table 6), which was confirmed in the papers by Korban et al. (1992), Toreegrasa et al. (2001) and Deng et al. (2009).

The adventitious shoots obtained in the present experiment were transferred to the ¾ MS + 1.722 µM IBA medium and they were rooted there. The plantlets thus obtained were acclimatized, transplanted to pots in the greenhouse. Their survival rate was as high as 90% suggesting that the whole experiment was successful.

In conclusion, we have found an optimal hormone combination and an ideal dark period for higher regeneration rate of explants of the grape 'Wink'. The shoots thus obtained displayed a high survival rate (Figure 1 J to L). Our findings might be of significance to the similar studies of other varieties of grapes, which is

Table 5. Callus induction and adventitious shoot regeneration response of the grape 'Wink' via *in vitro* internode explants as affected by thidiazuron (TDZ) and indole-3-butyric acid (IBA).

TDZ (µM)	IBA (µM)	Callus coverage (0-4)	Regeneration (%) (±SE)	Mean no. shoots (±SE)
4.55	0.05	4	10.23±1.25[fg]	1.31±0.07[e]
4.55	0.49	4	13.35±1.16[e]	1.50±0.04[e]
4.55	0.99	4	9.48±0.84[fg]	1.15±0.11[e]
4.55	1.48	4	0[i]	0[f]
4.55	2.47	4	0[i]	0[f]
9.10	0.05	4	25.48±2.77[b]	3.46±0.17[bc]
9.10	0.49	4	41.37±1.13[a]	4.74±0.64[b]
9.10	0.99	3-4	24.71±0.96[b]	5.15±0.61[a]
9.10	1.48	3-4	21.72±0.34[cd]	3.20±0.29[bc]
9.10	2.47	3-4	12.01±2.72[ef]	2.25±0.39[d]
13.65	0.05	3-4	23.30±1.40[bc]	2.95±0.16[cd]
13.65	0.49	3-4	25.81±1.57[b]	3.30±0.44[bc]
13.65	0.99	3-4	23.78±0.56[bc]	3.23±0.76[bc]
13.65	1.48	3-4	19.48±1.36[d]	2.37±0.09[d]
13.65	2.47	2-3	0[i]	0[f]
18.20	0.05	3-4	5.22±1.34[h]	1.18±0.27[e]
18.20	0.49	3-4	8.80±0.73[g]	1.16±0.13[e]
18.20	0.99	3-4	4.97±1.90[h]	1.25±0.37[e]
18.20	1.48	2-3	0[i]	0[f]
18.20	2.47	2-3	0[i]	0[f]

Classification of callus coverage on leaf explants: 0 = 0%, 1 = 25%, 2 = 50%, 3 = 75%, and 4 = 100%. Callus coverage observed on leaf explants for each plant growth regulator combination. Different letters in the same column denote significant differences at P<0.05 by Duncan's test.

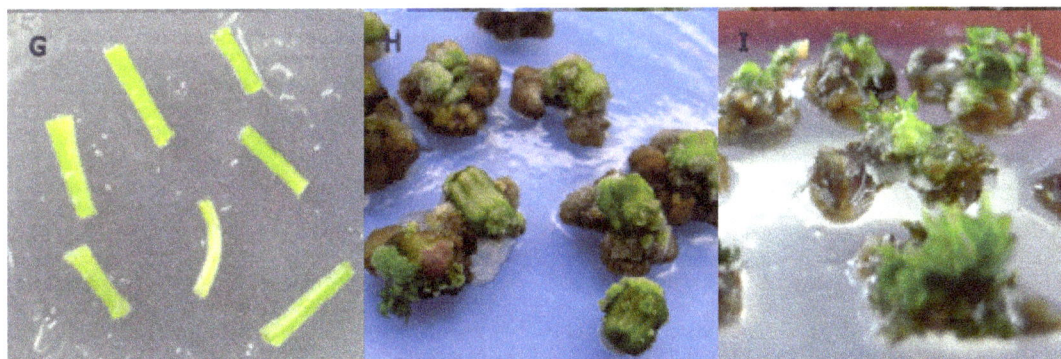

Figure 1 G to F. Adventitious shoot regeneration from in vitro explants of the grape 'Wink'. G to I: The process of regeneration of the internode explant with the adaxial side in contact with the MS medium supplemented with 9.10 µM TDZ and 0.49 µM IBA and in dark for two weeks.

Table 6. Effects of dark periods on shoot organogenesis from leaf explants of grape 'Wink'.

Days in darkness(week)	TDZ (µM)	IBA (µM)	Regeneration (%) (±SE)	Mean no. shoots (±SE)
0	18.20	0.49	0[d]	0[c]
1	18.20	0.49	67.16±0.92[c]	4.01±0.85[b]
2	18.20	0.49	78.68±1.38[a]	6.70±0.79[a]
3	18.20	0.49	78..66±0.96[a]	6.73±1.06[a]
4	18.20	0.49	72.36±2.23[b]	3.72±1.25[b]

Different letters in the same column denote significant differences at P<0.05 by Duncan's test.

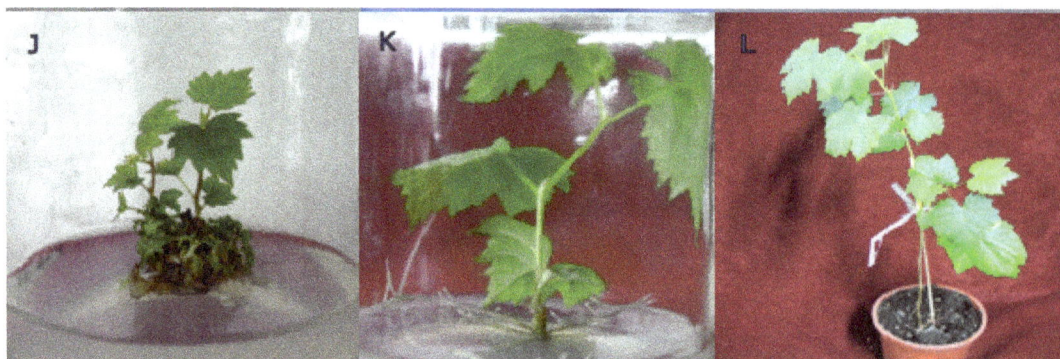

Figure 1 J to L. Adventitious shoot regeneration from *in vitro* explants of the grape 'Wink'. J to L: The process of transplant of the plantlets regenerated from the internode explant.

being confirmed by the study of the grape 'Manicure Finger' in our laboratory.

ACKNOWLEDGMENTS

We thank Prof. F. R. Gu (College of Foreign Studies, Nanjing Agricultural University) for language editing of the manuscript. This work was supported by The Technology System of National Grape Industry (No. CARS-30); the Innovation Fund of Jiangsu Agricultural Science and Technology, China [No. CX (10)110] and High-tech Project of Jiangsu Provincial Science and Technology Department, China (No. BE2009314-2).

REFERENCES

Araya S, Prieto H and Hinrichsen P (2008). An efficient buds culture method for the regeneration via somatic embryogenesis of table grapes 'Red Globe' and 'Flame Seedless'. Vitis, 47(4): 251-252.

Deng J, Liu X and Xue RG (2009). Induction of multiple shoots from stem explants of Chardonny grape. J. Fruit Sci., 26(2): 222-225.

Dhekney SA, Li ZT, Dutt M, Gray DJ (2008). *Agrobacterium*-mediated transformation of embryogenic cultures and plant regeneration in *Vitis rotundifolia* Michx. (muscadine grape). Plant Cell Rep., 27: 865-872.

Fang YL, Liu D and Song SR (2007). *In vitro* regenerating system of *Vitis rotundifolia* cv. 'Alachua'with its petioles. Acta Bot. Boreal. - Occident.Sin., 27(9): 1777-1781.

Jin WM, Dong J, Wang YH, Mao HL, Xiao Z, Chen MX (2009). Genetic fidelity of regeneration adventitious shoots in grape through organogenesis. Mole. Plant Breeding, 7(2): 375-379.

Korban SS, Oconnor PA and Elobeidy A (1992). Effects of thidiazuron naphthaleneacetic acid, dark incubation and genotype on shoot organogenesis from Malus leaves. Hort. Sci., 67(3): 341-349.

Li JF, Zhang Z, Zhuang ZM, Tong ZG, Tao JM (2007). Plant regeneration of grape rootstock '5BB' (*Vitis berlandieri* ×V. *riparia*) *in vitro*. Acta Bot. Boreal. –Occident. Sin., 27(7): 1323-1328.

Li Y, Feng H and Tian YT (2000). Progress in the Studies of Grape Regeneration System. Biotechnol. Inform., 2: 28-31.

Li Y, Feng H and Tian YT (2002). Developed *in vitro* regeneration system of *Vitis vinifera* cv. 'Red Globe' with its leaves and petioles. Acta. Hort. Sinica, 29(1): 60-62.

López-Pérez AJ, Velasco L, Pazos-Navarro M, Dabauza M (2008). Development of highly efficient genetic transformation protocols for table grape Sugraone and Crimson Seedless at low *Agrobacterium* density. Plant Cell Tiss. Organ Cult., 94: 189-199.

Mulwa RMS, Norton MA, Farrand SK and Skirvin RM (2007). *Agrobacterium*-mediated transformation and regeneration of transgenic 'Chancellor' wine grape plants expressing the *tfdA* gene. Vitis, 46(3): 110-115.

Pinto-Sintra AL (2007). Establishment of embryogenic cultures and plant regeneration in the Portuguese cultivar 'Touriga Nacional' of *Vitis vinifera* L. Plant Cell Tiss. Organ Cult., 88: 253-265.

Quan DL, Chang YY (2005). Study on direct regenerating adventitious buds from different organs of grape. J. Gansu Agric. Univ., 40(2): 173-177.

Stamp JA, Colby SM and Meredith CP (1990). Direct shoot organogenesis and plant regeneration from leaves of Grape (*vitis* spp.). Plant Cell Tiss. Organ Cult., 22: 127-133.

Tao JM, Zhang ZM, Zhang 7, Geng QF, Cai BH (2005). Study on plant regeneration from *in vitro* culture of *Vitis vinifera* cv. Manicure Finger. J. Fruit Sci., 22(5): 551-553.

Thomas MR, Locco P and Franks T (2000). Transgenic grapevines status and future. Acta Hort., 528: 279-287.

Toreegrasa L, Bouquet A and Goussard PG (2001). *In vitro* Culture and Propagation of Grapevine[M]. Netherland: Kluwer Acad. Publishers Mole. Biol. Biotechnol. Grapevine, pp. 281-326.

Tsvetkov IJ, Atanassov AI and Tsolova VM (2000). Gene transfer for stress resistance in grapes. Acta Hort., 528: 389-394.

Vidal JR, Rama J, Taboada L, Martin C, Ibañez M, Segura A, González-Benito ME (2009). Improved somatic embryogenesis of grapevine (*Vitis vinifera*) with focus on induction parameters and efficient plant regeneration. Plant Cell Tiss. Organ Cult., 96: 85-94.

Yang XM, An LZ, Wang YM, Li S (2006). Somatic Embryogenesis and Analysis of the Genetic Stability of Regeneration System in Grapevine 'Sinsaut'. Acta. Hort. Sinica, 33(6): 1317-1320.

Yuan WF, Xu K, Liao HY, Qian YM, Xu DC (2007). Adventitious bud regeneration from 'Pinot noir'grape. J. Anhui Agric. Univ., 34(1): 120-123.

Zhi YX Zhang JX and Wang YJ (2010). Studies on somatic embryogenesis and plant regeneration from anthers of Guangxi _2, a line of Chinese wild grape (*Vitis pseudoreticulata*). J. Fruit Sci., 27(1): 18-23.

19

Variation and association among characters genetically related to yield and yield stability in *Coffea canephora* genotypes

E. Anim-Kwapong*, G. J. Anim-Kwapong and B. Adomako

Cocoa Research Institute of Ghana, P. O. BOX 8, New Tafo-Akim, Ghana.

Water deficit stress is a main factor determining yield and yield stability in Robusta coffee (*Coffea canephora* P.). Studies were conducted to identify agronomic traits that offer genetic sources of drought tolerance associated with yields and yield stability. In a nine-year experiment, 18 genotypes were assessed for mean value and variation in three diverse environments, for nine vegetative traits, five reproductive traits and bean yield. Drought tolerance was expressed by a visual scale of leaf scorching from 0 to 5: tolerant – susceptible on the genotypes in 2000 when there was drought, and genetic associations between leaf scorching and the traits established. Significant interaction (P ≤ 0.05), and location (P ≤ 0.05) effects were observed for leaf scorching scores and all the traits, except stem diameter and diameter of primary branches. Significant (P ≤ 0.05) genotypic effects were also observed for all the traits, except fruits per node. Canopy diameter (Span), number of primary branches per plant (NPB), fruit-set (FS) and bean yield over seven years (MY1-7) were inversely and significantly (P ≤ 0.05) correlated with leaf scorching scores. Span, NPB and FS were also significantly (P ≤ 0.05) correlated with MY1-7. Span was highly correlated with stem diameter, length and diameter of primary branches. Eight genotypes each with high mean performance for (MY1-7) and fruit set (FS), seven for Span, and five for NPB, were among the top 10 genotypes which recorded the lowest leaf scorching scores. The results indicate that, Span and its associated traits, NPB, and FS could be exploited, through indirect selection for superior *Coffea canephora* genotypes, for direct utilisation or for breeding for adaptation to drought-stress.

Key words: Drought tolerance, leaf scorching, agronomic traits, yield, yield stability, genetic correlations, genetic variation, indirect selection, *Coffea canephora*.

INTRODUCTION

Coffea canephora production is confined to the inter-tropical zone, from 20 to 25°N and 24°S, mainly due to ecological factors related to temperature and humidity (Smith, 1989). Within this main production zone, the crop is affected adversely by different abiotic stresses such as extreme temperatures, salinity, fluctuations in incident light and drought (Andrea et al., 2003; DaMatta, 2004c; Partelli et al., 2009; Partelli et al., 2010; Batista-Santos et al., 2011). As drought episodes are much frequent than

the other stresses, drought-stress is considered the major environmental factor limiting coffee yield in most coffee growing areas (DaMatta and Ramalho, 2006). The effect on yield depends on the period of drought: before flowering, during fruit-setting, or fruit development. From the standpoint of breeding, the differential responses of genotypes to water-deficit stress have been identified as a main factor contributing to genotype-environment interactions, thus complicating selection for yield (Ehlers, 1994; Bidinger et al., 1996). In coffee, leaf-scorching has been attributed to sensitivity to drought (DaMatta and Rena, 2001; DaMatta, 2004a; DaMatta and Ramalho, 2006), with drought-tolerant cultivars showing delayed leaf wilting and shedding than drought-sensitive ones

---

*Corresponding author. E-mail: ekwapong06@yahoo.com.

(Orozco and Jaramillo, 1978). Visual scoring method of leaf-scorching has, therefore, been the main selection criterion for drought-tolerant genotypes at the vegetative phase (Cilas et al., 2003). In the absence of water-deficit stress, however, such method cannot be useful.

According to DaMatta and Ramalho (2006), cultivars more tolerant to drought generally differ morphologically and/or physiologically, with mechanisms allowing greater production under restricted water supply. Hence, understanding such mechanisms in genotypes naturally adapted to drought could help improve their agronomic performance. DaMatta (2004b) discusses some physiological mechanisms that have been shown to contribute to yield under drought conditions such as: gas exchange (Meinzer et al., 1992) carbon isotope discrimination (Gutiérrez and Meinzer, 1994), osmotic adjustment (Meinzer et al., 1990; DaMatta et al., 2003), and solute accumulation (Maestri et al., 1995). DaMatta (2004b), however, considered most of these methods not sufficiently convenient or effective mechanism of drought-tolerance in coffee.

Agronomic characteristics of most crops were found to be associated with their tolerance to drought (Zhong-hu and Rajaram, 1994; Omanya et al., 1996; Ehdaie et al., 2001). In coffee, root characteristics such as larger root mass (Ramos and Carvalho, 1977) and deeper root system (Pinheiro et al., 2005) were found to be associated with drought-tolerance. Methods based on root characteristics, however, are either not easy to measure, destructive, or too time consuming for plant breeders to evaluate large segregating populations. Hence, a complementary approach to improve Robusta coffee performance could involve the identification and selection of agronomic/morphological traits associated with tolerance to drought-stress that are relatively easy to measure. Correlation studies of agronomic traits, in this respect, should have the advantage of relying on measurement and count data that do not require sampling of tissue, and hence, non destructive. In coffee, agronomic traits associated with yields have been reported for both *C. canephora* (Leroy et al., 1994; Cilas et al., 2006; Anim-Kwapong and Adomako, 2010) and *C. arabica* (Walyaro and Van der Vossen, 1979; Cilas et al., 1998). However, little information exists on the agronomic traits that offer genetic sources of drought tolerance that are associated with yields and yield stability (Cilas et al., 2003).

The correlation between two traits plays an important role in breeding programmes as improvement of one trait causes simultaneous changes in the other trait (Falconer and Mackay, 1996). Genetic correlation, which is the proportion of variance that two traits share due to genetic causes, is useful in studying the associated genetic relationships among traits under selection. This correlated response among traits, forms the basis of indirect selection; hence, an understanding of the agronomic traits genetically correlated with drought-tolerance at the vegetative and reproductive phases of

the plant should provide an opportunity for exploiting such traits, when they occur in the population, for improving coffee yields and yield stability. The objectives of the present study were: (a) To assess variation among Robusta coffee genotypes for visual leaf-scorching, and for vegetative and reproductive traits, including yield. (b) To estimate the magnitude and direction of genetic associations between leaf-scorching and the traits and (c) To select traits that may be associated with tolerance to drought-stress at the vegetative and reproductive stages.

MATERIALS AND METHODS

Genetic material

Twelve clones extracted by individual selection, based on yield, from a population of three half-sib family groups introduced from Cote d'Ivoire, together with six clones introduced from Togo, were used for the study (Table 1).

Experimental sites and design

The study was conducted from 1996 to 2005 at three rain-fed sites at Tafo, Fumso and Bechem, representing a wide range of soil types and climatic regimes within the forest zone of Ghana where coffee is cultivated. The Tafo site is the research farm of the Cocoa Research Institute of Ghana (CRIG) situated at latitude 6° 13' N, longitude 0° 22' W, and altitude 220 m above sea level (asl) in the Eastern region of Ghana. Tafo has a mean total annual rainfall of approximately 1480 mm with a dry season from December to February recording mean monthly rainfall of approximately 44 mm. The soil is sandy loam classified as Haplic Luvisol, brown to yellowish red, well drained, developed in-situ from weathered materials of hornblende granodiorite (Adu and Asiamah, 1992)). Fumso (latitude 6° 6' N, longitude 1° 27' W, and altitude 122 m asl) has a total annual rainfall of approximately 1320 mm. The dry season at Fumso from December to February receives a monthly average rainfall of approximately 29 mm. The soils at the site (Fumso Cocoa Station) belong to the Kumasi Series and are classified as Ferric Luxisol, dark reddish brown to reddish brown, well drained and developed over coarsely-quartzose, biotite granodiorite and have a coarse sandy to fine gravelly topsoil (Ahn, 1961). Bechem (latitude 7° 5' N, longitude 2° 2' W, and altitude 259 m asl) is drier, with a mean total annual rainfall of about 1220 mm, and experiences a marked dry season from December to February receiving a monthly average rainfall of about 17 mm. The soils at the site belong to the Bechem series and of sedentary Forest Ochrosol-Rubrisol intergrades developed over non-micaceous hornblende granite. The topsoil is a dark brown, humus fine sandy loam to light clay, crumbly, loose and porous. It absorbs rainfall readily but liable to dry out during prolonged dry spell (Ahn, 1961). Mean average daily temperatures are about 26.8, 27.0 and 26.2℃ for Tafo, Fumso and Bechem, respectively.

Single-node cuttings of the clones used were rooted in propagators and cultured in nursery bags for six months. Thirty-two plants of each genotype were randomly assigned to each of the three environments. At each location, planting was done using a randomized complete block experimental design with four replications. Each plot, measuring 2.44 x 19.51 m, comprised of a single row of eight coffee trees of each genotype. Both inter-row and inter-plant spacings were 2.44 m giving a density of 1680 plants per hectare. *Gliricidia sepium* shade trees were planted in rows at 4.9 m spacing between each other in the trial plot. Planting at all three environments was done in June 1996. No fertilizer was

Table 1. Planting material and their sources of origin.

Clones from Cote d` Ivoire[†]	Clones from Togo[††]
A129	197
A115	149
A101	126
B170	181
B96	375
B36	107
B191	
E174	
E138	
E139	
E90	
E152	

[†]Selection based on 10-year yields. [††]Selection based on 3-year yields.

applied and crop-management practices were similar for all locations. In order to assess genetic differences in number of stems produced, the plants were allowed to grow on one or two stems developed from the single-node cuttings. Stems were capped at 18 months from field planting by removal of the terminal bud and subsequently capped to 1.8 m and maintained at that height. The first capping resulted in each main stem developing into two branches at the point of capping.

Data collected

Measurements of vegetative characteristics were taken three months after field planting on diameter of the main stem (girth), crown diameter (span) and number of primary branches, and repeated each year after field planting until the plants were 48 months in the field. Vegetative measurements taken when the plants were 48 months in the field, at the stage of maximum expansion, were used for this study. Four plants were randomly selected from each genotype per plot for assessment. Traits assessed included girth (taken at 10cm above the ground in mm), span (cm) taken as the width of the canopy measured at the widest portion of the tree canopy, number of stems, total number of primary branches counted per plant and per stem, and total number of secondary branches per plant. Length of primary branches (measured from the point of attachment to the main stem to the apex in cm), diameter of primary branches (10 cm from the main stem in mm) and number of nodes per primary branch were estimated as an average value of the six longest branches at the middle of the stem per plant. Where there were more than one stem, stem diameter was calculated according to Stewart and Salazar (1992), and span was taken for only the biggest stem.

At flowering and fruiting time in December 1998 to May 1999, two plants from each plot were randomly tagged. Three flowering primary branches at the middle of each plant were tagged for the determination of the number of flowering nodes per branch and number of flowers per node. Fruits that remained on the branches at six months from initial flowering were counted and used in estimating the number of fruiting nodes and fruits per node. Percent fruit-set (fruit-set) was estimated as the proportion of total flowers counted on the three flowering branches per tree that set fruit and remained on the branches at six months from flowering. Data recording was repeated the following season (December 1999 to May 2000), when there was severe drought at all three locations using three plants per plot, and data averaged across plots for the two seasons.

Assessment of plants for drought tolerance was done in February 2000 when coffee leaves appeared scorched due to severe moisture stress at all three sites. Leaf scoring of the plants was done based on dry leaves on a scale of 0 (no dry leaves) to 5 (virtually all dry leaves): drought resistant/tolerant – susceptible on all plants at all three sites. Yield was recorded on each tree for seven production years from Oct. to Jan. each year for the period 1998/1999 to 2004/2005. Transformation of cherry weight to average clean coffee yield per ha/yr was done for each of seven competitive stands per plot using the formula: wet cherry weight per stand (kg) x number of trees per ha x outturn. Outturn was estimated for each genotype as an average of weight of dry beans divided by weight of wet berries.

Statistical analysis

Analyses of variance and covariance were performed on the data using Minitab statistical software (MINITAB, 1997) to examine the presence of statistically significant differences among genotypes, locations and their interactions for these characters. The standard error of difference between means (SED) was estimated to identify genotypes that were significantly different from each other for the traits. The statistical model used for the combined analyses was:

$$Y_{ijk} = \mu + g_i + e_j + (ge)_{ij} + R_{jk} + \varepsilon_{ijk}$$

Where Y_{ijk} is the kth observation of any variable in the rth replications in environment j on genotype i; μ the general mean; g_i and e_j represent the effects of the ith genotype and the jth environment; $(ge)_{ij}$ is the interaction effect between the genotypes and the environment; R_{jk} is the effect of the k-th replication within the j-th location, ε_{ijk} is the random error associated with the kth observation on genotype i in environment j. i = 18; j = 3; r = 4. The effects g_i`s, $e_j$`s, $(ge)_{ij}$`s and $\varepsilon_{ijk}$`s are assumed independently and randomly distributed with zero means and variances σ^2_g, σ^2_l, σ^2_{gl} and σ^2_e respectively. The form of the analysis of variance and covariance with expectations of mean squares and cross products is presented in Table 2. Genotypic correlations (r_G) among the agronomic traits were computed from the variances and covariances as:

$$r_G = COV_{g(xy)} / \sqrt{\sigma^2_{g(x)} \sigma^2_{g(y)}}$$

Table 2. Form and generalized expectations of analysis of variance and covariance for two characters X and Y.

Source of Variation	Degrees of freedom[‡]	Mean square	Analysis of variance (X)	Analysis of variance (Y)	Analysis of covariance (XY)
Reps/locations	(r-1)L				
Locations (Loc)	L-1	M_l			
Genotypes(Gen)	N-1	M_g	$\sigma^2_{e(X)}+r\sigma^2_{gl(X)}+r\,L(\sigma^2_{g(X)})$	$\sigma^2_{e(Y)}+r\sigma^2_{gl(Y)}+rL(\sigma^2_{g(Y)})$	$\sigma_{e(XY)}+r\sigma_{gl(XY)}+rL(\sigma_{g(XY)})$
Gen x Loc	(N-1)(L-1)	M_i	$\sigma^2_{e(X)}+r\sigma^2_{gl(X)}$	$\sigma^2_{e(Y)}+r\sigma^2_{gl(Y)}$	$\sigma_{e(XY)}+r\sigma_{gl(XY)}$
Error	(N-1)(r-1)L	M_e	$\sigma^2_{e(X)}$	$\sigma^2_{e(Y)}$	$\sigma_{e(XY)}$

[‡] r = number of replications = 4; L = number of locations = 3; N = number of genotypes =18.

Table 3. Range and variance components of vegetative traits related to drought tolerance in 18 genotypes of Robusta coffee in three locations.

Traits	Range		Mean	Mean square values			
	Genotype	Location		Genotype	Location	Gen. x Loc.	Error
Girth (mm)	52 -76	63-66	64	493.4[***]	128.0	39.1	31.9
Span (cm)	190-229	206-217	210	1231.7[***]	2506.1[*]	226.7[*]	132.7
Length of prim. branch (cm)	108-132	111-121	117	650.53[***]	1701.62[*]	108.17[***]	47.69
Diam. of prim. branch (mm)	6.7-8.6	7.6-7.9	7.7	3.93[***]	1.96	0.20	0.18
No. of nodes/prim. branch	18.5-22.3	20.2-21.1	20.5	19.06[***]	15.05	3.76[***]	1.83
No. Stems/ plant	1.08-1.65	1.33-1.41	1.35	0.4118[***]	0.2613	0.0946[*]	0.0573
No. prim. branches /stem	84-125	73 – 145	103	1379.1[*]	102740.7[***]	663.1[***]	256
No. prim. branches /plant	105-157	97-186	132	2590.3[*]	164751.6[***]	1399[***]	570.4
No. Sec. branches /plant	7 – 55	17 – 28	22	1871.7[***]	2394.8[*]	348.1[***]	151.4
Leaf- scorching	1.02 - 2.92	1.64 - 2.35	1.99	2.76[***]	8.96[***]	0.90[***]	0.28
				df =17	df =2	df =34	df =153

[*] = $p < 0.05$, [**] = $p < 0.01$, [***] = $p < 0.001$.

Where $COV_{g(xy)}$ is the estimated genotypic covariance component for traits x and y, $\sigma^2_{g(x)}$ and $\sigma^2_{g(y)}$ are the genotypic variance components respectively for traits x and y (Falconer and Mackay, 1996). For correlations involving drought assessment scores, low vales for leaf scorching scores correspond to increased scores for drought tolerance. Test for significance of the correlations was by standard procedure (Steel et al., 1997).

RESULTS

Variance components

The analysis of variance revealed significant ($P \le 0.05$) interactions between the genotypes and the locations for leaf scorching and for all the vegetative traits except girth and diameter of primary branches. Highly significant ($P \le 0.001$) genotypic differences were also observed for leaf scorching and the vegetative traits except number of primary branches per plant and per stem, which showed lower variation ($P \le 0.05$). These observations indicate the presence of substantial variability among the tested genotypes (Table 3), and the possibility of selection for high performing genotypes for these traits. Genotypes with high values for span, girth, as well as length, number and diameter of primary branches had relatively lower leaf scorching compared with genotypes with less vigorous growth habits (Table 4). The lowest average leaf - scorching per tree was recorded from genotypes E90, E139, E138 and 149 compared to all other genotypes except B36, B96, 197 and E174. The locations vary significantly ($P \le 0.001$) for number of primary branches per plant and per stem, and for leaf scorching. Bechem had the highest leaf-scorching score with plants with relatively fewer numbers of primary branches the most affected. The locations also vary significantly ($P \le 0.05$) for span, length of primary branches and number of secondary branches. There were, however, no differences among the locations for girth, diameter and number of nodes of primary branches, and number of stems per plant.

Analysis of variance showed significant interactions ($P \le 0.001$) between the genotypes and the locations for all the reproductive traits observed, except number of

Table 4. Mean trait scores of vegetative traits related to drought tolerance in 18 genotypes of Robusta coffee in three locations.

Genotypes/ Locations	Girth (mm)	Span (cm)	Length of prim. branch (cm)	Diam. of prim. branch (mm)	No. nodes /prim. branch	No. stems /plant	No. prim branches/ stem	No. prim. branches /plant	No. sec. branches/ plant	Leaf scorching
Genotypes										
E138	76.4	219.4	119.9	8.2	21.9	1.2	119	139	7	1.31
E90	71.0	228.6	129.3	8.2	20.3	1.4	107	135	17	1.02
E139	65.0	216.0	119.1	7.9	20.9	1.6	100	152	20	1.15
E152	64.7	222.3	123.2	8.6	22.1	1.3	110	135	24	2.04
126	64.8	210.7	108.1	7.5	19.8	1.5	98	133	33	2.92
149	67.3	204.2	109.5	7.8	18.5	1.7	94	151	55	1.56
B36	64.7	219.9	132.3	7.4	20.9	1.4	97	136	16	1.85
197	58.5	196.9	108.7	6.9	19.7	1.5	114	157	14	1.98
B96	72.3	211.8	120.9	7.9	20.0	1.6	95	145	21	1.95
E174	66.0	221.2	118.2	8.2	21.8	1.2	101	113	8	1.98
A129	67.4	210.9	113.7	8.4	19.2	1.3	102	124	20	2.50
107	56.6	208.6	122.1	7.0	21.9	1.0	125	137	17	2.00
B191	52.7	195.7	112.3	6.7	18.5	1.3	91	111	32	2.33
181	66.3	203.6	107.9	7.6	21.2	1.7	84	133	10	2.40
A101	64.2	210.4	116.9	8.0	20.2	1.3	97	119	16	2.31
A115	58.0	210.9	116.2	7.8	22.3	1.2	93	105	15	2.19
B170	52.3	189.5	108.4	6.7	21.7	1.1	113	122	45	2.13
375	66.8	205.3	109.7	7.2	18.8	1.2	113	123	18	2.15
SED (153df)	2.3	4.7	2.8	0.2	0.6	0.1	7	10	5	0.22
Locations										
Bechem	63.7	216.9	121.0	7.9	21.1	1.4	73	97	28	2.35
Fumso	63.2	205.6	111.4	7.6	20.2	1.3	91	112	20	1.64
Tafo	65.7	208.4	117.0	7.6	20.3	1.3	145	186	17	1.97
SED (153df)	Ns	1.9	1.2	Ns	Ns	Ns	3	4	2	0.09

Table 5. Range and variance components of reproductive traits related to drought tolerance in 18 genotypes of Robusta coffee in three locations.

Traits	Range		Mean	Mean square values			
	Genotype	Location		Genotype	Location	Gen x Loc	Error
Fruit-set (%)	32.3-51.3	22.0-54.3	42.2	404.4***	22185.2***	98.3***	25.3
Number of fruits/node	12.7-17.6	10.6-18.6	15.6	28.1	1354.9***	17.3***	4.7
Number of fruiting nodes	8.7-11.3	6.2-12.3	10.2	8.8***	857.6***	3.0***	1.1
Number of flowers/node	28.7-41.8	31.7-39.0	35.2	155.0***	955.4***	17.3***	7.8
Number of flowering nodes	10.8-13.7	11.0-13.2	12.2	7.3***	88.9***	1.2	0.8
First 1- 3 years yield (kg/ha/yr.)	853-1635	445-1768	1324	512929***	41634417***	164807***	65813
Last 4-7 years yield (kg/ha/yr.)	1063-2582	1079-2451	1838	2278848***	35027941***	372520***	148158
Overall seven years yield (kg/ha/yr.)	1015-2178	808-2152	1625	1242185***	37096509***	194560***	76828
				df =17	df =2	df =34	df =153

* Significant at $P = 0.05$. ** Significant at $P = 0.01$, *** Significant at $P = 0.001$.

flowering nodes. Highly significant $(P \leq 0.001)$ differences among genotypes and locations were observed for all the traits, with the exception of fruits per node, for which the genotypes did not vary. There is, therefore, a substantial variability among tested genotypes and locations for these traits (Table 5). Genotypes with the lowest leaf-scorching score per tree generally also had the best average scores for fruit-set and yields (Table 6). Similarly, Bechem with the highest leaf- scorching score compared with Fumso and Tafo had

Table 6. Mean trait score of reproductive traits related to drought tolerance in 18 genotypes of Robusta coffee in three locations.

Genotypes/ Locations	Fruit-set (%)	No. fruits/ node	No. fruiting nodes	No. flowers/ node	No. flowering nodes	First 3-years yield (kg/ha/yr.)	Last 4-7 years yield (kg/ha/yr.)	Overall seven years yield (kg/ha/yr)
Genotypes								
E138	48.8	15.9	10.9	31.4	12.3	1618	2582	2178
E90	46.4	17.2	10.7	34.9	12.7	1635	2437	2109
E139	50.2	17.3	10.5	33.8	12.1	1629	2422	2073
E152	47.1	15.8	11.5	33.0	13.1	1610	2062	1876
126	43.0	13.3	9.2	28.7	12.0	1276	2235	1801
149	37.8	17.0	8.7	41.8	10.8	1444	1934	1786
B36	32.3	12.7	9.7	35.6	12.4	1283	2076	1772
197	48.9	17.6	10.7	34.9	11.6	1446	1827	1727
B96	44.9	15.3	10.2	31.4	12.4	1092	2060	1603
E174	51.3	17.0	11.3	31.6	12.7	1284	1833	1589
A129	40.6	17.1	9.3	40.0	11.4	1239	1862	1585
107	37.8	13.4	10.6	32.8	13.0	1230	1617	1460
B191	42.7	15.7	9.2	35.5	11.6	1287	1481	1404
181	35.6	15.3	9.6	40.0	11.5	1236	1556	1404
A101	38.8	16.3	10.2	39.9	12.2	1240	1480	1366
A115	34.9	13.5	11.0	35.6	13.4	1227	1443	1354
B170	35.9	15.4	11.1	38.5	13.7	1194	1123	1148
375	41.8	14.9	9.1	35.0	11.3	857	1063	1015
SED (153df)	2.05	Ns	0.42	1.14	0.36	104.7	157.1	113.2
Locations								
Bechem	22.0	10.6	6.2	39.0	11.0	445	1079	808
Fumso	50.1	18.6	12.4	35.1	13.2	1768	1986	1915
Tafo	54.3	17.5	12.0	31.7	12.5	1757	2451	2152
SED (153df)	0.84	0.36	0.17	0.47	0.15	42.8	64.2	46.2

the lowest average fruit-set, number of fruits per node and yields.

Substantial variability among genotypes and environments was also revealed by the wide range between the minimum and maximum values scored for the traits (Tables 3 and 5). For example, average leaf-scorching scores per tree ranged from 1.06 to 2.92 and span from 190 to 229 cm among the genotypes. Fruit-set also varied from 32.3 to 51.3% and average seven years yield from 1 015 to 2 178 kg/ha among the genotypes. Among the locations, drought reaction scores per tree varied from 1.64 to 2.35 and span from 206 to 217 cm. Fruit-set and average seven years yield also ranged from 22.0 to 54.3% and 808 to 2152 kg/ha, respectively. Genotypes with maximum values for the traits associated inversely with leaf scorching may, therefore, be utilised in the breeding programme for the improvement of Robusta coffee for adaptation to drought-stress.

Locations with minimum and maximum values for the traits associated inversely with leaf-scorching can also be noted as drought-stress and non-stress environments for testing genotypes for specific or broad adaptation to drought-stress.

Genetic correlations

Associations among leaf scorching and the vegetative traits are shown in Table 7. Almost all the vegetative traits studied were inversely correlated with leaf-scorching scores, except number of secondary branches, which was positively associated. Good prediction of leaf-scorching was by span $(r_G = -0.53; P \leq 0.05)$ and number of primary branches per plant. $(r_G = -0.47; P \leq 0.05)$.

Good prediction of average seven years yields was also by span $(r_G = 0.65; P \leq 0.01)$ and number of primary branches per plant $(r_G = 0.56; P \leq 0.05)$. Average seven years yield was also significantly correlated with girth and diameter of primary branches $(r_G = 0.52 - 0.55; P \leq 0.05)$. Span, however, showed highly significant genetic associations with girth, length and diameter of primary branches

Table 7. Genotypic correlation coefficients of leaf-scorching with vegetative traits and yield among 18 Robusta coffee clones in three environments.

S/N	Traits	1	2	3	4	5	6	7	8	9	10
1	Leaf-scorching	-									
2	Girth	-0.39									
3	Span	-0.53*	0.67**								
4	Length/prim. branch	-0.45	0.32	0.78***							
5	Diameter/prim. branch	-0.22	0.70***	0.77***	0.38						
6	No.nodes/prim. branch	-0.17	-0.05	0.35	0.39	0.04					
7	Number of stems/ plant	-0.15	0.36	0.00	-0.11	0.08	-0.41				
8	No. prim. branches/ stem	-0.28	-0.05	0.05	0.17	-0.15	0.28	-0.55*			
9	No. prim. branches/plant	-0.47*	0.35	0.08	0.09	0.01	-0.14	0.67**	0.22		
10	No. Sec. branches/ plant	0.11	-0.32	-0.48*	-0.39	-0.30	-0.43	0.18	-0.12	0.11	
11	Overall seven years yield	-0.64**	0.55*	0.65**	0.49*	0.52*	0.13	0.38	0.10	0.56*	-0.12

* Significant at $P = 0.05$. ** Significant at $P = 0.01$, *** Significant at $P = 0.001$.

Table 8. Genotypic correlations coefficients of leaf-scorching with yield and its components among 18 Robusta coffee clones in three environments.

S/N	Traits	1	2	3	4	5	6	7	8
1	Leaf –scorching	-							
2	Fruit-set	-0.52*							
3	Number of fruits /node	-0.49*	0.62**						
4	Number of fruiting nodes	-0.37	0.39	0.15					
5	Number of flowers/node	0.14	-0.56*	0.28	0.36				
6	Number of flowering nodes	-0.08	-0.05	-0.33	0.81***	-0.35			
7	First three years yield	-0.63**	0.48*	0.45	0.39	-0.13	0.11		
8	Last 4-7 years yield	-0.57**	0.53*	0.22	0.14	-0.45	-0.04	0.76***	
9	Overall seven years yield	-0.64**	0.54*	0.31	0.20	-0.36	-0.03	0.89***	0.98***

* Significant at $P = 0.05$. ** Significant at $P = 0.01$, *** Significant at $P = 0.001$.

$(r_G = 0.67 - 0.78; P \leq 0.01)$, but recorded significantly negative genetic correlation $(r_G = -0.48; P \leq 0.05)$ with number of secondary branches, a trait which was directly associated with leaf-scorching.

Among the reproductive traits (Table 8), good prediction of leaf-scorching was by fruit-set $(r_G = -0.52; P \leq 0.005)$, fruits per node $(r_G = -0.49; P \leq 0.05)$, and yield itself (first three years yield $(r_G = -0.63; P \leq 0.01)$, last 4 to 7 years yield $(r_G = -0.57; P \leq 0.01)$ and overall seven years $(r_G = -0.64 ; P \leq 0.01)$. Strong positive relationships were observed between fruit-set and yields.

DISCUSSION

In coffee, Orozco and Jaramillo (1978) reported that, drought tolerant cultivars showed delayed leaf wilting and shedding compared with drought sensitive ones. DaMatta and Rena (2001) also observed that, leaf senescence is a consequence of drought, in response to low plant water potential, and argued that leaf senescence in coffee is rather not a result of a survival mechanism to limit transpiration, as observed in some plants. Leaf senescence was, therefore, a reliable parameter for scoring the plants for drought tolerance.

The results showed that, plants with wide span recorded relatively lower leaf-scorching. But wide span is genetically strongly associated with girth, length and diameter of primary branches. Plants with wide span, therefore, had strong main stems and generally long and erect primary branches, implying that plants that showed lower leaf-scorching were more vigorous and larger than those that had higher rate of leaf-wilting in response to drought-stress. Such plants were likely to have more stored assimilates or stem reserves than the smaller plants. In general, plants with large span had higher fruit set, suggesting that, the likely higher stem reserves of these plants play a role in fruit-set of coffee under drought-stress. Stem reserves have been shown to play an important role as source of assimilate (translocated to other parts of the plant) during water-deficit stress in many plant studies (Bonnet and Incoll, 1992; Ehdaie and Waines, 1996; Blum et al., 1997; Ehdaie et al., 2006). Span and its associated traits could, therefore, be exploited for adaptation to drought-stress.

The significant negative genetic association of number of primary branches per plant with leaf-scorching also indicate the sensitivity of the primary branches to drought-stress. This is evident by the significantly lower values for number of primary branches recorded at Bechem, which experienced drier conditions during the study, than at Tafo. The reduced number of primary branches in genotypes with higher leaf-scorching scores as well as under drought-stress conditions at Bechem were likely to be a consequence of senescence of the leaves resulting in the death of the primary branches. The consequent smaller leaf area should lead to lower rate of carbon assimilation, which seemed to be more directly responsible for the decreased crop yield. Meinzer et al. (1992) observed that drought-sensitive coffee plants as well as drought-stressed plants, in addition to showing higher leaf-wilting and shedding, also have fewer and considerable smaller leaves than drought-tolerant plants. Barros et al. (1999) and DaMatta (2004a, b) associated die-back of the primary branches with soil and atmospheric water-deficit, high temperature and their combined effects. Rena and DaMatta (2002) reported die-back of the primary branches to be preceded by death of a large proportion of absorbing root following wilting of the leaves caused by water-deficit stress. This observation the authors argued, could affect yields negatively due to reduced assimilate production. Number of primary branches per plant can, therefore, be used to screen genotypes under drought-stress.

The very high negative genetic correlations observed in this study between leaf-scorching and fruit-set and yield (both at the early and late fruiting stages) show that fruit-set and cherry development of coffee plants are the reproductive traits most sensitive to moisture stress. Flowering of Robusta coffee in Ghana during the study period mostly coincided with the beginning of the dry season, in December, and lasted for about three months. In this study, mean total annual rainfall during the study period varied among the locations from 1220 mm at Bechem, 1320 mm at Fumso, to 1480 mm at Tafo. The effect of drought stress on coffee yield may, therefore, not be due to limiting total moisture availability during the growing period but particularly during the fruit-setting periods, or post-flowering drought stress. This stress was more severe at Bechem which recorded mean monthly rainfall of 17.4 mm during the dry season compared to 43.5 mm at Tafo over the same period. Hence, yields were lowest at Bechem and highest at Tafo. The coffee tree sheds a large number of pin-head berries in the first three months after flowering. Barros et al. (1999) attributed the fruit drop during the first month after flowering to fertilisation failure and that of the second to

third months to low carbohydrate supply and water deficit. In this study, up to 78% of flowers failed to develop into fruits at Bechem compared to 46% at Tafo, reflecting the relatively poorer yields recorded at Bechem than at Tafo. The differential ability of the genotypes to mobilise and translocate stem reserves to fruits could have resulted in the differences observed in fruit-set and yield among the genotypes. Wide ranges of physiological (Silva Ramos and Carvalho, 1997; DaMatta et al., 2003; Pinheiro et al., 2005) and biochemical (Lima et al., 2002; Pinheiro et al., 2004; Praxedes et al., 2006) attributes, including high plant water potential, relative water content, deep rooting, maintenance of leaf area, improved tolerance of oxidative stress, and ability for maintaining assimilate export, have been suggested as predictors of drought tolerance in *C. canephora* and *C. arabica.* Some of these attributes could be associated with the agronomic traits that are associated with drought tolerance in this study.

High negative genetic correlations observed between leaf-scorching and yield and the fact that mostly all the traits that confer reduced leaf-scorching also confer higher yields did not only imply that drought is an important environmental stress that determines coffee yields, but that, yield itself can be a criterion for selecting drought-tolerant Robusta coffee plants under stress as discussed for other crops (Mullet and Whitsitt, 1997). The agronomic implications of the association among leaf-scorching scores, early, late and overall seven years yields are that, adaptable genotypes could be selected at the early stages of bearing of the plant as well as from older plants in the later stages of the production cycle. Fruits per node was also inversely associated with leaf-scorching scores, but the comparatively stronger inverse associations between fruit-set and leaf-scorching, and the strong positive association between fruit-set and yields (both at the early and late bearing stages of the plants) clearly show that fruit-set is indeed the most important character, apart from yield, related to reproductive stage drought tolerance in Robusta coffee. From the analyses of variance, significant differences were observed among genotypes in leaf-scorching, span, number of primary branches, yields and fruit-set, implying that selection for good performing genotypes for these characters, which are assessed in a non-destructive manner, will help improve adaptation to drought-stress in this population and in future breeding programmes. Visual scoring of plants using leaf-scorching can only be possible in the presence of moisture stress. Moisture stress may, however, not be present at the juvenile stages of the plant when screening process of genotypes for most traits of agronomic importance may be advantageous for early selection. In this study, however, fruit-set, span and number of primary branches, the traits most related to reproductive and vegetative stage drought tolerance, and early three-year yields were estimated from data recorded on four to five and half-year old coffee plants under near optimum and moisture stress conditions in three diverse environments. They are, therefore, reliable methods of evaluation in near optimum and stress environments. Our results should, therefore, form an important base for a pre-selection index for high yielding and adaptable coffee genotypes in screening germplasm and segregating generations for characters of agronomic importance for improvement of the crop.

Conclusion

The present study revealed inverse genetic correlations between leaf-scorching and some vegetative and reproductive characters and positive associations among the characters. There was also substantial variation among the tested genotypes for these characters. Span and its associated traits, NPB, and FS could be exploited, through indirect selection for superior *C. canephora* genotypes. Genotypes with desirable high or low average values for these characters should be conserved for direct utilisation or for breeding for genetic improvement of the crop for adaptation to water-deficit stress.

ACKNOWLEDGEMENT

This paper, CRIG/09/2011/029/002 is published with the kind permission of the Executive Director of the Cocoa Research Institute of Ghana (CRIG).

REFERENCES

Adu SV, Asiamah RD (1992). Soils of the Ayensu-Densu Basin. Memoir No.9. Kumasi, Ghana: Soil Research Institute-Council for Scientific and Industrial Research. Advent Press, Accra, Ghana.

Ahn PA (1961). Soils of the Lower Tano Basin, South-Western Ghana.Memoir No.2. Soil and Land-use Survey Branch, Kumasi, Ghana. Ministry of Food and Agriculture. Government Printer, Accra, Ghana.

Andrea AL, Tina R Jonathan DGJ (2003). CDPK-mediated signalling pathways: specificity and cross talk. J. Exp. Bot., 395:181-188.

Anim-Kwapong E, Adomako B (2010). Genetic and environmental correlations between bean yield and agronomic traits in *Coffea canephora.* J Plant Breeding Crop Sci., 2(4): 64-72.

Barros RS, Maestri M, Rena AB (1999). Physiology of growth and production of the coffee tree – a review. J. Coffee Res., 27: 1- 54.

Batista-Santos P, Lidon FC, Fortunato A, Leitão AE, Lopes E, Partelli FL, Ribeiro AI, Ramalho JC (2011) The impact of cold on photosynthesis in genotypes of Coffea spp.-Photosystem sensitivity, photoprotective mechanisms and gene expression. Journal of Plant Physiology., 168: 792-806

Bidinger FR, Hammer GL, Muchow RC (1996). The physiological basis of Genotype by Environment Interaction in crop adaptation, In Plant Adaptation and Crop Improvement, edited by M Cooper and Hammer GL, Wallingford, UK: CAB International, 329-347.

Blum A, Golan G, Mayer J, Sinmena B (1997). The effect of dwarfing genes on sorghum grain filling from remobilized stem reserves under stress. Field Crops Res., 52: 43-54.

Bonnett G, Incoll LD (1992). The potential pre-anthesis and post-anthesis contributions of stem internodes to grain yield in crops of winter barley. Ann. Bot., 69: 219-225.

Cilas C, Bar-hen A, Montagnon C, Godon C (2006). Definition of architectural ideotypes for good yield capacity in *Coffea canephora.*

Ann. Bot., 97(3): 405-411.

Cilas C, Bouharmont P, Bar-Hen A (2003). Yield stability in *Coffea canephora* from diallel mating designs monitored for 14 years. Heredity, 91: 528-532.

Cilas C, Bouharmont P, Boccara M, Eskes AB, Baradat PH (1998). Prediction of genetic value for coffee production in *Coffea arabica* from a half-diallel with lines and hybrids. Euphytica, 104: 49-59.

DaMatta FM (2004a). Ecophysiological constraints on the production of shaded and unshaded coffee: a review. Field Crops Res., 86: 99 – 114.

DaMatta FM (2004b). Exploring drought tolerance in coffee: a physiological approach with some insights for plant breeders. Br. J. Plant Physiol., 16 (1): 1- 6.

DaMatta FM (2004c). Fisiologia do cafeeiro em sistemas arborizados. In: Matsumoto SN (ed), Arborização de Cafezais no Brasil, pp.85-118. Edições UEBS, Vitória da Conquista.

DaMatta FM, Chaves ARM, Pinheiro HA, Ducatti C, Loureiro ME (2003). Drought tolerance of two field-grown clones of *Coffea canephora.* Plant Sci., 164: 111-117.

DaMatta FM, Ramalho JDC (2006). Impact of drought and temperature stress on coffee physiology and production: a review. Br. J. Plant Physiol., 18(1): 55-81.

DaMatta FM, Rena AB (2001). Tolerancia do café à seca. In Tecnologias de Producao de Café com Qualidade, edited by Zambolina L, 65 – 100. Universidade Federal de Viçosa, Viçosa, Brazil.

Ehdaie B, Alloush GA, Madone MA, Waines JG (2006). Genotypic variation for stem reserves and mobilization in wheat. Crop Sci., 46: 735- 746.

Ehdaie B, Barnhart B, Waines JW (2001). Inheritance of root and shoot biomass in a bread wheat cross. J. Genet. Breeding, 55 (1): 1-10.

Ehdaie B, Waines JG (1996). The genetic variation for contribution of preanthesis assimilates to grain yield in spring wheat. J. Genet. Breeding, 50 (1): 47-55.

Ehlers JD (1994). Correlation of performance of sole crop and inter crop cowpeas with and without protection from insect pests. Field Crops Res., 36: 133-143.

Falconer DS, Mackay TFC (1996). Introduction to quantitative genetics. 4th ed. London: Longman Inc.

Gutiérrez MV, Meinzer FC (1994). Estimating water use and irrigation requirements of coffee in Hawaii. J. Am. Soc. Hort. Sci., 119: 652-657.

Leroy T, Montagnon C, Cilas C, Charrier A, Eskes AB (1994).Reciprocal recurrent selection applied to *Coffea canephora* Pierre. I. Characterization and evaluation of breeding populations and value of intergroup hybrids. Euphytica, 67: 113-125.

Lima AL, DaMatta FM, Pinheiro HA, Totola MR, Loureiro ME (2002). Photochemical responses and oxidative stress in two clones of *Coffea canephora* under water deficit conditions. Environ. Exp. Bot., 47: 239-247.

Maestri M, DaMatta FM, Regazzi AJ, Barros RS (1995). Accumulation of proline and quaternary ammonium compounds in mature leaves of water stressed coffee plants (*Coffea arabica* and *Coffea. canephora*). J. Hort. Sci., 70: 229-233.

Meinzer FC, Grantz DA, Goldstein G, Saliendra NZ (1990). Leaf water relations and maintenance of gas exchange in coffee cultivars grown in drying soil. Plant physiol., 94: 1781-1787.

Meinzer FC, Saliendra NS, Crisosto CH (1992). Carbon isotope discrimination and gas exchange in *Coffea arabica* during adjustment in different soil moisture regimes. Aust. J. Plant Physiol., 19: 171-184.

MINITAB (1997). Minitab Statistical Software. Release 12, Minitab Inc.,USA.

Mullet JE, Whitsitt MS (1997). Plant cellular responses to water deficit, In Drought Tolerance in Higher Plants: Genetical, Physiological and Molecular Biological Analysis, edited by Belhassen E, 41 – 46. The Netherlands: Kluwer Academic Publishers, Dordrecht.

Omanya GO, Ayiecho PO, Nyabundi JO (1996). Variation for adaptability to dry land conditions in Sorghum. Afr Crop Sci. J., 4 (2): 127-138.

Orozco-C FV, Jaramillo RA (1978). Coffea introductions behavior under conditions of moisture deficit in the soil. Cenicafé, 29: 61-93.

Partelli FL, Vieira HD, Silva MG, Ramalho JC (2010). Seasonal vegetative growth of different age branches of conilon coffee tree. Semina: Ciências Agrárias., 31: 619-626.

Partelli FL, Vieira HD, Viana AP, Batista-Santos P, Rodrigues AP, Leitão AE, Ramalho JC (2009). Low temperature impact on photosynthetic parameters of coffee genotypes. Pesquisa Agropecuária Brasileira, 44: 1404-14159.

Pinheiro HA, DaMatta FM, Chaves ARM, Fontes EPB, Loureiro ME (2004). Drought tolerance in relation to protection against oxidative stress in clones of *Coffea canephora* subjected to long-term drought. Plant Sci., 167: 1307-1314.

Pinheiro HA, DaMatta FM, Chaves ARM, Loureiro ME, Ducatti C (2005). Drought tolerance is associated with rooting depth and stomatal control of water use in clones of *Coffea canephora*. Ann. Bot., 96(1): 101-108.

Praxedes SC, DaMatta FM, Loureiro ME, Ferrão MAG, Cordeiro AT (2006). Effects of long-term soil drought on photosynthesis and carbohydrate metabolism in mature robusta coffee (*Coffea canephora* Pierre var. *kouillou*) leaves. Environ. Exp. Bot., 56(3): 263-273.

Ramos LS, Carvalho A (1977). Shoot and root evaluations on seedlings from *Coffea* genotypes. Bragantia, 56: 59-68.

Rena AB, DaMatta FM (2002). O sistema radicular do cafeeiro: estrutura e ecofisiologia, In O Estado da Arte de Tecnologias na Produção de Café, edited by Zambolin L, 11 – 92. Universidade Federal de Viçosa, Viçosa, Brazil.

Silva Ramos LC, Carvalho da A (1997). Shoot and root evaluations on seedlings from *Coffea* genotypes. Bragantia, 56 (1): 58-68.

Smith AW (1989). Introduction. In: Clarke RJ, Macrae R (eds), Coffee, Vol.1 – Chemistry, pp. 1-41. Elsevier, London.

Steel RGD, Torrie JM, Dickey DA (1997). Principles and Procedures of Statistics: A biometrical approach, 3rd ed. New York: McGraw-Hill Co. Inc.

Stewart JL, Salazar R (1992). A review of measurement options for multipurpose trees. Agrofor. Syst., 19: 173- 183.

Walyaro DJ, Van der Vossen HAM (1979). Early determination of yield potential in Arabica coffee by applying index selection. Euphytica, 28 (2): 465-472.

Zhong-hu H, Rajaram S (1994). Differential responses of bread wheat characters to high temperature. Euphytica, 72: 197-203.

20

Anti-nutritional factors as screening criteria for some diseases resistance in sesame (*Sesamum indicum* L.) genotypes

Mohamed Abd El-Himed Sayed Ahmad El-Bramawy

Agronomy Department, Faculty of Agriculture, Suez Canal Univ., 41522 Ismailia, Egypt. E-mail: el_bramawy71@hotmail.com.

Since sesame genotypes differed significantly in the levels of anti-nutritional factors (phytate, trypsin inhibitor and tannins), these anti-nutritional factors were examined as screening criteria for diseases resistance. Forty eight sesame genotypes from different sources were grown in soil infested by *Fusarium oxysporum* (FOS) and *Macrophomina phaseolina* (MPH) at Experimental Farm, Faculty of Agriculture, Suez Canal University, Egypt through two seasons (2009 and 2010) and approved for the above mentioned aim. The anti-nutritional factors of the tested sesame genotypes were associated significantly with both of infection pathogens percentages and seed yield. Most tested sesame genotypes were gathered in the scale of resistance (R) by scoring 48.90% and moderate resistance (MR) by scoring 40.7% of all the tested genotypes. On the other hand, 9.90% and 0.50% of the tested sesame genotypes were gathered in the scale of moderate susceptible (MS) and susceptible(S), respectively. Sesame genotype groups (resistant, moderately resistant, moderately susceptible and susceptible) were observed differenced significantly in their contents of the anti-nutritional factors. Most of the relationships among sesame genotype groups or among the anti-nutritional factors and among themselves were showed significant correlations in both seasons (2009 and 2010) over the two fungal pathogens (FOS and MPH) together. The resistance (R) class possessed the highest values of the anti-nutritional factors, while the lowest values were related to the susceptible (S) class of sesame group. However, the other two classes, the moderately resistance and moderately susceptible possessed the medium values of phytic acid, trypsin inhibitor, and tannins over an average of the two seasons. Hence, these anti-nutritional factors could have played a clear role in the range and magnitude of resistance or susceptibility to fungal pathogens, *F. oxysporum* and *M. phaseolina* in sesame genotype groups considered in this work.

Key words: Disease resistance, *Fusarium oxysporum*, infection percentage, *Macrophomina phaseolina*, phytate, screening, seed yield, sesame, tannins, trypsin inhibitor.

INTRODUCTION

Sesame is a plant species of *Sesamum indicum* L., and herbaceous annual plant belonging to the Pedaliaceae family (Sugano et al., 1990; Sugano and Akinmoto, 1993). Sesame seed is one of the world's important and oldest oilseed crops known according to some archaeological findings (Nayar and Sastry, 1990; Elleuch et al., 2010). Its cultivation went back to 2130 B.C. (Weiss, 1983). However, its recorded history in Egypt returned to 1300 B.C. (Burkill, 1953). Despite the fact that 100% of the world's sesame area is found in the developing countries (Ashri, 1998), it is common in many countries over the world with different names according to the region of production. For example, in some areas, it is known as Sesamum (China, Mexico, South and Central America), gingelly (South India, Burma),

Abbreviations: FOS, *Fusarium oxysporum*; **MPH,** *Macrophomina phaseolina*.

benniseed (Sierra Leone, Guinea in West Africa), sim-sim (Middle East) and till (East and North Africa via Egypt). China, India, Sudan, Mexico and Burma are the major producers of sesame seeds in the world by contributing to approximately 60% of its total world production (Kemal and Yalçin, 2008). In Egypt, it represents about 0.37% of the total cultivated area (Anonymous, 2010), and approximately 80 to 90% of the sesame crop is used in the production of a popular food (Halawa Tahnia), while the remainder amount of about 10 to 20% is used in the production of Tehineh, the bakeries and confectionery (Serry and Satour, 1981).

However, sesame has been cultivated for centuries, particularly in Asia and Africa, for its high content of edible oil and protein (Salunkhe et al., 1991). It is an important source of food worldwide and constitutes an inexpensive source of protein, fat, minerals and vitamins in the diets of rural populations, especially children (Namiki, 1995; Shahidi et al., 2006). The chemical composition of sesame shows that the seed is an important source of oil (48 to 60%), protein (18 to 23.5%). It was also reported to have carbohydrate (13.5%), ash (5.3%) and moisture (5.2%) (Obiajunwa et al., 2005; Kahyaoglu and Kaya, 2006). Moreover, sesame oil is known as the king of oil seeds due to continuous antioxidative constituents that is, sesamolin, sesamin and sesamol contents, compared to other oil seeds (Brar and Ahuja, 1979).

On the other hand, other chemical components such as anti-nutritional factors such as phytate, trypsin, a-amylase inhibitors, lectin, and tannins are in existence in sesame seeds and can limit their utilization in food system (Moran et al., 1969; Wu and Inglett, 1974). However, these anti-nutritional factors may be related considering disease resistance and since sesame plants are susceptible to fungal diseases, they are liable to attack at least by eight economically important fungal diseases (Kolte, 1985). The most serious fungal diseases, which infect sesame plants, are Fusarium wilt disease, caused by Fusarium oxysporum f. sp. sesami (FOS) and charcoal root rot disease caused by Macrophomina phaseolina (MPH). Since sesame plants are so affective with both diseases, it differed greatly in their response and had heavy losses in the yield crop. F. oxysporum (FOS) pathogen causes vascular wilt disease and sudden death of sesame plants. Therefore, these soil-borne pathogens may cause heavy yield losses in sesame ranging from 50 to 100% and survive in soil for several years (Gaber et al., 1998; Khaleifa, 2003; El-Bramawy, 2006a, b; Bayoumi and El-Bramawy, 2007; El-Bramawy and Shaban, 2007). While, M. phaseolina (MPH) infects sesame plants at all stages and damage can result in poor seedling establishment, pre and post emergence damping off and reduced vigor and productivity of older plants. Major symptoms on infected plants are stunting, chlorosis, premature defoliation, early maturity and death (Abawi and Corrales, 1989; El-Bramawy and Abd Al-Wahid, 2007; El-Bramawy

and Shaban, 2007; El-Bramawy, 2008). However, the control of the main diseases that infect sesame in Egypt is a problem and is difficult to overcome completely, because most of them are caused by ground pathogenic inhabitants, which require a special treatment for their eradication or diminution. Application of some agronomic practices and improvement of diseases resistance in current genotypes have been proposed as the most effective strategies of solving the disease problems in sesame. Therefore, the improvement of current sesame genotypes to be more diseases resistant is an alternative option to solving disease problems in sesame (Zahra, 1990; Ziedan, 1993; Zhang et al., 2001; Ragab et al., 2002; El-Fiki et al., 2004; El-Bramawy, 2006, 2008; El-Shahkess and Khaleifa, 2007).

Moreover, there is a wide diversity in agronomic traits, namely: plant height, branches number, capsules number, capsule length, length of fruiting zone, maturity date, seed index, seed yield, oil content and seed color among sesame species and genotypes within a species (Li et al., 1991). Seed chemical components such as phytate, trypsin, a-amylase inhibitors, lectin, and tannins are also assorted among sesame accessions and germoplasm within accessions (Chang et al., 2002; Kanu et al., 2009). This biodiversity in agronomic traits or/and chemical components characteristic could be a valuable tool for screening and breeding a new germplasm for higher diseases resistance. El-Bramawy et al. (2009a) found that maturity date, branches number per plant and seed color traits may successfully be used to predict the resistance of sesame genotypes to Fusarium wilt (caused by F. oxysporum) and charcoal rot (caused by M. phaseolina) diseases without conducting tedious crop experiments. In the same connect, but with other field crops, Jalali and Chand (1992) observed that most practical and cost-efficient method for management of Fusarium wilt of chickpea is the use of resistant cultivars. Also, Gupta and Sharma (2006) reported that evaluation and exploitation of wild accessions for different morphological and agronomical traits will help in tapping the unexplored variability in cultivated lentil. Introgression of related wild lens taxa into the cultivated lentil will help in the flow of useful genes into the cultivated lentil and thus including these wild species in the common gene pool (Ladizinsky, 1993).

Numerous studies have shown positive and negative correlations between agronomic traits and disease resistance. For example, Dubin et al. (1998) found that heading date and plant height in wheat showed a negative significant correlation with area under disease progress curve (AUDPC). Duveiller et al. (1998 a, b) also found that plant height is expected to influence spot blotch severity in wheat. Through their studies for the relationship of plant height and days to maturity in homozygous wheat lines with resistance to spot blotch, Joshi et al. (2002) reported that resistance to spot blotch severity was independent of plant height and days to maturity in progenies from wheat crosses. These lines showed a significant difference for plant height and days to maturity, but not for AUDPC values of spot blotch disease. Several reports have indicated that seed colors in various crops are associated with diseases resistance. It has also been reported that polyphenolics and tannins are subsets involved in seed color expression and are often associated with plant resistance to pathogens or insects (Li et al., 1991; Pastor-Corrales et al., 1998; Islam et al., 2003). In general, more total phenols in common bean plants have been linked with greater resistance to root-rot disease (Statler, 1970). Also, tannin is beneficial in the field because its presence provides resistance to fungi. Therefore, it is important to develop an effective evaluation approach for screening disease resistant genotypes that would be reliable, quick, easy, practical and economical. Whereas, screening under natural field conditions is not feasible due to high degree of heterogeneity as well as fungal competition (Harris and Burns, 1973). Therefore, the main aim of the study was assessment the validity of the anti-nutritional factors (phytate, trypsin inhibitor, tannins) as screening criteria of 48 sesame germplasm for resistance to fungal pathogens (FOS and MPH). Sup-objective was to investigate the predictive screening parameters by judging the relationships among these anti-nutritional factors of sesame seeds in relation to the diseases resistance levels.

MATERIALS AND METHODS

Plant materials

Forty eight sesame genotypes were used as plant materials in this work. They include two mutants (these two mutants achieved by Nuclear Research Center, Atomic Energy Authority in Egypt), eight local cultivars, three introduced varieties, fifteen promising lines originated from hybridization and selections through a breeding program in the Agronomy Department, and twenty land races collected from Egypt district over Egypt Governorates. The names, origin, pedigree and sources of these sesame genotypes, which were considered in this work are listed in Table 1.

Field treatments

At the Experimental Plant Breeding Farm, Faculty of Agriculture, Suez Canal University, Egypt, field treatments were established in two growing seasons (2009 and 2010). These field treatments were screened for assessment of the resistance levels of sesame germplasm with their seed yield potential against isolate of the two fungal pathogens [F. oxysporum (FOS) and M. phaseolina (MPH)].

Soil experimental analysis and characteristics

The experimental soil was classified as a sandy soil (94.46% sand, 2.50% silt and 3.04 clay, in season 2008 and 94.66% sand, 2.54% silt and 3.24 clay in season, 2009) according to the method outlined by Kilmer and Alexander (1949). This soil possessed pH values of 7.62 and 7.56, contained 3.14 and 3.45 ppm available N, 1.81 and

1.85 ppm available P, 11.64 and 11.92 ppm available K, and 0.049 and 0.058% organic matter in the two growing seasons, respectively.

Experimental design and inoculation of soil

A randomized complete block design (RCBD) with three replicates was used in this work. Two field experiments were conducted at Experimental Farm, Faculty of Agriculture, on infected field by the two fungal pathogens [F. oxysporum (FOS) and M. phaseolina (MPH)] during 2008 and 2009 growing seasons. Each sesame entry was planted on a plot containing two ridges that were 60 cm apart and 4 m in length (4.8 m^2). The seeds were planted on the upper third of the ridges in hills with 10 cm between hills. Sowing was done in wilt-sick and charcoal rot-sick plot in soil known to have a high inoculums density of both pathogenic fungi from a long-term sesame research field that was naturally infested with FOS and MPH (El-Bramawy, 1997, 2003, 2006 and 2008; El-Bramawy and Shaban, 2007; El-Bramawy and Abd Al-Wahid, 2009; El-Shazly et al., 1999; Bayoumi and El-Bramawy, 2007; El-Bramawy et al., 2009a).

Measurements of infection percentage

The sesame genotypes were examined weekly and noted for any wilt and charcoal root rot symptoms for 10 weeks of sowing. The infected sesame plants by Fusarium wilt were characterized by the internal vesicular discoloration, appearance of wilt on plants and might die and fell down (considered wilted). However, the charcoal root rot infection was expressed as root discoloration, black stem rot and pronounced reduction in root system of the infected plants (Smith and Carvil, 1997). Re-isolation of the wilt (F. oxysporum) and charcoal root rot (M. phaseolina) pathogens from the diseased plant were then done and compared with the original isolates to assure the existence of a relation between inoculation pathogen and the development of diseases.

Scoring of resistance levels for sesame genotypes

To evaluate the resistance levels of sesame genotypes to both fungal pathogens, 1- 6 scales of Kavak and Boydak (2006) were formed based on the infection percentage of each sesame genotype by each pathogen. This Scale presented in Table 2 was based on twenty plants selected randomly per plot. The wilted and rotted sesame plants were counted and calculated as percentage of infected plants, then transformed to Arcsine values and prepared for statistical analysis.

Determination of Anti-nutritional factors

The anti-nutritional factors, that is, phytic acid, and trypsin inhibitor, and tannins were determined in Laboratory of Food Technology Department. These estimates were done on four sesame groups [Resistant (R), Moderately Resistant (MR), Moderately Susceptible (MS) and Susceptible (S)] depending on their behavior towards the fungal pathogens (FOS and MPH). Five gram seeds of the different four groups among all sesame genotypes were used as samples for determining the anti-nutritional factors as follows:

Phytic acid: Phytic acid was determined according to the method described by Latta and Eskin (1980), and later modified by Vaintraub and Lapteva (1988). One gram of dried sample was

Table 1. Serial number, names, origin and pedigree of 148 sesame genotypes used in the study.

S/N	Genotype name	Origin	Sources	S/N	Genotype name	Origin	Sources
1	Mutants 48	Egypt	Giza 24 D 20 M 3 R-10	25	S 2	Egypt	Selection from local lines under breeding program
2	Mutants 8	Egypt	Giza 24 D 20 M 6 R -12	26	S 3	Egypt	Selection from local lines under breeding program
3	Toshka 1	Egypt	M2 A1 B11 (CAN 114 x Type 29) x NA413	27	S 4	Egypt	Selection from local lines under breeding program
4	Toshka 2	Egypt	M2 A2 B11 (CAN 114 x Type 29) x NA413	28	S 5	Egypt	Selection from local lines under breeding program
5	Toshka 3	Egypt	M2 A3 B11 (CAN 114 x Type 29) x NA413	29	L.R1	Egypt	Land race collected from Ismailia Governorate
6	Taka 1	Egypt	Not available	30	L.R2	Egypt	Land race collected from El-Sharkia Governorate
7	Taka 2	Egypt	Not available	31	L.R3	Egypt	Land race collected from El-Mina Governorate
8	Taka 3	Egypt	Not available	32	L.R4	Egypt	Land race collected from El-Mina Governorate
9	Giza 25	Egypt	Giza white x Type 9	33	L.R56	Egypt	Land race collected from Assuite Governorate
10	Giza 32	Egypt	B 32 (CAN 114 x Type29)	34	L.R6	Egypt	Land race collected from Assuite Governorate
11	U.N. A 130	U.S.A	Unknown	35	L.R7	Egypt	Land race collected from Sohag Governorate
12	G N. A 574	Greece	Unknown	36	L.R8	Egypt	Land race collected from Ismailia Governorate
13	K N. A 592	Korea	Unknown	37	L.R9	Egypt	Land race collected from El-Fayum Governorate
14	H 1	Egypt	Ismailia line 10 x Neu H.B	38	L.R10	Egypt	Land race collected from El-Sharkia Governorate
15	H 2	Egypt	Oro x Local line 274	39	L.R11	Egypt	Land race collected from El-Sharkia Governorate
16	H 3	Egypt	U C. R 10 x Giza 32	40	L.R12	Egypt	Land race collected from Ismailia Governorate
17	H 4	Egypt	U. C. R 11 x Giza 25	41	L.R13	Egypt	Land race collected from El-Fayum Governorate
18	H 5	Egypt	Ismailia line 8 x Ismailia line 20	42	L.R14	Egypt	Land race collected from El-Fayum Governorate
19	H 6	Egypt	A cross between two local lines	43	L.R15	Egypt	Land race collected from Ismailia Governorate
20	H 7	Egypt	A cross between two local lines	44	L.R16	Egypt	Land race collected from Bani sueif Governorate
21	H 8	Egypt	A cross between two local lines	45	L.R17	Egypt	Land race collected from Bani sueif Governorate
22	H 9	Egypt	A cross between two local lines	46	L.R18	Egypt	Land race collected from Wady El-Gaded Govenorate
23	H 10	Egypt	A cross between two local lines	47	L.R19	Egypt	Land race collected from Wady El-Gaded Govenorate
24	S 1	Egypt	Selection from local lines under breeding program	48	L.R20	Egypt	Land race collected from El-Sharkia Governorate

Table 2. The scale used in this investigation.

Scale number	Infection percentage (%)	Category
1	0.00	Immune (I)
2	0.1 – 20	Resistant (R)
3	20.1 – 40	Moderately resistant (MR)
4	40.1 – 60	Moderately susceptible (MS)
5	60.1 – 80	Susceptible (S)
6	80.1 – 100	Highly susceptible (HS)

extracted with 50 mL of 2.4% HCl for 1 h at an ambient temperature and centrifuged (3000 xg/30 min). The clear supernatant was used for the phytate estimation. One milliliter of Wade reagent (0.03% solution of $FeC13.6H2O$ containing 0.3% sulfosalicylic acid in water) was added to 3mL of the sample solution and the mixture was centrifuged. The absorbance at 500nm was measured using spectrophotometer. The phytate concentration was calculated from the difference between the absorbance of the control (3 mL of water+1 mL Wade reagent) and that of assayed sample. The concentration of phytate was calculated using phytic acid standard curve and results were expressed as phytic acids in g per 100 g dry weight.

Trypsin inhibitor: Trypsin inhibitor activities were determined using the procedure of Kakade et al. (1974). One gram of defatted seed flour was mixed with 100 ml of 0.009 HCl. The mixture was shaken at an ambient temperature for 2 h and centrifuged (10000xg/ 20 min) and the supernatant was used for inhibitor estimation. The diluted extract (1ml) was incubated with 1 ml of trypsin solution at 37°C for 10 min. A 2.5 ml of prewarmed substrate (BAPNA) was added and after exactly 10 min at 37°C the reaction was stopped with 0.5 ml of acetic acid (30%, v/v). The absorbance was measured at 410 nm against a blank using a spectrophotometer.

Tannins: Total tannins were determined as described in the A.O.A.C. (1990).

Statistical analysis

Data obtained were subjected to statistical analysis using analysis of variance for randomized complete block design, with sesame genotypes as a whole plot and replicates as blocks. Statistical analysis was done for sesame genotypes with their pathogens as well as anti-nutritional factors were estimated using CoStat Version 6.311 (CoHort software, Berkeley, CA 94701). Treatments' means were compared using Duncan's Multiple Range Test (Steel and Torrie, 1980). Probability levels lower than 0.05 or 0.01 were held to be significant.

The relationships between the four sesame crop [resistant (R), moderately resistant (MR), moderately susceptible (MS) and susceptible (S)] due to the interaction with two fungal pathogens (F. oxysporum and M. phaseolina) and the anti-nutritional factors (phytic acid, a-amylase inhibitor and trypsin inhibitor, and tannins were analyzed by regressions. The best equations to fit the relations were chosen by regression procedures with selection of forward, backward and stepwise methods. The relationships were analyzed comparing the slops of the linear regressions using a covariance analysis (Antunez et al., 2001).

RESULTS

The infection percentage and seed yield (kg/ha) of sesame genotypes infected with both fungal pathogens F. oxysporum (FOS) and M. phaseolina (MHP) in the two growing seasons (2009 and 2010) are presented in Tables 3 and 4, respectively. Highly significant variations were observed among the evaluated sesame genotypes based on the infection percentages and seed yield at infection by F. oxysporum (Tables 3) and M. phaseolina (Table 4). The infection percentages recorded for both pathogens varied among sesame genotypes from 1.7% (S_2) with rank of resistance (R) to 61.6% (H_4) with rank of susceptible (S) for FOS (Table 3) and from 2.2% (H_9)

with rank of resistance (R) to 54.2% (LR_{17}) with rank of moderately susceptibility (MS) for MPH (Table 4) in the first season (2009). In the second season (2010) however, it ranged from 1.4% (H_1) as resistance to 54.2% as a moderately susceptibility (FOS, Table 3) and from 3.7% (Toshka 2) as resistance to 55.1% as moderately susceptibility (MPH, Table 3). The seed yield (kg/ha) ranged from 416.8 to 886.4 kg ha^{-1} (Table 3) and 443.2 to 917.5 kg ha^{-1} (Table 4), in 2009 under infection of F. oxysporum (FOS) and M. phaseolina (MHP), respectively, while in 2010, the values obtained ranged from 374.4 to 873.1 kg ha^{-1} for FOS and 409.5 to 884.8 kg ha^{-1} for MHP in Tables 3 and 4, respectively.

The most evaluated sesame genotypes were gathered in the scale of resistance (R) and moderate resistance (MR), while neither immune (I) genotypes nor highly susceptibility (HS) genotypes among all tested sesame genotypes were observed for both fungal pathogens (F. oxysporum and M. phaseolina) in the two seasons (2009 and 2010). On averaged over the two seasons, 49.0 and 38% at the infection of F. oxysporum (FOS) and 49.0 and 44% at the infection of M. phaseolina (MPH) were gathered in the scale of resistance (R) and moderate resistance (MR), respectively. However, only 13 and 7% of the tested sesame genotypes (averaged over the two seasons) at infection with F. oxysporum and M. phaseolina, respectively were grouped in the scale of moderately susceptible (MS) and susceptible (S). Among all tested sesame genotypes only one genotype tagged no.17 (U C. R 11 x Giza 25) possessed the rank of susceptible (S) with infection rate (61.00) by F. oxysporum only in the first season (2009), while this changed to the rank of moderately susceptible (MS) by infection rate of 44.00 in the second season (2010). On the other hand, the same genotype (U C. R 11 x Giza 25) under M. phaseolina infection, possessed the rank of moderately susceptible (MS) with infection rate of 32.50 and 21.40 in seasons 2009 and 2010, respectively (Table 4). It was noted also that the sesame genotypes such as Toshka 2, U N. A 130, H1, H 8, H9, S1, S 2 and S5 had the resistance (R) rank under infection by F. oxysporum or/and M. phaseolina through both seasons (2009 and 2010).

For this instance, it is very interesting to report that, some sesame genotypes such as Toshka 2, H1, H9 and L.R15 kept their resistance characteristic over the two fungal pathogens. Other genotypes possessed the resistance (HR) or moderately resistance rank (MR) to one or two diseases/ fungal pathogens and changed their behavior to moderately susceptible (MS) or susceptible (S) with the other pathogen such as H 4 and H 10 for F. oxysporum (FOS) and L.R7 for M. phaseolina (MPH). However, it should be taken into account that by contrast, some genotypes lost their susceptible (S) or moderately susceptible (MS) rank with pathogen to be more resistance (R) or, moderately resistance (MR) to other fungal pathogen.

By viewing to the data obtained, it can conclude that fascinatingly, 68, 57 and 44% in season 2009 and 78, 73

Table 3. Infection percentage and seed yield (kg/ha) of 48 sesame genotypes studied for the fungal pathogens (*F. oxysporum*) in 2009 and 2010.

	Fusarium oxysporum f. sp. *sesami*, FOS					
	Season, 2009			Season, 2010		
S/N	Infection (%)	Rank	Yield (kg/ha)	Infection (%)	Rank	Yield (kg/ha)
1	25.40	MR	665.3	9.60	R	817.6
2	16.00	R	763.8	12.90	R	817.1
3	21.70	MR	717.7	16.20	R	756.5
4	19.20	R	606.8	20.00	R	814.1
5	17.10	R	737.1	23.30	MR	680.0
6	12.10	R	787.4	13.80	R	861.8
7	3.40	R	886.4	3.20	R	575.6
8	58.90	MS	598.3	47.70	MS	661.8
9	27.10	MR	635.1	14.20	R	784.4
10	42.70	MS	512.9	19.80	R	727.5
11	31.10	MR	619.4	28.00	MR	678.4
12	13.10	R	812.9	6.80	R	875.6
13	18.50	R	709.7	9.30	R	530.5
14	9.20	R	820.2	1.40	R	608.4
15	20.6	MR	774.3	14.90	R	797.7
16	15.80	R	773.1	14.90	R	783.9
17	61.60	S	416.8	44.00	MS	519.5
18	27.70	MR	696.4	12.80	R	774.1
19	1.80	R	855.0	2.30	R	830.1
20	23.40	MR	683.3	10.80	R	822.5
21	18.80	R	775.2	6.70	R	561.6
22	3.90	R	831.4	7.90	R	810.8
23	41.50	MS	599.0	23.60	MR	374.4
24	7.50	R	869.3	9.40	R	886.6
LSD (0.05)						

	Fusarium oxysporum f. sp. *sesami*, FOS					
	Season, 2009			Season, 2010		
S/N	Infection (%)	Rank	Yield (kg/ha)	Infection (%)	Rank	Yield (kg/ha)
25	1.70	R	873.1	2.70	R	896.7
26	16.70	R	802.2	20.50	MR	710.9
27	20.00	R	717.9	22.70	MR	694.3
28	6.90	R	865.3	10.10	R	815.3
29	15.10	R	749.5	21.10	MR	801.0
30	25.50	MR	690.5	33.50	MR	737.3
31	30.30	MR	679.1	32.30	MR	679.1
32	30.20	MR	678.8	42.20	MS	636.7
33	19.00	R	749.5	21.00	MR	702.7
34	20.20	MR	751.6	26.20	MR	845.2
35	50.20	MS	433.1	54.20	MS	456.5
36	45.60	MS	515.3	38.90	MR	506.4
37	22.90	MR	678.8	16.90	R	721.0
38	38.20	MR	644.0	30.20	MR	620.6
39	40.70	MS	492.1	48.70	MS	450.0
40	15.80	R	854.3	23.80	MR	821.6
41	16.00	R	751.8	20.00	R	798.6
42	30.10	MR	655.7	24.10	MR	697.8
43	15.10	R	819.7	11.10	R	838.4
44	21.20	MR	797.0	17.20	R	801.7
45	20.10	MR	702.2	20.10	MR	800.5
46	25.10	MR	609.6	31.10	MR	647.0
47	22.20	MR	702.7	18.20	R	707.4
48	45.30	MS	643.7	37.30	MR	671.8
LSD (0.05)	2.7		59.90	2.6		31.82

R = Resistance, MR= moderately resistance, MS = moderately susceptibility, S =susceptible and HS = highly susceptibility.

and 65% in season 2010 of resistance genotypes for FOS and 68, 78 and 61% in 2009 and 70, 76 and 56% in 2010 of resistance genotypes for MPH had a high of anti-nutritional factors, and are medium anti-nutritional factors and having low anti-nutritional factors, respectively. However, 22, 17 and 11% in 2009 and 14, 10 and 8% in 2010 of resistance genotypes to FOS and 25, 13 and 8 % in 2009 and 17, 14 and 5% in 2010 of resistance genotypes to MPH had high seed yield kg/ha, medium seed yield kg/ha and low seed yield kg/ha, respectively.

Analysis of variance was performed for the anti-nutritional factors (phytic acid, trypsin inhibitor and tannins) among the four sesame groups (resistant,

Table 4. Infection percentage and seed yield (kg/ha) of 48 sesame genotypes studied for the fungal pathogens (*M. phaseolina*) in 2009 and 2010.

Macrophomina phaseolina, MPH

S/N	Season 2009 Infection (%)	Season 2009 Rank	Season 2009 Yield (kg/ha)	Season 2010 Infection (%)	Season 2010 Rank	Season 2010 Yield (kg/ha)
1	20.40	MR	691.2	14.40	R	837.0
2	26.20	MR	686.8	16.50	R	762.4
3	24.20	MR	692.4	16.30	R	803.3
4	18.00	R	747.2	3.70	R	890.4
5	15.30	R	738.7	3.70	R	892.5
6	15.60	R	761.2	16.10	R	781.1
7	9.00	R	824.6	7.90	R	444.6
8	26.10	MR	777.8	20.10	MR	825.6
9	30.10	MR	463.6	11.60	R	848.7
10	23.50	MR	758.6	20.10	MR	698.7
11	14.20	R	631.1	14.20	R	814.8
12	33.40	MR	524.2	25.60	MR	662.9
13	31.90	MR	565.1	13.90	R	670.9
14	10.80	R	794.0	7.10	R	526.5
15	30.40	MR	649.4	17.60	R	790.0
16	13.20	R	819.2	21.70	MR	698.0
17	32.50	MR	648.4	21.40	MR	727.0
18	33.10	MR	624.8	17.90	R	764.7
19	36.80	MR	502.2	23.70	MR	636.2
20	19.30	R	711.6	9.80	R	819.9
21	19.20	R	724.5	9.80	R	518.8
22	2.20	R	892.9	5.80	R	834.2
23	52.20	MS	443.2	42.30	MS	409.5
24	20.00	R	703.2	6.10	R	856.7
25	4.10	R	887.3	3.80	R	853.2
26	2.60	R	917.5	4.20	R	893.4
27	16.20	R	770.6	21.80	MR	702.0
28	10.30	R	830.2	10.00	R	832.6
29	16.10	R	749.0	22.10	MR	795.8
30	16.10	R	702.2	13.80	R	744.4
31	11.00	R	795.8	17.00	R	884.8
32	20.10	MR	714.2	20.10	MR	686.1
33	20.10	MR	772.9	30.10	MR	730.8
34	20.30	MR	796.1	22.30	MR	772.7
35	45.10	MS	562.1	55.10	MS	604.2
36	20.30	MR	669.0	22.30	MR	734.5
37	25.10	MR	690.8	31.10	MR	714.2
38	23.30	MR	702.2	19.30	R	833.3
39	32.20	MR	749.5	30.20	MR	702.7
40	32.10	MR	778.3	26.10	MR	764.2
41	25.30	MR	690.5	29.30	MR	704.6
42	16.50	R	808.5	20.50	MR	827.2
43	25.20	MR	714.2	33.20	MR	681.4
44	18.10	R	807.5	20.10	MR	830.9
45	19.20	R	750.2	19.20	R	754.9
46	17.00	R	797.7	21.00	MR	839.8
47	23.40	MR	807.8	19.40	MR	784.4
48	40.20	MS	632.3	48.20	MS	672.0
LSD (0.05)	2.1		41.42	2.3		28.08

R = Resistance, MR= moderately resistance, MS = moderately susceptibility, S =susceptible and HS = Highly susceptibility.

moderately resistant, moderately susceptible and susceptible). According to the results, highly significant differences were noted among the sesame groups for phytic acid, a-amylase inhibitor and trypsin inhibitor, lectin, and tannins (Table 5),

indicating high variability among genotypes. Mean values of phytic acid, trypsin inhibitor, and tannins in the four sesame genotype groups are presented in Table 6. Significant differences were observed among the resistant, moderately

resistant, moderately susceptible and susceptible sesame genotype groups for their anti-nutritional factors. Moreover, the phytic acid, trypsin inhibitor and tannins as a anti-nutritional there were a significant variations among them within each of

Table 5. Mean squares of phytic acid, trypsin inhibitor and tannins in four sesame genotype groups through average of the two seasons (2009/2010).

S.O.V.	Degree of freedom	Anti-nutritional factors		
		Phytic acid	Trypsin inhibitor	Tannins
Replication	2	n.s	n.s	n.s
Sesame genotype groups	3	6.18 **	19.78**	491.01**
Error	9	1.69	4.01	25.98

*Significant at $P<0.05$, **Significant at $P<0.01$, NS= Non-significant.

Table 6. Mean values of phytic acid, and trypsin inhibitor and tannins in four sesame genotype groups through average of the two seasons (2009/2010).

Sesame genotype groups	Anti-nutritional factors		
	Phytic acid (%)	Trypsin inhibitor (TIU/ ml)	Tannins (mg/100 g)
Resistant (R)	6.07	12.79	413.67
Moderately Resistant (MR)	5.02	9.13	381
Moderately Susceptible (MS)	3.14	5.27	334
Susceptible (S)	2.94	4.40	218.67
L.S.D., 0.05	1.11	3.09	30.67

sesame group, whereas the phytic acid content was differed significantly among the all sesame groups with high value (6.07) in seed of the resistance (R) sesame crop, but decreased to score latest one (2.94) in seed of susceptible (S) group.

Trypsin inhibitor showed significant differences through all sesame groups, where the two sesame groups (resistant and moderately resistant) differed significantly among themselves as well as with the other two groups (moderately susceptible and susceptible). However, these two sesame groups moderately susceptible (MS) and susceptible (S) did not differ significantly from each other. These four sesame groups possessed 12.79, 9.13, 5.27, and 4.40 as trypsin inhibitor values, respectively (Table 6).

Tannins content in resistant sesame group (R) was the highest (413.67) and differed significantly from those of moderately resistant (381), moderately susceptible (334), and susceptible (218.67) groups. However, both sesame groups (moderately resistant and moderately susceptible) did not show any significant difference from each other, but differed significantly from the susceptible group. Therefore, it was apparent that the resistant (R) class possessed the highest values of the anti-nutritional factors, while the lowest values were related to the susceptible (S) class of sesame group. On the other hand, the two classes of the moderately resistant and moderately susceptible sesame genotypes possessed the medium values of phytic acid, trypsin inhibitor and tannins over average of the two seasons (Table 6).

In order to evaluate and select the predictive and reliable chemical or physiological traits that can be used as screening criteria for resistance to Fusarium wilt

(F. oxysporum) and charcoal root rot (M. phaseolina) diseases among the sesame genotype groups, an objective measure based on the relationship among phytic acid, trypsin inhibitor and tannins as well as the four groups of sesame genotypes (resistant, moderately resistant, moderately susceptible and susceptible) and seed yield were adopted. These relationships were explored according to the class resistance or susceptibility of sesame genotypes containing anti-nutritional factors and their seed yield through 2009 and 2010 seasons (Table 7). If the coefficient of determination (R^2) is significant, these traits may reflect the degree of genotypic variation of sesame genotype groups. The significant value of the determination coefficient (R^2) may indicate to be a useful criteria for evaluation and assessment the content of phytic acid, trypsin inhibitor, tannins in sesame seeds genotypes with their relationship of the resistance or susceptible rank to the Fusarium wilt (F. oxysporum) disease and charcoal root rot disease (M. phaseolina). A quadratic regression equation based on stepwise analysis best fit the relationship. In general, most of these relationships showed significant correlations in both seasons (2009 and 2010) for the two fungal pathogens (FOS and MPH) together (Table 7).

A linear regression, through a covariance analysis, of sesame groups under both fungal pathogens together was used in studying the relationship between the anti-nutritional factors and the four sesame groups infected by both pathogens (FOS and MPH). Also, seed yield was considered in this relationship studied. Anti-nutritional factors were considered as the dependent variables (Y), with the four sesame genotype groups as independent variables (X).

Table 7. Regression equations and correlation coefficients between phytic acid, trypsin inhibitor and tannins sesame genotypes group through the seasons 2009 and 2010.

Sesame genotypes group	Anti-nutritional factors	
	2009	2010
	Regression equation (R^2)	Regression equation (R^2)
Phytic acid		
Resistant (R)	0.62**	0.57**
Moderately resistant (MR)	0.11^{ns}	0.04^{ns}
Moderately susceptible (MS)	0.31**	0.29**
Susceptible (MS)	0.19*	0.12*
Trypsin inhibitor		
Resistant (R)	0.22*	0.20*
Moderately resistant (MR)	0.21*	0.01^{ns}
Moderately susceptible (MS)	0.45**	0.39**
Susceptible (MS)	0.14*	0.16*
Tannins		
Resistant (R)	0.52**	0.60**
Moderately resistant (MR)	0.25 *	0.21*
Moderately susceptible (MS)	0.61**	0.55**
Susceptible (MS)	0.42**	0.34**

For sesame groups, all sesame genotypes (independent variables) were classified into four groups *via* resistant, moderately resistant, moderately susceptible and susceptible according their interaction with FOS and MPH, while the anti-nutritional factors (dependent variables) *via* phytic acid, trypsin inhibitor, and tannins. These two variables (independent and dependent), represented by 24 equivalent equations and fitted as they followed:

Phytic acid, (Equations 1 - 8)

Results of the equations (Equations 1, 2, 3, 4, 5, 6, 7 and 8), indicated that the relationship between sesame groups (resistant, moderately resistant, moderately susceptible and susceptible) and the phytic acid, which were analyzed simultaneously, showed significant correlation in all groups, except moderately resistant group (Table 7). However, the sesame genotypes that had high phytic acid were the only covariate which is significantly correlated with the resistance (R) rank at low infection by FOS and MPH as well as high potential seed yield (kg/ha) in both seasons (Equations 1 and 2), while low phytic acid showed correlation with the susceptible group and seed yield (Equations 7 and 8). The moderately resistant (Equations 3 and 4) and moderately susceptible (Equations 5 and 6) sesame groups were medium in their correlations with phytic acid. The results in the first season (2009) showed similar trend with the

results in the second season (2010).

$$Y = 34.6 - 11.32X_1 + 3.29X_2$$
$$\underset{(t=-1.91)^{0.05}}{} \quad \underset{(t=-0.72)^{0.54}}{} \quad \text{(2009)}$$
$$(1)$$

$$Y = 31.16 - 10.33X_1 - 2.91X_2$$
$$\underset{(t=-1.93)^{0.05}}{} \quad \underset{(t=-0.52)^{0.54}}{} \quad \text{(2010)}$$
$$(2)$$

Where; Y, X_1 and X_2 are, the phytic acid, the resistant (R) sesame group and seed yield (kg/ha), respectively.

$$Y = 21.6 - 10.25X_1 - 1.99X_2$$
$$\underset{(t=-1.73)^{0.05}}{} \quad \underset{(t=-0.52)^{0.54}}{} \quad \text{(2009)}$$
$$(3)$$

$$Y = 19.16 - 9.13X_1 - 1.79X_2$$
$$\underset{(t=-1.90)^{0.25}}{} \quad \underset{(t=-0.62)^{0.50}}{} \quad \text{(2010)}$$
$$(4)$$

Where; Y, X_1 and X_2 are, phytic acid, the moderately resistant (MR) sesame group seed yield (kg/ha), respectively.

$$Y = 23.12 - 0.94X_1 + 3.04X_2$$
$$\underset{(t=-0.21)^{0.84}}{} \quad \underset{(t=0.52)^{0.61}}{} \quad \text{(2009)}$$
$$(5)$$

$$Y = 22.71 - 1.32X_1 - 2.11X_2$$
$$(t=-1.7)^{0.10} \quad (t=0.41)^{0.69}$$

(2010)
(6)

Where; Y, X_1 and X_2 are, the phytic acid, the moderately susceptible (MS) sesame group and seed yield (kg/ha), respectively.

$$Y = 18.55 - 10.20X_1 - 1.91X_2$$
$$(t=-1.03)^{0.05} \quad (t=-0.72)^{0.51}$$

(2009)
(7)

$$Y = 15.63 - 8.39X_1 - 1.82X_2$$
$$(t=-1.03)^{0.04} \quad (t=-0.22)^{0.14}$$

(2010)
(8)

Where; Y, X_1 and X_2 are, the phytic acid, the susceptible (S) sesame group and seed yield (kg/ha), respectively.

Trypsin inhibitor, (Equations 9 - 16)

In respect of trypsin inhibitor, R^2 ranged from highly significant to significant values in all cases in both seasons, except moderately resistant (MR) group only in the second season (Table 7). However, Equations from 17 to 24, inferred that the values of dependent variable (trypsin) increase by increasing that of in dependent variables gradually from susceptible (S) sesame group through resistance (R) sesame group to seed yield (kg/ha).

$$Y = 31.36 - 11.23X_1 - 3.92X_2$$
$$(t=-1.93)^{0.05} \quad (t=-0.62)^{0.54}$$

(2009)
(9)

$$Y = 28.62 - 12.12X_1 - 2.19X_2$$
$$(t=-1.93)^{0.05} \quad (t=-0.62)^{0.54}$$

(2010)
(10)

Where; Y, X_1 and X_2 are, trypsin inhibitor, the resistant (R) sesame group, and seed yield (kg/ha), respectively.

$$Y = 14.23 - 1.98X_1 - 3.44X_2$$
$$(t=-1.13)^{0.02} \quad (t=-0.32)^{0.14}$$

(2009)
(11)

$$Y = 11.26 - 1.34X_1 - 2.49X_2$$
$$(t=-1.73)^{0.15} \quad (t=-0.64)^{0.44}$$

(2010)
(12)

Where; Y, X_1 and X_2 are, trypsin inhibitor, the moderately

resistant (MR) sesame group and seed yield (kg/ha), respectively.

$$Y = 13.16 - 4.13X_1 - 1.49X_2$$
$$(t=-1.43)^{0.45} \quad (t=-0.12)^{0.64}$$

(2009)
(13)

$$Y = 12.11 - 3.34X_1 - 1.19X_2$$
$$(t=-1.90)^{0.15} \quad (t=-0.22)^{0.14}$$

(2010)
(14)

Where; Y, $X1$ and X_2 are, trypsin inhibitor, the moderately susceptible (MS) sesame group, and seed yield (kg/ha), respectively.

$$Y = 11.66 - 10.30X_1 - 0.91X_2$$
$$(t=-2.03)^{0.25} \quad (t=-0.64)^{0.94}$$

(2009)
(15)

$$Y = 10.46 - 9.13X_1 - 0.79X_2$$
$$(t=-2.93)^{0.04} \quad (t=-0.42)^{0.14}$$

(2010)
(16)

Where; Y, X_1 and X_2 are, trypsin inhibitor, the susceptible (S) sesame group and seed yield (kg/ha), in respectively.

Tannins, (Equations 17 - 24)

Significant to highly significant values of R^2 were shown in each of the seasons (2009 and 2010) among the four sesame groups for tannins (Table 7). The dependent variable tannins (Y) increased gradually in the sesame group from susceptible class to resistant class, including the moderately susceptible class as well as moderately resistant class (Equations 33 to 40). This increase indicates that high level of tannins was related to resistance of sesame group, while the low level of tannins was related to susceptible class. The seed yield (kg/ha) also increased gradually from low tannins containing sesame group to high tannin containing one.

$$Y = 9.96 - 10.11X_1 + 3.41X_2$$
$$(t=-2.93)^{0.04} \quad (t=-0.42)^{0.50}$$

(2009)

(17)

$$Y = 9.66 - 10.01X_1 - 3.01X_2$$
$$(t=-0.13)^{0.05} \quad (t=-0.62)^{0.14}$$

(2010)
(18)

Where; Y, X_1 and X_2 are, tannins, the resistant (R) sesame group, and seed yield (kg/ha), respectively.

$$Y = 8.01 - \underset{(t=-1.03)^{0.25}}{6.63X_1} - \underset{(t=-0.22)^{0.24}}{1.94X_2}$$

(2009)
(19)

$$Y = 7.52 - \underset{(t=-1.53)^{0.55}}{2.93X_1} - \underset{(t=-0.52)^{0.54}}{0.98X_2}$$

(2010)
(20)

Where; Y, X_1 and X_2 are, tannins, the moderately resistant (MR) sesame group, and seed yield (kg/ha), respectively.

$$Y = 10.16 - \underset{(t=-3.13)^{0.15}}{2.63X_1} - \underset{(t=-1.02)^{0.04}}{1.19X_2}$$

(2009)
(21)

$$Y = 8.77 - \underset{(t=-1.13)^{0.05}}{2.43X_1} - \underset{(t=-1.02)^{0.50}}{2.01X_2}$$

(2010)
(22)

Where; Y, X_1 and X_2 are, tannins, the moderately susceptible (MS) sesame group, and seed yield (kg/ha), respectively.

$$Y = 7.76 - \underset{(t=-1.90)^{0.05}}{2.93X_1} - \underset{(t=-0.62)^{0.54}}{1.32X_2}$$

(2009)
(23)

$$Y = 6.66 - \underset{(t=-3.13)^{0.15}}{2.39X_1} - \underset{(t=-1.62)^{0.04}}{1.9X_2}$$

(2010)
(24)

Where; Y, X_1 and X_2 are, tannins, the susceptible (S) sesame group, and seed yield (kg/ha), respectively.

DISCUSSION

Various diseases cause considerable damage to crop plants and reduce their yield. Successful breeding for disease resistance has occurred in many different crop types, including vegetables, fruits, field crops and ornamentals. Because field crops are considered as low value crops compared with fruits and vegetables, control costs must be minimized and it is in these crops that host plant resistance breeding has had the most attention and success. Therefore, success of improving the disease resistance of genotypes requires an effective and reliable screening attitude of characters in breeding programs. However, since field study of plant reaction to the fungal pathogens is difficult, laborious and time consuming, breeders often search for easily and rapidly evaluated

characters that are correlated with resistance to diseases. The highly significant values, which were observed among sesame genotypes or sesame groups in resistance to fungal diseases (FOS and MPH) and seed yield, indicated the presence of sufficient variability for all investigated sesame genotypes, which could be valuable in a further breeding program for wilt and charcoal root rot diseases management with high seed yield. Also, the significant differences in the anti-nutritional factors (phytic acid, trypsin inhibitor and tannins) of the sesame genotypes / sesame groups in both seasons (2009/2010), confirmed the presence of fair amount of genetic variability considered adequate for further bio-anti-nutritional factors assessment in sesame genotypes considered in this work. These findings are more or less in agreement with those reported by several investigators (El-Shakhess, 1998; Chattopadhyay and Kalpana, 2000; Dinakaran and Mohammed, 2001; Philip et al., 2007; Nzikou et al., 2009; El-Bramawy et al., 2009 a, b; Akande et al., 2010) who worked with different sesame populations under various conditions.

The characterization of the four sesame genotypes groups (resistant, moderately resistant, moderately susceptible and susceptible), possessed differences significant, base on presence or absence of anti-nutritional factors in their seeds. Anti-nutritional factors (phytic acid, trypsin inhibitor and tannins) seeds estimated were highly variable in sesame seeds. Similarity coefficients among these four sesame genotype groups, showed a wide variability range, where the relationships among these (Anti-nutritional factors) can be observed clearly (Tables 5 and 6). This indicates that, there is a great adjustment to the sesame genotypes used (four sesame group) in the similarity behavior, based on the high value of the phytate, trypsin, a-amylase inhibitors, lectin and tannins in sesame seeds.

According to these findings presented in Table 5, there is a very close relationship among sesame group and anti-nutritional factors, with incidence of Fusarium wilt (caused by F. oxysporum) and charcoal root rot (caused by M. phaseolina) fungal pathogens on sesame. Susceptible sesame group showed the lowest values for phytate, trypsin, a-amylase inhibitors, lectin and tannins (Table 6). Laurentin et al. (2003) found similar results when they evaluated the same six genotypes during two years in Turén, Venezuela. Bergvinson et al. (1997) indicated that, the behavior of some diseases and insects can be influenced by the nutritional quality of plants on which they infect or feed.

The resistance (R) and moderately resistance (MR) to both fungal pathogens (FOS and MPH) are the famous cases among all classes, indicating the high ability of sesame genotypes for resistance to the pathogens and keeping a good, fairly high seed yield potential. This also explains the absence of highly susceptible (HS) genotypes among all tested sesame genotypes through the two seasons (2009 and 2010). These results are in harmony with those reported by Gaber et al. (1998), Khaleifa (2003), El-Bramawy (2006a, b), Bayoumi and El-Bramawy (2007), and El-Bramawy and Shaban (2008). Also, the nonexistence of immune (I) sesame genotypes was reported before by Li et al. (1991), El-Bramawy, (2003) and El-Bramawy et al. (2009a, b).

Sesame genotypes, which kept their resistant (R) characteristic over the two fungal pathogens such as Toshka 2, H1, H9 and L.R15, exhibited their stability level to both pathogens. While, the genotypes, which possessed the high resistance (HR) or moderately resistance rank (MR) to one or two diseases/ fungal pathogens and changed their behavior to be moderately susceptible (MS) or susceptible (S) to the other pathogen such as H 4 and H 10 for *F. oxysporum* (FOS) and L.R7 for *M. phaseolina* (MPH), indicated non stability or variable stability according to conditions of their interaction with the pathogens. The stability and non-stability of sesame genotypes with fungal pathogens were detected before (El-Shakhess, 1998; El-Bramawy, 1997, 2003 and 2006; Bayoumi and El-Bramawy, 2007; El-Bramawy and Shaban, 2008; El-Bramawy et al., 2009a, b). However, the genotypes, which lost their susceptible (S) or moderately susceptible (MS) rank to pathogen to be more resistance (R) or, moderately resistance (MR) fungal pathogens indicated improvement in their resistant characteristics to diseases causing fungal pathogens.

Plant resistance to diseases has reportedly been associated with fungal pathogens, mainly *F. oxysporum* and *M. phaseolina* density in field crops (Ragab et al., 2002; Baydar, 2005; Sarwar and Haq, 2006; El-Bramawy et al., 2009a, b; El-Bramawy, 2010). However, very few papers have reported biochemical compounds of plants conferring resistance to *F. oxysporum* (FOS) and *M. phaseolinan* (MPH). Highly significant differences were detected among the sesame groups for phytic acid, a-amylase inhibitor and trypsin inhibitor, lectin, and tannins (Table 5), indicating high variability among sesame genotypes groups, which varied in their content of the anti-nutritional factors, hence varied in their reaction with fungal pathogens (FOS and MPH). These variations could be confirmed with the selection for the high content of phytic acid, trypsin inhibitor and tannins, which could also confer more resistance on sesame genotypes against each of *F. oxysporum* (FOS) and *M. phaseolina* (MPH). However, the significant differences, which were observed among the resistant, moderately resistant, moderately susceptible and susceptible sesame genotype groups for their content of the anti-nutritional factors (Table 6), confirmed the enhancement in the feasibility of selection for the high levels of phytic acid, trypsin inhibitor and tannins, which were more related to the resistant characters. While, the moderate to low values of these anti-nutritional factors were so related to the moderate to low susceptible characters respectively.

These findings are in harmony with the results reported by Li et al. (1991) and El-Bramawy et al. (2009 b).

Due to the presence of the significant relationship, which noted between the anti-nutritional factors as dependent variables (phytic acid, trypsin inhibitor and tannins) and sesame groups (resistant, moderately resistant, moderately resistant and susceptible) as independent variables, it is possible to put our hands on the positive relationship between the level of resistance in sesame genotype seeds with the pathogenic fungi (FOS and MPH) and their contents of phytic acid, or trypsin inhibitor or tannins as a typical chemical in the sesame seeds. hese results were more or less agreements with the results reported by Hatano et al. (1989), Tazawa et al. (2000), Oliveira et al. (2002), Leandro et al. (2008), El-Bramawy et al. (2009a) through their working on some field crops.

Since sesame genotypes differed in their response to the fungal pathogen (FOS and MPH) as well as in their chemical components in seeds, the anti-nutritional factors could be adopted as screening criteria in building a breeding program for resistance to *F. oxysporum* and *M. phaseolina*, which cause Fusarium wilt and charcoal root rot diseases, respectively. Similar findings were reported by several authors (El-Fiki et al., 2004; El-Bramawy, 2006 and 2008; El-Bramawy and Shaban, 2007; El-Bramawy and Abd Al-Wahid, 2009; El-Shakhess and Khalifa, 2007 Bayoumi and El-Bramawy, 2007; El-Bramawy et al., 2009 a, b).

Wu et al. (2000) reported that, the selection for chemical characters may be significant in the development of Amaranthus for resistance to root rot disease. Furthermore, Lee and Choi (1986) reported that the germplasm resources of sesame could be classified based on level of chemical components and these traits are controlled by one gene (nb) (Brar and Ahuja, 1979). These indicate that the trait of chemical components (anti-nutritional factors) in seeds may be possibly used as screening criteria, if it was associated with the infection percentage caused by the fungal pathogens (FOS and MPH). However, the results found also showed that all sesame genotypes were significantly correlated with the infection percentage by FOS and also with the infection percentage by MPH (Table 5). This result may be due to the fungus of the Fusarium wilt penetrating sesame plant roots and spreading up into the stem through the water conducting vessels which causes plants to wilt from the top down or rot from down (roots) and exhibits plant vessels that are plugged and damaged.

The concentrations of phytic acid, or a-amylase inhibitor or trypsin inhibitor or lectin, or tannins phenols and tannins in sesame seeds were considered an average of its group (data not shown). The findings of this study were in agreement with those reported by Islam et al. (2003), who found that Polyphenolics, of which tannins are a subset, are involved in seed character expression and are often associated with plant resistance to pathogens or insects. Statler (1970) also reported that the more total tannins and phenols in common bean plants causing the greater resistance to root-rot disease. Harris and Burns (1973) mentioned that, tannin is beneficial in the field due to the fact that its presence provide resistance to fungi and seed vivipary. El-Fiki et al. (2004) observed that the amounts of total tannins with phenols were obviously higher in the sesame entries that were classified as highly resistant and resistant than did those classified as susceptible and high susceptible sesame entries. Li et al. (1991) found in 2992 accessions of sesame depending on seed color, that the most resistance of sesame genotypes to MPH had white seed colors, while black or grey seeded genotypes tended to be susceptible, and yellow or brown seeded ones intermediate. El-Bramawy et al. (2009a) used the phenotypical traits (tannins and phenols contents) in forty-eight sesame accessions as selection criteria for Fusarium wilt and charcoal root rot diseases resistance for a breeding program. The detected desirable phenological trait could be transfer to high-yielding cultivars by using conventional hybridization, biotechnological and bridge techniques for the introgression of both diseases resistance genes. Therefore, it indicate that the sesame seed content of phytic acid, trypsin inhibitor and tannins could successfully be used to predict the resistance of sesame genotypes to Fusarium wilt (FOS) and charcoal root rot (MPH) diseases without conducting tedious field crop experiments. In addition, we suggest that this desirable phenological trait could be transfer to high-yielding cultivars by using conventional hybridization, biotechnological and bridge techniques for the introgression of wilt and charcoal rot diseases resistance genes.

For sesame groups, all sesame genotypes (independent variables) were classified to four groups via resistant (R), moderately resistant (MR), moderately susceptible (MS) and susceptible (S) according their interaction with FOS and MPH, while the anti-nutritional factors and infection percentage by *Fusarium oxysporum* f. sp. *sesami* (FOS) and/or *Macrophomina phaseolina* (MPH), might be a suitable attitudes for direct selection among the sesame genotypes for resistance to the Fusarium wilt (FOS) and/or charcoal root rot (MPH) diseases. Therefore, it can be concluded that the resistance (R) class possessed the highest values of the anti-nutritional factors, while the lowest values were related to the susceptible (S) class of sesame group. On the other hand, the two class of the moderately resistance and moderately susceptible possessed the medium values of phytic acid, trypsin inhibitor and tannins over average of the two seasons. Hence, these anti-nutritional factors could play a clear role and magnitude in the range of resistance or susceptible to fungal pathogens, *F. oxysporum* and *M. phaseolina* in sesame genotypes groups considering this work.

According to the regression equations and correlation coefficients between contents of phytic acid, or a-amylase inhibitor or trypsin inhibitor or lectin, or tannins, sesame

genotype groups through the seasons 2009 and 2010 were listed in equations 1 to 40 and collected together in Table 7. Our results found that the sesame genotypes had medium or low phytic acid, a-amylase inhibitor and trypsin inhibitor and therefore had higher infection by FOS and MHP than did genotypes that had high contents of the above (phytic acid, trypsin inhibitor and tannins). The high rate of the phytic acid and trypsin inhibitor contents in the resistant sesame seeds genotypes may increase the levels of resistance to both of FOS and MPH, while low contents of them increase the susceptibility to infection by these fungal pathogens (Equations 1 to 16).

Because of a high contents of phytic acid, a-amylase inhibitor and trypsin inhibitor in sesame seeds, it can become a good favorable conditions for spreading the fungal pathogens, hence it cause the disease. Furthermore, as infection spreads, the water feeding system becomes blocked, therefore the water uptake has not been commensurate with the water transpiration so that the plants becomes more susceptible to fungi infection and increases symptom expression.

The results of our study found that the Fusarium wilt (FOS) or charcoal root rot (MPH) incidence were not correlated with some cases of resistance range (R and MR) or susceptible rank (MS and S) as noted in Table 7 in relation to their contents of phytic acid, a-amylase inhibitor and trypsin inhibitor. It means that the selection criteria by this way may be not suitable approach for direct selection among the sesame genotypes for resistance to Fusarium wilt and charcoal root rot diseases. These finding could be due to the fact that sesame plants can gets the infestation by the both pathogens at any stage of sesame crop development. In this regard, Songa et al. (1997) found that the sesame lines did not seem to influence or affect the susceptibility or resistance to MPH, which cause the charcoal root rot diseases of the various bean accessions.

Our results also found that the sesame genotypes having high level of tannins was generally more resistance to Fusarium wilt (FOS) and charcoal root rot (MPH) diseases than genotypes having low or moderate level (Table 7). These results are confirmed from the linear regression, through a covariance analysis between the infection percentages by FOS and MPH, and the four groups of sesame seed contents (Equations 17, 18, 19, 20, 21, 22, 23 and 24). The increase in resistance of genotypes to both fungal pathogens may be due to the increase in concentrations of total phenols and tannins in their seeds. Therefore, this increase indicates that high level of tannins was related to resistance in sesame groups, while the low level of tannins was related to susceptible class. The seed yield (kg/ha) was also increased gradually from low tannins contents to high tannins contents. These findings were in harmony with the results published before by El-Bramawy et al. (2009a, b).

Conclusion

The anti-nutritional factors in sesame seed genotypes were used as screening criteria for diseases resistance. Significant differences were obtained among sesame genotype groups (resistant, moderately resistant, moderately susceptible and susceptible) for their contents of the anti-nutritional factors. Resistant (R) class possessed the highest values of the anti-nutritional factors, while the lowest values were related to the susceptible (S) class of sesame group. While the moderately resistant and moderately susceptible classes were associated with the medium values of phytic acid, trypsin inhibitor, and tannins over the two seasons of assessment. Hence, these anti-nutritional factors could have played a clear role in the magnitude and range of resistance or susceptibility to fungal pathogens, FOS and MPH in sesame genotype groups considered in this work.

REFERENCES

A.O.A.C. (1990). Official Methods of Analysis. Association of Official Analytical Chemists. 15th ed., K. Helrich (Ed), Arlington, VA.

Abawi GS, Corrales MA (1989). Ashy stem blight screening procedures and virulence of Macrophomina phaseolina isolates on dry edible beans. Turrialba, 39: 200-2007.

Akande FB, Adejumo OA, Adamade CA, Bodunde J (2010). Processing of locust bean fruits: Challenges and prospects. Afr. J. Agric. Res., 5(17): 2268-2271.

Anonymous (2010). Annual Statistical Book of Agricultural in Egypt (In Arabic), p. 219.

Antunez I, Retamosa EC, Villar R (2001). Relative growth rate in phylogenetically related deciduous and evergreen woody species. Oecologia, 128: 172-180.

Ashri A (1998). Plant Breeding Reviews. Volume 16, edited jules Janick. ISBN 0-471-25446-0. ©1998 John Wiley and Sons. Inc.

Baydar H (2005). Breeding for the improvement of the ideal plant type of sesame. Plant Breed., 124: 263-267.

Bayoumi TY, El-Bramawy MAS (2007). Genetic analyses of some quantitative characters and Fusarium wilt disease resistance in sesame. African Crop Science Conference Proceedings, 8: 2198-2204.

Bergvinson D, Arnason JT, Mihm JA, Jewell DC (1997). Phytochemical basis for multiple borer resistance in maize. pp. 82-90. In: J.A. Mihm (ed.) Insect resistance maize: recent advances and utilization; proceeding of an international symposium held at CIMMYT, Mexico D.F., Mexico, 27 Nov.-3 Dec., 1994.

Brar GS, Ahuja KL (1979). Sesame, its culture, genetics breeding and biochemistry. Annu. Rev. Plant Sci. 1: 245-313. Kalyani Puble., New Delhi.

Burkill IH (1953). Habits of man and the origins of cultivated species of the Old World. Linn. Soc. (London) Proc., 164: 12-41.

Chang T-H, Liu X-Y, Zhang X-H, Wang H-L (2002). Effects of dl-praeruptorin A on interleukin-6 level and Fas, bax, bcl-2 protein expression in ischemia-reperfusion myocardium.

Chattopadhyay C, Kalpana Sastry R (2000). Methods for Screening Against Sesame Stem-root Rot Disease.

Dinakaran D, Mohammed N (2001). Identification of resistant sources to root rot of sesame caused by Macrophomina phaseolina (Tassi.) Goid. Sesame and Safflower Newsletter, 16: 68-71.

Dubin HJ, Arun B, Begum SN, Bhatta M, Dhari R, Goel LB, Joshi AK, Khanna BM, Malaker PK, Pokhrel DR, Rahman MM, Saha NK, Shaheed MA, Sharma RC, Singh AK, Singh RM, Singh RV, Vargas M, Verma PC (1998). Results of the South Asia regional Helminthosporium leaf blight and yield experiments, 1993-1994. In: E. Duveiller, H.J. Dubin, J. Reeves and A., McNab (Eds.),

Helminthosporium Blights of Wheat: Spot Blotch.

Duveiller E, Garcia J, Franco J, Toledo J, Crossa F, Lopez G (1998a). Evaluating spot blotch resistance of wheat: Improving disease assessment under controlled conditions and in the field. In: E. Duveiller, H.J. Dubin, J. Reeves and A. McNab (Eds.), Helminthosporium Blights of Wheat: Spot Blotch and Tan Spot, CIMMYT, Mexico, D.F., pp. 171-181.

Duveiller E, van Ginkel M, Dubin HJ (1998b). Helminthosporium diseases of wheat: Summary of group discussions and recommendations. In: E. Duveiller, H.J. Dubin, J. Reeves and A. McNab (Eds.), Helminthosporium Blights of wheat: Spot blotch and Tan Spot, CIMMYT, Mexico, D.F., pp. 1-5.

El-Bramawy MAS (2006 a). Inheritance of resistance to Fusarium wilt in some crosses under field conditions. Plant Protect. Sci., 42(2): 99-105.

El-Bramawy MAS (2006b). Inheritance of Fusarium wilt disease resistance caused by Fusarium oxysporum f sp. sesami in some crosses under field conditions. Sesame and Safflower Newsletters Sesame and Safflower Newsletters, 21: 1-8.

El-Bramawy MAS (2008). Assessment of resistance to Macrophomina phaseolina (Tassi) Goid. in some sesame collections. Egypt J. Appl. Sci., 23(2B): 476-485.

El-Bramawy MAS (2010). Genetic analysis of yield components and diseases resistance in sesame (Sesamum indicum L.) using two progenies of diallel. Res. J. Agron., 4(3): 44-56.

El-Bramawy MAS, Abd Al-Wahid OA (2007). Identification of genetic resources for resistance to Fusarium wilt, charcoal root rot and Rhizocotonia root rot among sesame (Sesamum indicum L.) gemplasm. African Crop Science Conference Proceedings, 8: 1893-1900.

El-Bramawy MAS, Abd Al-Wahid OA (2009). Evaluation of Resistance of Selected Sesame (Sesamum indicum L.) Genotypes to Fusarium Wilt Disease Caused by Fusarium oxysporum f. sp. sesami. Tunisian J. Plant Protect., 4(1): 29-39.

El-Bramawy MAS, El-Hendawy SE, Shaban WI (2009a). Assessing the suitability of morphological and phenological traits to screen Sesame genotypes for Fusarium wilt and charcoal rot disease resistance. J. Plant Protect. Res., 48(4): 397-410.

El-Bramawy MAS, El-Hendawy SE, Shaban WI (2009b). Assessing the suitability of morphological and phenotypical traits to screen sesame accessions for resistance to Fusarium wilt and charcoal rot diseases. Plant Protect. Sci., 45(2): 49-58.

El-Bramawy MAS, Shaban IS (2007). Nature of Gene Action for Yield, Yield Components and Major Diseases Resistance in Sesame (Sesamum indicum L.).

El-Bramawy MAS, Shaban IS (2008). Short communication. Inheritance of yield, yield components and resistance to major diseases in Sesamum indicum L. Spanish J. Agric. Res., 6(4): 623-628.

El-Fiki AI, El-Deeb F, Mohamed FG, Khalifa MMA (2004). Controlling sesame charcoal rot incidence by Macrophomina phaseolina under field conditions by using the resistant cultivars and some seed and soil treatments. Egypt. J. Phytopathol., 32(1-2): 103-118.

Elleuch M, Bedigian D, Zitoun A, Zouari N (2010). Sesame (Sesamum indicum L.) seeds in food, nutrition and health. In V.R. Preedy, R.R. Watson and V.B. Patel, Eds. Nuts and Seeds in Health and Disease Prevention. Elsevier.

El-Shakhess SAM (1998). Inheritance of some economic characters and disease reaction in some sesame (Sesamum indicum L.). Ph.D. Thesis. Argon. Dep. Fac. of Agric., Cario Univ., Cario, Egypt.

El-Shakhess SAM, Khalifa MMA (2007). Combining ability and heterosis for yield, yield components, charcoal-rot and Fusarium wilt diseases in sesame. Egypt J. Plant Breed., 11(1): 351-371.

El-Shazly MS, Abd Al-Wahid OA, El-Ashry MA, Ammar SM, El-Bramawy MAS (1999). Evaluation of resistance to Fusarium wilt disease in sesame germplasm. Int. J. Pest Manage., 45(3): 207-210.

Gaber MR, Hussein NA, Saleh OI, Khalil MA (1998). Susceptibility of certain varieties and genotypes and control of wilt and root rot diseases of sesame attributed to Fusarium oxysporum f. sp sesami and Macrophomina phaseolina. Egypt. J. Microbiol., 33(3): 403-428.

Gupta D, Sharma SK (2006). Evaluation of wild Lens taxa for agro-morphological traits, fungal diseases and moisture stress in North Western Indian Hills. Genet. Resour. Crop Evol., 53: 1233-1241.

Harris HB, Burns RE (1973). Relationship between tannin content of sorghum grain and preharvest seed molding. Agron. J., 65: 957-959.

Hatano T, Edamatsu R, Hiramatsu M, Mori A, Fujita Y, Yasuhara T, Yoshida T, Okuda T (1989). Effect of interaction of tannins with co-existing substances VI. Effect of tannins and related polyphenols on superoxide anion radical and on DPPH radical. Chem. Pharm. Bull., 37: 2016-2021.

Islam FM, Rengifo AJ, Redden RJ, Basford KE, Beebe SE (2003). Association Between Seed Coat Polyphenolics (Tannins) and Disease Resistance in Common Bean. Plant Foods Hum. Nutr., 58: 285-297.

Jalali BL, Chand H (1992). Chickpea wilt. In: Singh US, Mukhopadhayay AN, Kumar J, Chambe HS, eds. Plant Diseases of Cereals and Pulses. Englewood Cliffs, NJ: Prentice Hall, pp. 429-44.

Joshi AK, Chand R, Arun B (2002). Relationship of plant height and days to maturity with resistance to spot blotch in wheat. Euphytica, 123: 221-228.

Kahyaoglu T, Kaya S (2006). Modelling of moisture color and texture changes in sesame seeds during the conventional roasting. J. Food Eng., 75(2): 167-177.

Kakade ML, Rackis JL, McGhee JE, Puski G (1974). Determination of trypsin inhibitor activity of soy bean products: a collaborative analysis of an improved procedure. Cereal Chem., 51: 376-382.

Kanu PJ, Kanu JB, Sandy EH, Kandeh JBA, Mornya PMP, Huiming Z (2009). Optimization of enzymatic hydrolysis of defatted sesame flour by different proteases and their effect on the functional properties of the resulting protein hydrolysate. Am. J. Food Technol., 4: 226-240.

Kavak H, Boydak E (2006). Screeing of the risistance levels of 26 sesame breeding lines of fusarium wilt disease. Plant Pathol. J., 5: 157-160.

Kemal ÜM, Yalçin H (2008). Proximate composition of Turkish sesame seeds and characterization of their oils. Fats oils, 59(1): 23-26.

Khaleifa MMA (2003). Pathological studies on charcoal rot disease of sesame. Ph.D. Thesis. Argon. Dep. Fac. of Agric., Moshtohor, Zagazig Univ., Benha, branch, Egypt, p. 295.

Kilmer VG, Alexander LT (1949). Methods of Machanical analysis of soils. Soils Sci., 68: 15.

Kolte SJ (1985). Diseases of annual edible oil seed crops. Volume II: Rapeseed, Mustard, Safflower and Sesame diseases. CRC Press inc. Boca Raton, Florida.

Ladizinsky G (1993). Wild lentils. Crit. Rev. Plant Sci., 12: 169–184.

Latta M, Eskin M (1980). A simple and rapid colorimetric method for phytate determination. J. Agric. Food Chem., 28: 1313-1315.

Laurentin H, Pereira C, Sanabria M (2003). Phytochemical characterization of six sesame (Sesamum indicum L.) genotypes and their relationships with resistance against whitefly (Bemisia tabaci Gennadius). Agron. J., 95(6): 1577-1582.

Leandro M, Lu GD, José Alberto P, Leticia E (2008). Antioxidant properties, total phenols and pollen analysis of propolis samples from Portugal. Food Chem. Toxicol., 46(2008): 3482-3485.

Lee CH, Choi KY (1986). Genetic studies on quantities characters for sesame breeding (in Korean, English Abstr.) Korean J. Breed., 18: 242-248.

Li LL, Wang SY, Fang XP, Hung ZH, Wang ST, Li ML, Cui MQ (1991). Identification of Macrophmina phaseolina resistant germoplasm of sesame in China. Oil Crops China, 1(3-6): 23.

Moran ET, Pepper WF, Summers JD (1969). Processed Feather and Hog Hair Meals as Sources of Dietary Protein for the Laying Hen with Emphasis on Their Use in Meeting Maintenance Needs. Poult. Sci., 48: 1245-1251. doi:10.3382/ps.0481245.

Namiki M (1995). The chemistry and physiological functions of sesame. Food Rev. Int., 11: 281-329.

Nayar MP, Sastry ARK (1990). Red Data Book of Indian Plants. Vol. 3 Botanical Survey of Indian, Calcutta, Indian.

Nzikou L, Matos G, Bouanga-Kalou CB, Ndangui NPG, Pambou-Tobi A, Kimbonguila TH, Silou LM, Desobry S (2009). Chemical Composition on the Seeds and Oil of Sesame (Sesamum indicum L.). Grown in Congo-Brazzaville.

Obiajunwa EI, Adebiyi FM, Omode PE (2005). Determination of Essential Minerals and Trace Elements in Nigerian Sesame Seeds, Using TXRF Technique. Pak. J. Nutr., 4(6): 393-395.

Oliveira NJF, Melo MM, Araujo MS (2002). Clinical characteristics in crossbred cattle fed citrus pulp pellets. Arq. Bras. Med. Vet. Zootec., 54(3).

Pastor-Corrales MA, Jara C, Singh SP (1998). Pathogenic variation in sources of and breeding for resistance to Phaeoisariopsis griseola causing angular leaf spot in common bean. Euph. 103: 161-171.

Philip JK, Kerui Z, Jestina BK, Huiming Z, Haifeng Q, Kexue Z (2007). Retraction - Biologically active components and nutraceuticals in sesame and related products: a review and prospect. Trends Food Sci. Technol., 19(4): 221.

Ragab AI, Kassem M, El-Deeb AA (2002). New varieties of sesame Taka 1, Taka 2 and Taka 3. 3-Evaluation of these variety reaction against infection by the major fungal disease problems in Egypt and its effects on yield components. Egypt J. Appl. Sci., 17 (6): 167-183.

Salunkhe DK, Chavan JK, Adsule RN, Kadam SS (1991). World Oilseeds: Chemistry, Technology and Utilisation. New York, Van Nostrand Reinhold, p. 554.

Sarwar G, Haq MA (2006). Evaluation of sesame germplasm for genetic parameters and disease resistance. J. Agric. Res., 44(2): 89-96.

Serry M, Satour M (1981). Major diseases of sesame and sources of resistance in Egypt. FOA. Plant Prod. Prot. Paper, 29: 71-72.

Shahidi F, Liyana-Pathirana CM, Wall DS (2006). Antioxidant activity of white and black sesame seeds and their hull fractions. Food Chem., 99: 478-483.

Smith GS, Carvil ON (1997). Field screening of commercial and experimental soybean cultivars for their reaction to Macrophomina phaseolina. Plant Dis., 81: 363-368.

Songa W, Hillocks RJ, Mwango-mbe AW, Buruchara R, Ronno WK (1997). Screening common bean accessions for resistance to charcoal rot Macrophomina phaseolina in Eastern Kenya. Experimental-Agriculture (UK). 33: 459-468.

Statler GD (1970). Resistance of bean plants to Fusarium Solani F Phaseoli. Plant Dis. Rep., 54: 698-699.

Steel RGD, Torrie JH (1980). Principles and Procedures of Statistics, 2nd ed. McGraw-Hill, New York.

Sugano A, Inoue T, Koba K, Yoshida K, Hirose N, Akimoto Y, Shinmen Y, Amachi T (1990). Influence of sesame lignans on various lipid parameters in rats. Agric. Biol. Chem., 54: 2669-2673.

Sugano M, Akinmoto KA (1993). Multifunctional gift from nature. J. Chin. Soc. Anim. Sci., 18: 1-11.

Tazawa HA, Moriya K, Gefen E, Pearson JT (2000). Embryonic heart rate measurements during artificial incubation of emu eggs. Br. Poult. Sci., 41: 89-93.

Vaintraub IA, Lapteva NA (1988). Colorimetric determination of phytate in unpurified extracts of seeds and the products of their processing. Anal. Biochem., 175(1): 227-230.

Weiss E A (1983). Oil Seed Crops. Longman, London, New York, p. 660.

Wu H, Sun M, Yue S, Sun H, Cai Y, Huang R, Brenner D, Corke H (2000). Field evaluation of an Amaranthus genetic resource collection in China. Genet. Resour. Crop Evolut., 47: 43-53.

Wu YV, Inglett GE (1974). Denaturation of plant proteins related to functionality and food applications. J. Food Sci., 39: 218.

Zahra AM (1990). Studies on wilt disease of (Sesamum indicum L.) in upper Egypt. Ph. D. Thesis. Argon. Dep. Fac. Agric., Assuit Univ., Assuit, Egypt.

Zhang X, Cheng Y, Liu SY, Feng X (2001). Evaluation of sesame germplasms resistant to Macrophmina phaseolina and Fusarium oxysporum. Chin. J. Oil Crops Sci., 23(4): 47: 23-27.

Ziedan EHE (1993). Studies on Fusarium wilt disease of seasme in A. R. E. (Egypt). M Sc. Thesis Plant Patho. Dep., Fac. of Agric., Ain-Shams, Univ., Cario, Egypt.

21

Segregation analysis of *cry7Aa1* gene in F₁ progenies of transgenic and non-transgenic sweetpotato crosses

Rukarwa R. J.[1], Mukasa S. B.[1], Sefasi A.[2], Ssemakula G.[3], Mwanga R. O. M.[4] and Ghislain M.[5]

[1]School of Agricultural Sciences, Makerere University, P.O. Box 7062, Kampala, Uganda.
[2]International Potato Center, Applied Biotechnology Laboratory, P. O. Box 1558 Lima, Peru.
[3]National Crop Resources Research Institute, P. O. Box 7084, Uganda.
[4]International Potato Center – Kampala, P. O. Box 7065, Kampala, Uganda.
[5]International Potato Center, Nairobi, P. O. Box 25171, Nairobi 00603, Kenya.

Sweetpotato weevils are the most devastating pests of sweetpotato causing yield losses ranging from 60 to 100%. Their cryptic nature, where the larvae is found within plant tissues render them difficult to manage especially using chemicals control. Development of weevil resistant sweetpotato was conducted by crossing a transgenic event CIP410008.7 as a female parent with three Ugandan cultivars as male parents. Crossing event CIP410008.7 with New Kawogo, Tanzania and NASPOT1 gave 57, 32 and 19 seeds respectively. A total of 86 F1 progenies were analysed for the presence of *cry7Aa1* using polymerase chain reaction (PCR). Expected 608 bp bands were amplified in progenies that contained the cry gene. The gene was integrated at different frequencies in the F1 progenies of different families: CIP410008.7 x New Kawogo (47.2%), CIP410008.7 x Tanzania (52%) and CIP410008.7 x NASPOT1 (44.4%). Chi-square test showed that all the three families followed a 1:1 segregation *cry7Aa1* gene ratio. This study shows the transfer of a transgene from genetically modified event into elite sweetpotato lines.

Key words: Weevil resistance, transgenic plants, sweetpotato, *cry7Aa1*.

INTRODUCTION

The African sweetpotato weevils, *Cylas puncticollis* B. and *C. brunneus* F. are highly destructive pests of sweetpotato throughout sub-Saharan Africa (Fuglie, 2007; Andrade et al., 2009). They attack sweetpotato both in the field and during storage causing substantial losses of up to 60 to 100% (Sorensen, 2009). Stem damage to the vascular system caused by larval feeding and tunnelling reduces the size and number of roots. Weevil feeding on storage roots induces terpenoid production that makes even slightly damaged roots unpalatable (Stathers et al., 2003).

The use of resistant varieties is the most appropriate way to control sweetpotato weevils because it is more effective and safe for the environment. However, cultivars with reliable levels of resistance are not yet available (Stevenson et al., 2009). Progress in breeding weevil-resistant cultivars has been slow due to inconsistent resistance displayed by the genotypes across different areas. Additionally, the polyploidy nature of sweetpotato (2n = 6x = 90), outcrossing behavior, and numerous mating incompatibilities, make conventional breeding difficult. Given the lack of progress in conventional breeding, exploring the option of genetic engineering using *Bacillus thuringiensis* (Bt) could provide the resource

Figure 1. Map of genetic elements in pCIP78 transferred into sweetpotato for expression of the Btprotein, *Cry7Aa1*.

poor farmers with a better alternative to sweetpotato weevil control.

Significant progress has been made in the recent past in transferring Bt cry genes into the American sweetpotato cultivar Jewel (Kreuze et al., 2009). However, this cultivar is not well adapted to sub-Saharan Africa. Transformation of cultivars adapted to Africa using the protocol for Jewel would thus be a suitable option. Unfortunately, the response of sweetpotato to *in vitro* regeneration and transformation conditions is cultivar-specific. Farmer preferred African sweetpotato cultivars have been reported to be recalcitrant and hardly regenerate transformed shoots from calli (Sefasi et al., 2012). As a result, there is limited development of African sweetpotato cultivars expressing agronomically important traits such as weevil resistance. Crossing adapted African cultivars with characterised transgenic events carrying cry genes can circumvent the problem of *in vitro* recalcitrance of elite cultivars and facilitate the introgression of the transgene. Furthermore, use of these transgenic cultivars as parents in breeding programs has some advantages over transformation of existing elite African cultivars which requires several years of evaluation to ensure that introduced gene is stably expressed and no other detrimental phenotypic effects are induced by transformation process. In this case, large numbers of transgenic progeny can be produced from crosses involving transgenic parents, and these can be evaluated following conventional selection programmes. However, in order for transgenic crops to be acceptable for commercial exploitation it is essential that transgenes should be predictable and stably incorporated.

Thus the aim of this study was to introduce a weevil resistant gene (*cry7Aa1*) from transgenic event CIP410008.7 of cultivar Jewel into African cultivars New Kawogo, Tanzania and NASPOT1, and verifying the integration and segregation in F1 progenies of different families.

MATERIALS AND METHODS

Parental plants

The transgenic sweetpotato from Cv. Jewel expressing *cry7Aa1* gene for partial resistance to sweetpotato weevil was produced as described by Luo et al. (2006). The cultivar was transformed using *Agrobacterium tumefaciens* strain EHA105 with the pCIP78 plasmid

construct containing the *cry7Aa1* gene under the control of a β-amylase promoter. The cassette also contained the *nptII* as a marker gene under the control of the Nopaline synthase (*Pnos*) promoter (Figure 1). One of the resulting transformed sweetpotato events, CIP410008.7, was selected for this study after undergoing several cycles of vegetative propagation and phenotypic evaluation. Transgene integration patterns and copy number for this event were also determined from Southern Blot analysis and transgene expression was determined by DAS ELISA. The event CIP410008.7 was used as a female parent in crosses with three Ugandan cultivars; Tanzania, New Kawogo and NASPOT 1. Cultivars Tanzania and New Kawogo are ranked among the five superior and farmer preferred cultivars grown in Uganda because of their wide adaptability and high dry matter content (Mwanga et al., 2001), whereas cv. NASPOT 1 dominates in the central and eastern parts of Uganda because of its high dry matter content and moderate resistance to sweetpotato virus disease (Mwanga et al., 2003). Cultivar New Kawogo is rated as one of the most tolerant cultivars to weevil infestation while Tanzania and NASPOT 1 have been ranked as very susceptible to sweetpotato weevils.

Progeny generation and management

Crossing the sweetpotato was done in a screenhouse at National Crops Resources Research Institute (NaCRRI). The transgenic Jewel was used as female parent and the Ugandan cultivars as male parents in bi-parental crossings. In this experiment the transgenic plants were maintained in a screenhouse and subsequently grafted to *Ipomoea setosa* to induce flowering. The plants were fertilized and irrigated according to standard regimes for screenhouse sweetpotato propagation. Pollen of the male parents was collected from a crossing block outside the facility. Hand pollination of event CIP410008.7 was carried out in the morning before 10.00 am to minimise the rate of flower abortion. Four to five weeks after pollination the ripe seed capsules were harvested and dried for a week. Seeds were removed from capsules followed by soaking in concentrated sulfuric acid (98% H_2SO_4) for 45 min to break dormancy. This was followed by several rinsings with water for 5 min and subsequently sown in nursery boxes.

PCR analysis of *cry7Aa1* gene

Genomic DNA was isolated from all sweetpotato leaf samples using the CTAB method (Stacey et al., 2000). Polymerase chain reaction (PCR) amplification of the 608 bp DNA fragment of the *cry7Aa1* gene was carried out in a Multigene Thermocycler (Labnet International, Inc. NJ, USA) using 5'-ACAACTCATCACCATACCAAAC-3' and 5'-AAGAGCAAGATGCAAGTTTG-3' as forward and reverse primers, respectively. All the PCR experiments took place in a total volume of 25 μl containing 50 ng of total plant DNA made as follows 12.5 μl of Ready mix Taq® DNA polymerase (Sigma-Aldrich, Poole, UK)

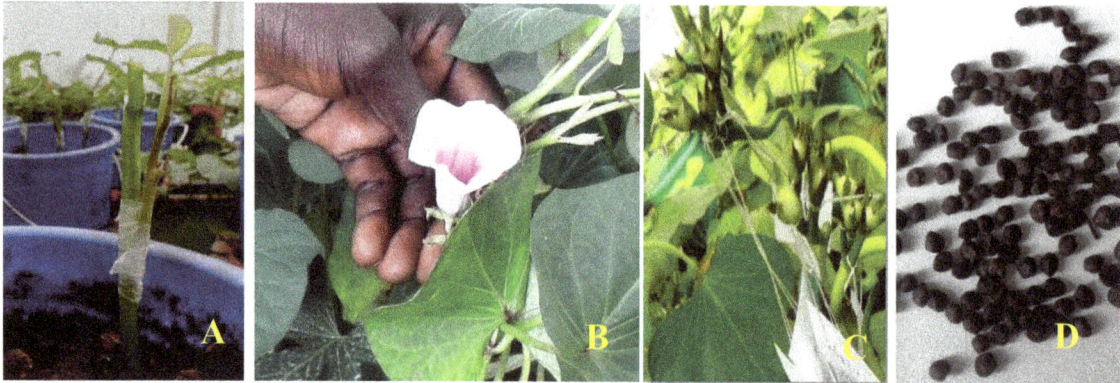

Figure 2. Sweetpotato breeding process. (A) sweetpotato scion grafted on *I. setosa* root stock; (B) sweetpotato fully bloomed flower (C); seed capsules of sweetpotato produced from the crosses; (D) sweetpotato seed.

and primer F (0.5 µl) and primer R (0.5 µl) and adjusted to 25 µl with nuclease free water. PCR conditions were as follows: initial denaturation of DNA at 93°C for 2 min and then amplified through 35 thermal cycles of 93°C for 15 s, 55°C for 30 s, 72°C for two minutes and ending by a final extension step at 72°C for 7 min. The PCR amplicons were separated by electrophoresis on 1% agrose gel in Tris-EDTA (TE) buffer stained with ethidium bromide and visualized under UV light. The presence or absence of the transgenes in the hybrids was used to confirm the transgenic and non transgenic progeny. The presence or absence of the transgenes in the F1 progeny was used to confirm the segregation in transgenic lines. The segregation data were analysed by the χ^2 test at P < 0.05 to determine if the observed segregation ratios of *cry7Aa1* fit the expected Mendelian 1:1 phenotypic ratio.

RESULTS AND DISCUSSION

Sweetpotato progeny development

In this study, all sweetpotato cultivars failed to flower under screenhouse conditions five months after planting. The failure to flower could be due to the uncontrolled environmental conditions in which the plants were exposed to in the screenhouse. The success of flowering initiation and flower development is influenced by environmental conditions such as light intensity, relative humidity and temperature (Rossel et al., 2008). A photoperiod of 11.5 h day length or less, temperatures of 20 to 25°C and relative humidity of 75% usually promote flowering in sweetpotato. Flowering in the transgenic event was only achieved after grafting sweetpotato to its wild relative *I. setosa* (Figure 2A). Similarly, the Ugandan cultivars failed to flower under screenhouse conditions even after grafting; pollen was obtained from a field crossing block. Pollination was done on flowers of the female parent which were fully bloomed (Figure 2B). After pollination the initial fruit grew and produced round shaped green capsules with a diameter of 1 to 2cm (Figure 2C). Seeds that were removed from the capsules were dark brown, small with a diameter of 3 to 4mm (Figure 2D).

Not all sweetpotato flowers which were pollinated produced capsules that could be harvested. When ovaries of the transgenic event were crossed with the pollen from the Ugandan cultivars, the flowers had a tendency to abort. Probably, various degenerative processes occurred during the flower and ovule development stage that resulted in low capacities of the capsule and seed formation. Serious physiological problems, occurring primarily as post pollen barriers to fertility, often impede seed production in sweetpotato. Problems of incompatibility and sterility impede controlled pollination in this crop. Besides difficulties in flower setting in sweetpotato, the flower that developed readily developed seed upon fertilization despite previous reports of complex incompatibilities and sterility mechanisms (Murata and Matsuda, 2003).

The polyploidy nature of sweetpotato (2n = 6x = 90) and numerous mating incompatibilities, make crossing this crop difficult (Murata and Matsuda, 2003). Crosses between event CIP410008.7 and Ugandan cultivars, however, managed to produce capsules and seeds in this study. The number of capsules produced from a cross between CIP410008.7 and New Kawogo were higher than NASPOT 1 and Tanzania. The average number of seeds per capsule ranged from 1.10 to 1.68 (Table 1). Some crosses were not possible and practically all crosses produced much less than the potential four seed per capsule. Each sweetpotato cross in this study produced 1 to 2 seeds per capsule. After scarification with sulfuric acid to break the seed dormancy and floating in water, all the seeds planted germinated. Seeds that were light or deformed were discarded.

Segregation and integration of cry gene

Due to the heterozygous nature of sweetpotato, crosses among the sweetpotato cultivars resulted in high levels of segregation at F1 generation that enable evaluation or selection. The F1 progenies produced in this study were

Table 1. Sweetpotato seed capsule and seed production from crosses between CIP410008.7 and three Ugandan cultivars.

Cross	Number of seed capsules collected	Total seeds collected	Mean number of seed per capsule
CIP 410008.7 x NASPOT 1	13	19	1.46
CIP 410008.7 x New Kawogo	52	57	1.10
CIP 410008.7 x Tanzania	19	32	1.68

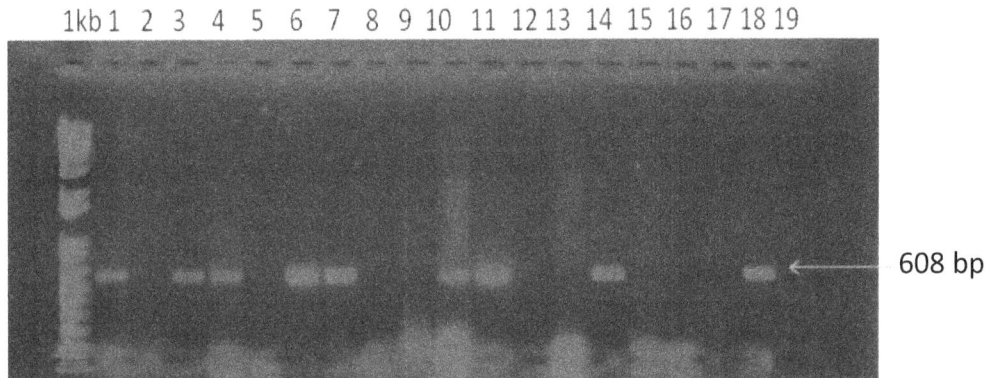

Figure 3. PCR analysis for the F1 plants resulting from crosses between CIP410008.7 and Ugandan cultivars. The pair of primers used amplified the cry7Aa1 gene in plants yielding a 608-bp fragment. 1kb ladder, 1-17 plant numbers, Positive control (18) and negative control (19).

evaluated using PCR to check the integration of the weevil resistance gene. PCR analysis of *cry7Aa1* using the gene-specific primers in some samples of progenies from crosses between CIP410008.7 and Ugandan cultivars are shown in Figure 3. In some sweetpotato samples evaluated, the amplification produced a DNA band sized 608 bp and some did not show any amplification. This suggests that some of the evaluated progenies positively contained the cry gene that was integrated from transgenic CIP410008.7 which is the weevil resistant parent in the crosses.

Sweetpotato is considered an autohexaploid with hexasomic inheritance (Kumagai et al., 1990; Anwar et al., 2009). The segregation pattern of traits in sweetpotato is more complex because there more than two homologous chromosomes that can pair during meiosis (Kumagai et al., 1990). A transgene present in just one parent is expected to segregate in a 1:1 ratio if present as a single copy (simplex) or a 4:1 ratio if present as a double copy (duplex). Southern blot analysis has shown that the cry gene was integrated with one copy in the genome of transgenic event CIP410008.7 (Ghislain M, Pers comm). The cry gene dominant allele controlling the resistance was expressed as a simplex configuration (Rrrrrr) in the genome of transgenic sweetpotato. The Ugandan cultivars that were susceptible to the weevil had an allellic composition at the locus as rrrrrrr which is recessive for weevil resistance. The presence of the cry gene in this study followed a simplex hexasomic model of inheritance in all the three families.

In all the 86 plants tested, dominance effects of the cry gene were observed because at least 50% of all progenies from the crosses carried cry7Aa1 gene. Crosses of CIP410008.7 x New Kawogo, CIP410008.7 x Tanzania and CIP410008.7 x NASPOT 1 produced 25, 13 and 8 progenies carrying the cry gene derived from the transgenic parent respectively (Table 2). Based on the Chi square (x^2) test of the three families, the segregation pattern obtained was in accordance to the expected Mendelian segregation ratio of 1:1 (transgenic to non transgenic plants). Similar results were reported by Okada et al. (2002), who evaluated the offspring from crosses between transgenic and non transgenic sweetpotato and concluded that *hpt* and *CP* gene were transmitted in a Mendalian pattern.

Crosses between commercial sweetpotato cultivars would theoretically produce a high diversity for many characters. Each F1 progeny has the potential to become a new variety. In the case of crosses between the event and Ugandan cultivars, it is expected that superior characteristics which are farmer preferred can be collected from the parents. The sweetpotato lines produced from this study that positively contained *cry7Aa1* gene need to be evaluated further for their resistance to the weevil both in the laboratory and under field conditions as well as further evaluation of various agronomic traits.

Table 2. Segregation of cry7Aa1 gene by PCR analysis on the F1 progenies of sweetpotato.

Cross	Number tested	Observed negative	Observed negative	χ^2 (1 : 1)	P-value
CIP 410008.7 x NASPOT 1	18	10	8	0.22	0.637
CIP 410008.7 x New Kawogo	53	28	25	0.17	0.680
CIP 410008.7 x Tanzania	25	12	13	0.04	0.841

In practice, a large numbers of progenies are required for the breeder to have the opportunity to select for other characteristics other than the presence of the transgene.

Conclusion

Our data illustrates the feasibility of crossing transgenic and non transgenic sweetpotato to obtain transgenic progeny following the established segregation ratios. The F1 sweetpotato progenies that produced DNA bands sized 608 bp positively contained cry gene derived from transgenic event CIP410008.7. The sweetpotato lines that contained the cry gene need to be tested further for their resistance to the sweetpotato weevil in the laboratory and selected further for their famer preferred morphological characteristics.The purpose of this work is to provide plant breeders with some direction for future research in transgenic breeding.

ACKNOWLEDGEMENTS

We acknowledge Mr Moses Mwondha for taking care of transgenic plants in the screenhouse, and the International Potato Center and the Regional Universities Forum for Capacity Building in Agriculture (RUFORUM) for funding this work.

REFERENCES

Andrade M, Barker I, Cole D, Dapaah H, Elliott H, Fuentes S, Grüneberg W, Kapinga R, Kroschel J, Labarta R, Lemaga B, Loechl C, Low J, Lynam J, Mwanga R, Ortiz O, Oswald A, Thiele G (2009). Unleashing the potential of sweetpotato in Sub- Saharan Africa: Current challenges and way forward. Working Paper 2009. Lima, Peru: International Potato Center (CIP).

Anwar N, Kikuchi A, Kamugai T, Watanabe KN (2009). Nucleotide sequence variation associated with β-amylase in the sweetpotato Ipomoea batatas (L.) Lam. Breed. Sci. 59:209-216.

Fuglie KO (2007). Priorities for sweetpotato research in developing Countries: Results of a survey. Hort. Sci. 42:1200-1206.

Kreuze KF, Valkonen JPT, Ghislain M (2009). Genetic Engineering. In Loebenstein G, Thottappilly G (eds.), The Sweetpotato. Springer-Verlag New York Inc.

Kumagai T, Umemura Y, Baba T, Iwanaga M (1990). The inheritance of beta-amylase null in storage roots of sweet potato, Ipomoea batatas (L.) Lam. Theor. Appl. Genet. 79(3):369-376.

Luo HR, Santa Maria M, Benavides J, Zhang DP, Zhang YZ, Ghislain M (2006). Rapid genetic transformation of sweetpotato (Ipomoea batatas (L.) Lam) via organogenesis. Afr. J. Biotechnol. 5:1851-1857.

Murata T, Matsuda Y (2003). Histological studies on the relationship between the process from fertilisation to embryogenesis and the low seed set of sweetpotato, Ipomoea batatas (L.) Lam. Breed. Sci. 53:41-49.

Mwanga ROM, Odongo B, P'Obwoya CO, Gibson RW, Smit NEJM, Carey EE (2001) Release of five sweet potato cultivars in Uganda. Hort. Sci. 36:385-386.

Mwanga, ROM, Odongo B, Turyamureeba GM, Alajo A, Yencho GC, Gibson RW, Smit NEJM, Carey E E (2003). Release of six Sweet potato cultivars (NASPOT1 to NASPOT 6) in Uganda. Hort. Sci. 38:475-476.

Okada Y, Nishiguchi M, Saito A, Kimura T, Mori M, Hanada K, Sakai J, Matsuda Y, Murata T (2002). Inheritance and stability of the virus-resistant gene in the progeny of transgenic sweetpotato. Plant Breed. 121:249-253.

Rossel G, Espinoza C, Javier M, Tay D (2008). Regeneration guidelines: sweet potato. In: Dulloo ME, Thormann I, Jorge MA, Hanson J (eds). Crop specific regeneration guidelines CGIAR System-wide Genetic Resource Programme, Rome, Italy.

Sefasi A, Kreuze J, Ghislain M, Manrique S, Kiggundu A, Ssemakula G, Mukasa SB (2012). Induction of somatic embryogenesis in recalcitrant sweetpotato (Ipomoea batatas L.) cultivars. Afr. J. Biotechnol (In press).

Sorensen KA (2009). Sweetpotato Insects: Identification, Biology and Management. In: Loebenstein G, Thottappilly G (eds.) The Sweetpotato. Springer-Verlag New York Inc.

Stacey J, Isaac PG, Rapley R (2000). Isolation and Purification of DNA from Plants. The Nucleic Acid Protocols Handbook. Humana Press.

Stathers TE, Rees D, Nyango A, Kiozya H, Mbilinyi L, Jeremiah S, Kabi S, Smit N (2003). Sweetpotato infestation by Cylas spp. in East Africa: II. Investigating the role of root characteristics. Int. J. Pest Manag. 49:141-146.

Stevenson PC, Muyinza H, Hall DR, Porter EA, Farman D, Talwana H, Mwanga, ROM (2009). Chemical basis for resistance in sweetpotato Ipomoea batatas to the sweetpotato weevil Cylas puncticollis. Pure Appl. Chem. 81:141-151.

22

Evaluation of various fungicides and soil solarization practices for the management of common bean anthracnose (*Colletotrichum lindemuthianum*) and seed yield and loss in Hararghe Highlands of Ethiopia

Mohammed Amin[1], Ayalew Amare[2] and Thangavel Selvaraj[1]

[1]Department of Plant Sciences and Horticulture, College of Agriculture and Veterinary Sciences, Ambo University, P.O. Box No 19 Ambo, Ethiopia.
[2]School of Plant Sciences, College of Agriculture, Haramaya University, P. O. Box No: 138, Dire Dawa, Ethiopia.

The study was undertaken to evaluate the effect of integrated management of bean anthracnose through soil solarization and various fungicides applications on epidemics of the disease and to determine the effect of integrated management of bean anthracnose on seed yield and loss of common bean variety Mexican-142. Field experiments were conducted at Haramaya and Hirna in 2010 main cropping season of Ethiopia. Soil solarization was integrated with mancozeb and carbendazim seed treatment and with foliar sprays of carbendazim at the rate of 0.5 kg/ha at 10 and 20 days intervals. The experiment was arranged in 2 × 3 × 3 split-split plot design with three replications. A total of 18 treatments were evaluated. There was a significant difference in the anthracnose incidence, severity, seed yield and yield loss. At Haramaya, severe epidemics of anthracnose developed. Seed treatments, foliar sprays and soil solarization alone as well as their interactions did not significantly affect pods per plant and seeds per pod at both locations. The combinations of solarized soil + mancozeb seed treatment + carbendazim foliar spray at 10 days intervals produced seed yield of 3.8 t ha^{-1} at Haramaya and 3.6 t ha^{-1} at Hirna over the control. Integrations of soil solarization, seed treatments and foliar spray were found to be effective in reducing bean anthracnose epidemics and increasing yield.

Key words: Carbendazim, foliar spray, Mancozeb, *Phaseolus vulgaris,* seed treatment.

INTRODUCTION

Common bean (*Phaseolus vulgaris* L.) is an important legume crop in the daily diet of more than 300 million of the world's population and an important food and cash legume in the East and the Great Lakes region of Africa (Hadi et al., 2006). In Ethiopia, common bean is mainly grown in Hararghe highlands (eastern), southern, south-western, and the Rift Valley regions (Habtu et al., 1996; CSA, 2005). In Ethiopia, the national average yield of common beans is low ranging from 0.615 to 1.487

tons/ha between 2004 and 2010 (CSA, 2010), which is far below the corresponding yield recorded at research sites (2.5 to 3 tons ha^{-1}) using improved varieties (EPPA, 2004). The major biotic constraints to common bean production in Ethiopia are anthracnose, rust, bacterial blight, angular leaf spot, ascochyta blight and bean common mosaic virus (Habtu, 1987). Anthracnose caused by *Colletotrichum lindemuthianum* (Sacc. and Magnus) Lams-Scrib is one of the most devastating seed-borne diseases of common bean in Ethiopia (Schwartz et al., 1983). The crop is vulnerable to the attack of the pathogen at all the stages, from seedling to maturity, depending on the prevalence of favorable environmental conditions that are essential for the initiation and further development of the disease and causes an estimated yield loss of 63% in Ethiopia (Beshir, 1997).

The management of primary inoculum sources is key strategy in controlling bean anthracnose (Hall, 1994). The use of seed treatment is an important tactic for disease control in general and for anthracnose management in particular (Freeman et al., 1997). However, seed treatment alone could be inefficient and would often require follow-up applications of contact or systemic foliar fungicides (Koch, 1996; Tesfaye and Pretorius, 2005). Furthermore, chemical disease control should form part of integrated disease management system including physical practices and manipulation of environmental conditions. Soil solarization has been identified as hydrothermal process and its success depends on moisture for maximum heat transfer to soil-borne organisms; it is a function of time and temperature relationships. The thermal decline of soil-borne organisms during solarization depends on both the soil temperature and exposure time, which are inversely related. The effectiveness of solarization for disinfesting soil and increasing plant growth and development depends on soil colour and structure, air temperature, soil moisture, length of day, intensity of sunlight, and the thickness and light transmittance of the plastic film (Stapleton and DeVay, 1982).

Although bean anthracnose is rampant in *Phaseolus* beans of Hararghe highlands, its economic impact has not been evaluated. In addition to assessment of various management methods like chemical and use of physical practices are urgently needed in order to provide an alternative to other control methods. However, management of bean anthracnose through the effect of soil solarization has not been studied so far in Ethiopia. Much work still needs to be done in Ethiopia, particularly in Hararghe highlands, especially with the bean anthracnose through various fungicides and the effect of soil solarization, since the disease is still causing much devastation on the crop. Therefore, the present study was undertaken to evaluate the various fungicides and the effect of soil solarization for the integrated management of common bean anthracnose and on seed yield loss in Hararghe highlands of Ethiopia under field condition.

MATERIALS AND METHODS

Description of the study sites

The study was conducted at Haramaya University experimental field in the main campus and Hirna Sub-Research Station. Haramaya is located at 9°26'N latitude, 42°30'E longitude and at an altitude of 1980 m. a. s. l. The site receives mean annual rainfall of 780 mm, with mean minimum and maximum temperatures of 8.25 and 24.4°C, respectively. On the other hand, Hirna is located at 9°12'N latitude, 41°4'E longitude and at an altitude of 1870 m. a. s. l. The site receives a mean annual rainfall of 990 to 1010 mm with an average temperature of 24°C.

Experimental design and materials

A split-split plot design with three replications was used at both experimental sites. Total of 18 treatment combinations were evaluated. Main plots were assigned to soil solarization, sub-plots to frequencies of foliar sprays and sub-sub plots to seed treatments. Each sub-sub-plot (2.4 × 2 m) consisted of six rows, and distance between rows and between plants were 40 and 10 cm, respectively. There were 20 plants per row and the four central rows were harvested for determining yield. There were 0.8 m space between sub-sub plots, 1 m between sub-plots and 1 m between main plots. The bean variety Mexican-142, which is susceptible to anthracnose (Beshir, 1997) was used for both experiments; in addition, mancozeb was used for seed treatment at 3 g/kg while carbendazim was used for seed treatment at 2 g/kg and as foliar spray at 0.5 kg/ha.

Soil solarization treatment was performed as follows: After ploughing the plots, the area to be solarized was leveled and made free from weeds, debris, or large clods that could raise the plastic-off ground. The soil was turned over by hand and raked smooth to provide an even surface. Transparent plastic sheeting that was 0.0254 mm thick laid by hand (that is, close to the soil), and anchored to the soil by burying the edges in trench around the treated area. The plastic was kept in place from last week of May to last week of June for 1 month to allow the soil to heat to the greatest depth as possible.

Foliar spray with carbendazim (Bavistin DF) at the rate of 0.5 kg ha[-1] (CIAT, 1988), and at three frequencies (including control) was used as the third component of the experimental treatments. Frequencies consisted of spraying at 10 day-interval as of disease onset, spraying at 20 day-interval and no spray. Spraying of fungicide was started 42 days after sowing (DAS) at Haramaya and 50 DAS at Hirna, that is, when the first symptoms of the diseases appeared. The experiments relied entirely on natural infection, because both sites are hot spot areas for the disease. Two seed treatments, namely mancozeb (contact) and carbendazim (systemic) were tested at doses of 3 and 2 g/kg of seeds, respectively. Seeds were treated with mancozeb or carbendazim 24 hours before sowing and untreated seeds served as control. Planting was done on July 5, 2010 at Haramaya University and July 13, 2010 at Hirna Sub Research Station on previous plots sown of common bean to get increased disease pressure.

Data collection and analysis

Anthracnose incidence was assessed as of disease onset at 10 day-intervals at both locations. Using 10 randomly pre-tagged bean plants in the four central rows, severity was rated using standard disease scales of 1-9 (CIAT, 1988), where, 1= no visible disease symptoms and 9= more than 25% of leaf surface area with large coalescing and generally necrotic lesions resulting in defoliation. The severity grades were converted into percentage severity index for analysis (Wheeler, 1969).

$$PSI = \frac{Sum\ of\ numerical\ ratings \times 100}{No.of\ plants\ scored \times maximum\ score\ on\ scale}$$

Seed yield was obtained from the four central rows of each

Table 1. Average temperatures of soil during four weeks of soil solarization at different depth at Haramaya and Hirna.

Weeks	Depth (cm)	Haramaya (°C)		Hirna (°C)	
		Solarized	Non solarized	Solarized	Non solarized
1	5	58.2	31.6	61.8	33.2
	15	58.0	31.2	59.1	32.9
2	5	54.7	29.7	57.9	31.1
	15	52.5	28.9	55.3	30.5
3	5	53.5	26.1	56.7	28.8
	15	53.9	27.6	55.2	26.7
4	5	48.2	25.1	53.1	25.2
	15	45.7	24.7	52.4	23.3

experimental plot; it was converted into tons per hectare (after adjusting to 10% seed moisture content using moisture meter). In addition, soil temperature for periods of soil solarization was recorded using digital soil thermometer by inserting into the soil at 5 and 15 cm depths (Table 1).

All data were subjected to analysis of variance using General Linear Model (GLM) procedure of SAS statistical version 9.2 software (SAS, 2009) except mean separation for significant interaction effects which was carried out using GenStat version 12.1 software (GenStat, 2009). Least Significant Difference (LSD) was used to separate means.

RESULTS AND DISCUSSION

Solarization was undertaken from last week of May to last week of June and soil temperature for period of solarization was recorded using soil thermometer. Average temperatures of soil during four weeks of soil solarization at different depth at both sites were described in Table 1. Average temperatures of soil were variable across sites, solarization and depth. Soil temperatures were warmer at Hirna than at Haramaya, because there was a cloud and rainfall was more variable across sites during period of solarization. According to Stapleton (1997), soil solarization is based on utilizing the solar energy for heating soil mulched with a transparent sheet, reaching a level of 40 to 55°C in the upper soil layer. There is a gradient of temperatures from the upper to lower soil layer during the appropriate season. The temperature elevation is facilitated by wetting the soil before and/or during mulching with the transparent sheet. The main factor involved in the pest control process is the physical mechanism of thermal killing.

Incidence of anthracnose

Incidence of bean anthracnose showed significant difference between solarized and non-solarized plots at 52 days after sowing (DAS) (P<0.05) and highly significant

difference at 62, 72, 82 and 92 DAS (P<0.01) at Haramaya and all four successive disease assessments at Hirna (P<0.01). In these experiments, disease incidence in solarized soil decreased as compared with non-solarized soil, this might be due to anthracnose pathogen population in the soil were reduced by heat. As reported by Dillard and Cobb (1993), in areas where beans are consecutively cropped, over seasonal inoculum can initiate epidemics of bean anthracnose. The primary inoculum from infested debris was relatively more damaging than other inoculum source, causing early epidemic development and yield reduction (Fininsa and Tefera, 2002). The different seed treatments showed highly significant (P<0.01) difference in terms of anthracnose incidence at 52, 62 DAS at Haramaya, 60 days after plantin at Hirna, and showed significant difference (P<0.05) 72 DAS at Haramaya, 70, 80 and 90 DAS at Hirna (Table 2). At Hirna, the maximum incidence was recorded from control. While, lower levels of incidence were encountered for seed treatments with mancozeb and carbendazim, respectively at final date of incidence assessment. This might be due to the above fungicides able to reduce primary source of anthracnose inoculum on/in the seeds.

There was no significant variation in disease incidence between frequencies of carbendazim foliar sprays at 62, 72, 82 and 92 DAS dates of disease assessment at Haramaya except initial date 52 DAS (P<0.01), while significant differences were observed at all dates of assessment at Hirna (P<0.01) (Figure 2). Lower mean incidence was calculated from carbendazim foliar sprays at 10 (16.1%) and 20 (17.2%) days' intervals at initial date of incidence assessment 60 DAS. The difference in incidence might be due to the fact that frequent fungicides spraying could reduce spreading of secondary inoculum between beans. Pastor-Corrales and Tu (1989) reported that preventive spraying of foliage at flower initiation, late flowering and pod-filling with protective systemic fungicides like maneb, zineb, benomyl, carbendazim, and fentin hydroxide have been used to

Table 2. Effect of soil solarization, seed treatment and foliar sprays on incidence of common bean anthracnose at Haramaya and Hirna during 2010 main cropping season.

Treatments	Incidence at Haramaya (%)		Incidence at Hirna (%)	
	Initial (52 DAS)	Final (92 DAS)	Initial (60 DAS)	Final (90 DAS)
Soil solarization				
Solarized	19.6	60.7	15.9	45.6
Non-solarized	24.4	72.2	21.5	50.7
CV (%)	9.1	9.6	7.3	11.9
LSD (5%)	3.6	6.5	3.1	3.7
Foliar spray				
CFS10	18.9	63.9	16.1	45.6
CFS20	19.4	66.1	17.2	47.2
Control	24.4	72.2	22.8	51.7
CV (%)	15.7	13.0	16.1	12.4
LSD (5%)	4.4	7.9	3.8	4.4
Seed treatment				
MST	18.9	63.3	15.6	43.3
CST	22.8	66.7	17.8	47.2
Control	24.4	72.2	22.8	53.9
CV (%)	14.2	13.4	11.8	15.8
LSD (5%)	4.5	7.9	3.8	4.5

MST=Mancozeb seed treatment; CST=Carbendazim seed treatment; CFS10 and; DAS=Days after sowing; 20=Carbendazim foliar spray at 10 and 20 days interval; LSD=Least significant difference.

control bean anthracnose. As reported by Sindhan and Bose (1981), carbendazim was effective as seed dressing and foliar sprays, and these treatments increase seed germination and seed yield, while reducing disease incidence. Study conducted by Gwary (2008) support present experiment in that sorghum anthracnose can best be managed on the early susceptible variety by integrating seed dressing treatment supplemented with foliar sprays.

Severity of anthracnose

Percent severity index was higher in non-solarized soil as compared to solarized soil at both locations. There were significant interaction effects (*P<0.01*) between levels of soil solarization and seed treatment in reducing PSI at 52, 62 72 DAS at Haramaya and non significant difference at 82, 92 DAS at Haramaya and all dates of severity assessment at Hirna. There were significant interaction effects between soil solarization × foliar sprays (*P<0.01*) at 52, 62 and 92 DAS at Haramaya and (*P<0.05*) 70, 80 and 90 DAS at Hirna. Percentage severity index was not significantly different at the remaining dates of severity assessment at both locations. At final date of assessment, the combination of solarized soil with carbendazim sprays at 10 and 20 days intervals minimum over the control at Haramaya. Soil solarization alone or with low dosages of a fungicide, biocide and bio-

agent resulted in complete reduction of the pathogen; soil solarization integrated with applications of *T. harzianum*, carbendazim and neem was the most effective treatment (Chakraborty et al., 2008).

There were significant interaction effects between seed treatment × foliar sprays in reducing severity (*P<0.01*) at Haramaya which were no significant at all dates of severity assessment at Hirna. The combination of mancozeb seed treatment with spray frequencies of carbendazim at 10 and 20 day intervals showed much lower over the control at final date of disease assessment at Haramaya. Likewise, the interactions of carbendazim seed treatment with carbendazim spray at intervals of 10 and 20 days much minimum over the control on the final date of severity assessment at Haramaya (Table 3). In connection with this, Jenny (2010) reported that the seed treatment alone provide effective control for a maximum of four to six weeks after sowing, but do not provide absolute control. Current study also showed that combination of seed treatment with foliar spray was more effective in reducing severity than seed treatment or foliar applications of fungicides alone.

There were significant interaction effects among soil solarization × seed treatment × foliar sprays (*P<0.01*) at 52, 72 and 82 DAS at Haramaya and non significant at the remaining dates at Haramaya and all dates disease assessment at Hirna. The interactions of solarized soil × mancozeb seed treatment × carbendazim foliar sprays at 10 and 20 days gaps reduced PSI at initial date of

Table 3. Effect of soil solarization x seed treatment, soil solarization × foliar sprays and seed treatment × foliar sprays interactions on percentage severity index (PSI) of anthracnose at Haramaya and Hirna during 2010 main cropping season.

Treatments interaction		PSI at Haramaya (%)		PSI at Hirna (%)	
		Initial (52 DAS)	Final (92 DAS)	Initial (60 DAS)	Final (90 DAS)
Soil solarization	**Seed treatment**				
Solarized	MST	13.8	28.7	16.3	25.2
	CST	13.1	29.1	16.6	25.7
	US	20.7	38.5	15.6	26.2
Non-solarized	MST	13.9	33.1	20.3	29.4
	CST	14.3	34.8	20.9	31.4
	US	26.9	44.7	21.9	32.1
CV (%)		13.0	12.3	9.6	9.7
LSD at 5% for SS × ST		4.4	NS	NS	NS
Soil solarization	**Foliar spray**				
Solarized	CFS10	13.3	30.5	19.0	21.9
	CFS20	15.1	31.5	15.1	25.2
	UF	19.3	34.4	14.3	29.9
Non-solarized	CFS10	15.1	32.6	21.2	25.2
	CFS20	17.9	32.7	17.5	31.8
	UF	22.1	47.4	24.4	35.8
CV (%)		14.7	12.1	25.8	21.1
LSD at 5% for SS x FS		2.5	4.3	NS	14
Seed treatment	**Foliar spray**				
MST	CFS10	12.2	28.2	20.0	29.6
	CFS20	13.4	29.3	16.3	24.8
	UF	15.9	35.3	18.5	27.4
CST	CFS10	14.5	31.1	21.1	31.8
	CFS20	13.6	30.3	15.9	25.2
	UF	12.9	34.6	19.2	28.5
US	CFS10	15.8	35.4	19.3	31.1
	CFS20	19.8	36.7	16.7	25.6
	UF	33.2	52.8	20.4	30.7
CV (%)		13.0	12.3	9.6	9.7
LSD at 5% for ST x FS		3.6	5.3	NS	NS

MST=Mancozeb seed treatment; CST=Carbendazim seed treatment; CFS10 and 20=Carbendazim foliar spray at 10 and 20 days interval; UF=Unsprayed foliar; DAS =Days after sowing; US=Untreated seed; LSD=Least significant difference; FS=Foliar spray; ST=Seed treatment; SS=Soil solarization; NS=Non significant difference.

anthracnose disease assessment. Similarly, the interaction of carbendazim seed treatment with the same fungicide foliar sprays frequencies at 10 and 20 days intervals reduced disease severity at initial (52 DAS) date of severity assessment.

The primary inoculum from infested debris was relatively more damaging, causing early epidemic development and yield reduction, and the difference among severity levels could be influenced by concentration of inoculum harbored and which survived the temperature, relative humidity, and rain during over seasoning and the location of debris could have an effect (Fininsa and Tefera, 2002). Under favorable temperature and relative humidity, the pathogen leads to reduction of the grain quality due to occurrence of the tags, besides to decrease in the productive potential of the common bean

Table 4. Effect of soil solarization x foliar spray x seed treatment interactions on seed yield of common bean during 2010 main cropping season at Haramaya and Hirna.

Treatments interactions			Yield (tons/ha)	
			Haramaya	Hirna
Soil solarization	Foliar spray	Seed treatment		
	CFS10	CST	3.6	3.5
	CFS10	MST	3.8	3.6
	CFS10	US	2.5	2.8
	CFS20	CST	2.9	3.1
Solarized	CFS20	MST	3.2	3.1
	CFS20	US	2.3	2.6
	UF	CST	2.6	2.9
	UF	MST	2.8	2.9
	UF	US	2.2	2.5
	CFS10	CST	3.2	3.2
	CFS10	MST	3.4	3.5
	CFS10	US	2.4	2.5
	CFS20	CST	2.7	2.8
Non-solarized	CFS20	MST	3.0	3.1
	CFS20	US	2.2	2.4
	UF	CST	2.4	2.6
	UF	MST	2.5	2.8
	UF	US	1.9	2.3
CV (%)			17.1	12.5
LSD at 5% for SS × FS × ST			0.88	0.68

MST=Mancozeb seed treatment; CST=Carbendazim seed treatment; CFS10 and 20=Carbendazim foliar spray at 10 and 20 days interval; UF=Unsprayed foliar; US=Untreated seed; LSD=Least significant difference; SS=Soil solarization; FS=Foliar spray; ST=Seed treatment.

(Pastor-Corrales et al., 1995). Generally, variation in severity between locations might have been due to the differences in environmental factors, inoculum load and time of infection. At both study areas, the magnitude of disease severity was significantly different between levels of solarization, it was higher for non-solarized than solarized soil (Figure 1).

Effect of integrated management of bean anthracnose on seed yield

There were significant (P<0.05) differences in interactions among the levels of solarization × foliar sprays × seed treatments on seed yield both at Haramaya and Hirna. While, the lowest seed yields were recorded on the combinations of non-solarized soil × unsprayed foliar × seed treatments at both study sites (Table 4). Low application rates of fungicide, fumigant, or herbicide have been successfully combined with soil solarization to achieve better pest control (Hartz et al., 1993). However, the interaction effects of soil solarization × seed treatments as well as seed treatments × foliar sprays were no

significant both at Haramaya and Hirna.

Yield components

Hundred seed weight was varied significantly (P<0.01) between each levels of soil solarization at Hirna and showed no significant differences at Haramaya (Tables 5 and 6). Maximum 100 seed weight was calculated from solarized soil as compared with the non-solarized control at Hirna. Solarization levels had no significant effect on the number of pods per plant and seeds per pod at both locations. Plants often grew faster and produced yields of increased quantity and quality (size and appearance) when grown in solarized as compared to non-solarized soil. Soil solarization was found to be effective against a wide range of soil-borne pathogens that affect many agricultural crops in various regions of the world were reported by Katan et al. (1976), Pullmann et al. (1979) and Camporota et al. (1986). Soil solarization has been used effectively to reduce populations of soil-borne pathogens (Katan et al., 1980). Seed treatments and foliar sprays alone had no significant effect on pods per

Table 5. Effect of soil solarization x foliar sprays interaction on yield and its components of common bean during 2010 main cropping season at Haramaya.

Treatments interaction		Pods/plant	Seeds/pod	100 seed weight (g)	Yield (tons/ha)
Soil solarization	Foliar sprays				
	CFS10	27.4	3.5	16.8	3.4
Solarized	CFS20	24.1	3.8	16.0	2.8
	UF	21.8	3.7	16.6	2.5
	CFS10	21.4	3.7	16.1	3.5
Non-solarized	CFS20	20.6	3.9	16.6	2.4
	UF	21.0	3.8	15.6	2.3
CV (%)		38.4	17.3	7.3	25.0
LSD at 5% for SS × FS		NS	NS	0.99	0.55

CFS10 and 20=Carbendazim foliar spray at 10 and 20 days interval; UF=Unsprayed foliar; LSD=Least significant difference; NS=Non significant difference; SS=Soil solarization; FS=Foliar spray.

Table 6. Effect of anthracnose on yield and its components of common bean under soil solarization, fungicide seed treatments using carbendazim and mancozeb, and foliar applications of carbendazim during 2010 main cropping season at Hirna.

Treatments	Pods/plant	Seeds/pod	100 seed weight (g)	Yield (tons/ha)
Soil solarization				
Solarized	18.5	4.9	16.0	3.0
Non-solarized	18.2	4.8	15.4	2.8
CV (%)	4.4	2.7	1.6	4.1
LSD (5%)	NS	NS	0.41	NS
Foliar sprays				
CFS10	17.6	4.8	15.6	3.2
CFS20	18.1	4.9	15.6	3.0
Control	19.3	4.8	15.9	2.6
CV (%)	21.1	6.0	6.7	15.5
LSD (5%)	NS	NS	NS	0.25
Seed treatments				
MST	17.2	4.8	16.0	3.1
CST	18.6	4.9	15.4	2.7
Control	19.2	4.7	15.7	2.9
CV (%)	19.1	6.9	4.5	12.5
LSD (5%)	NS	NS	NS	NS

MST=Mancozeb seed treatment; CST=Carbendazim seed treatment; CFS10 and 20=Carbendazim foliar spray at 10 and 20 days interval; LSD=Least significant difference; NS=Non significant difference.

plant, seeds per pod and 100 seed weight both at Haramaya and Hirna. Interaction between soil solarization × seed treatments was significant ($P<0.05$) on 100 seed weight at Hirna and non significant at Haramaya (Tables 7). The difference between soil solarization × seed treatments effects was not significant on the pods per plant and seeds per pod at both study areas. According to Beshir (1997), seed treatment with fungicides resulted good disease control which resulted in the high seed weight and low seed yield losses and effective in reducing the severity of anthracnose.

Relative yield loss in seed yield

Estimates of yield loss were made on the basis of field experiments at Haramaya and Hirna, in which fungicides application, particularly, in combination, promoted seed yield gain in each levels of solarization. Yield losses differed among plots treated with different fungicides and spray frequencies in each solarization levels. Highest yield losses were recorded on the control plots of non-solarized and followed by fungicides untreated plots in the solarized soil at both locations (Table 8).

Table 7. Effect of soil solarization x foliar sprays, soil solarization × seed treatments interactions on yield and yield components of common bean during 2010 main cropping season at Hirna.

Treatments interaction		Pods/plant	Seeds/pod	100 seed weight (g)	Yield (tons/ha)
Soil solarization	Seed treatments				
Solarized	MST	16.8	4.7	15.5	3.1
	CST	18.6	4.8	14.9	2.8
	US	19.9	4.8	15.8	3.1
Non-solarized	MST	17.5	4.7	16.5	3.2
	CST	18.5	5.0	15.9	2.8
	US	18.4	4.7	15.6	2.7
]CV (%)		19.1	6.9	4.5	12.5
LSD at 5% for SS × ST		NS	NS	0.64	NS
Soil solarization	Foliar sprays				
Solarized	CFS10	17.9	4.7	15.3	3.4
	CFS20	17.8	4.8	15.3	2.9
	UF	19.6	4.8	15.6	2.6
Non-solarized	CFS10	17.3	4.8	15.8	3.5
	CFS20	18.3	5.0	15.9	2.6
	UF	18.9	4.7	16.3	2.5
CV (%)		21.1	6.0	6.7	15.5
LSD at 5% for SS x FS		NS	NS	NS	0.47

MST=Mancozeb seed treatment; CST=Carbendazim seed treatment; CFS10 and 20=Carbendazim foliar spray at 10 and 20 days interval; UF=Unsprayed foliar; LSD=Least significant difference; NS=Non significant; SS=Soil solarization; FS=Foliar spray; ST=Seed treatment

Table 8. Relative seed yield losses caused by bean anthracnose under soil solarization, fungicide seed treatments using mancozeb and carbendazim and carbendazim foliar spray during 2010 main cropping season at Haramaya and Hirna.

Treatments		Haramaya		Hirna	
Soil solarization	Fungicides	Yield (tons/ha)	RYL (%)	Yield (tons/ha)	RYL (%)
Solarized	CST+UF	2.6	25.7	2.9	17.1
	MST+UF	2.8	20.0	2.9	17.1
	MST+CFS20	3.2	8.6	3.1	11.4
	CST+CFS20	2.9	17.1	3.1	11.4
	US+CFS20	2.3	34.3	2.6	25.7
	MST+CFS10	3.8	-8.6	3.6	-2.9
	CST+CFS10	3.6	-2.9	3.5	0.0
	US+CFS10	2.5	28.5	2.8	20.0
	Control	2.2	37.1	2.5	28.6
Non-solarized	MST+CFS20	3.0	9.1	3.1	8.8
	US+CFS20	2.2	33.3	2.4	29.4
	CST+CFS20	2.7	18.2	2.8	17.7
	MST+CFS10	3.4	-3.0	3.5	-2.9
	CST+CFS10	3.2	3.0	3.2	5.9
	US+CFS10	2.4	27.3	2.5	26.5
	MST+UF	2.5	24.2	2.8	17.7
	CST+UF	2.4	27.3	2.6	23.5
	Control	1.9	42.4	2.3	32.4

MST=Mancozeb seed treatment; CST=Carbendazim seed treatment; CST10 and 20=Carbendazim foliar spray at 10 and 20 days interval; UF=Unsprayed foliar; US=Untreated seed; RYL=Relative yield loss.

Figure 1. Infected pods by *C. lindemuthianum* on control plant at Haramaya.

Figure 2. Healthy plants on treated plot with MST+CFS10 at Hirna.

Conclusions

Differences in epidemics occurred within each level of solarization due to fungicide applications under natural infection. Plots treated with combinations of mancozeb seed treatment and carbendazim spray frequencies at 10 days' intervals obtained the best protection in terms of disease severity and increased seed yield at bothlocations. Therefore, this study contributes to integrated bean anthracnose management options using soil solarization seed treatment and foliar spray is justified. Bean growers should be encouraged to use seed treatment and other cultural practices applicable in the area against anthracnose when growing susceptible bean cultivars. And the use of resistant varieties and producing pathogen free seeds would be a cheap means

of controlling the anthracnose disease and need to be explored.

REFERENCES

Beshir T (1997). Loss assessment study on haricot bean due to Anthracnose. Pest Manage. J. Ethiopia 1:69-72.

Camporota P, Baudrand M, Taussig C and Drame A (1986). La de´sinfection solaire du sol contre la pourriture basale de la laitue. Phytoma 3:41–42.

Central Statistical Authority (CSA) (2005). Agricultural Sample Survey 2005/2006. Report on forecast of area and production of major crops, *Statistical Bulletin, No.* 171, Addis Ababa, Ethiopia.

Central Statistical Authority (CSA) (2010). Area under production of major crops. Statistical Bulletin. 245, Addis Ababa, Ethiopia.

Chakraborty N, Chatterjee C, Quimio TH (2008). Integrated management of fusarial wilt of egg plant (*Solanum melongena*) with soil solarization. *Micologia* Aplicada Int. 21(1):25-36.

CIAT (1988). Inform annual (1988): program de frijol. Documento de Trabajo 72. CIAT, Cal, Colombia. CIAT African Workshop Series, No.4. pp. 110-120.

Dillard HR, Cobb AC (1993). Survival of *Colletotrichum lindemuthianum* in bean debris in New York State. Plant Dis. 77:1233-1238.

Ethiopian Pulses Profile Agency (EPPA) (2004). Ethiopian Export Promotion Agency Product Development & Market Research Directorate Ethiopia.

Fininsa C, Tefera T (2002). Effect of primarily inoculum source of bean anthracnose and their effect on epidemics and yield. Trop. Sci. 42:30-34.

Freeman S, Nizanl Y, Dotan S, Even S, Sando T (1997). Control of *Colletotrichum acutatum* in strawberry under laboratory, green house and field conditions. Plant Dis. 81:749-752.

GenStat (2009). GenStat Release 12.1. PC/windows XP/ 12th edition 06 January. Copy right 2009 VSN International. Ltd.

Gwary DM (2008). Integrated Management of Sorghum Anthracnose Through the Use of Fungicides, Crop Varieties and Manipulation of Sowing Dates in Sudan Savanna of Nigeria. Int. J. Agric. Biol. 10(6):661-664

Habtu A (1987). Haricot bean diseases and their importance in Ethiopia. Eth. J. Agric. Sci. 9:968-973.

Habtu A, Sache I, Zadoks JC (1996). A survey of cropping practices and foliar diseases II. Severity-damage relationship. Pest Manage. J. Ethiopia 2:113.

Hadi H, Kazem GG, Rahimzadeh KF, Mostafa V, Reza SM (2006). Response of common bean to different levels of shade. J. Agron. 5:595-599.

Hall R (1994). Compendium of bean diseases. 2nd edition. APS Press, the American Phytopathological Society, St. Paul, Minnesota, P. 73.

Hartz T, DeVay J, Elmore C (1993). Solarization is an effective soil disinfestations technique for strawberry production. Hort. Sci. 28(2):104-106.

Jenny D (2010). Pulse seed treatments and foliar fungicides 7th edition South Australian Research and Development Institute, (08) 8303 9389; Kurt Lindbeck, Industry and Investment New South Wales (02) 6938 1608.

Katan J, Greenberger A, Alon H, Grinstein A (1976). Solar heating by polyethylene mulching for the control of diseases caused by soil-borne pathogens. Phytopathology 76:683-688.

Katan J, Rotem I, Finkel Y, Daniel J (1980). Solar heating of the soil for the control of pink root and other soil-borne diseases on onions. *Phytoparasitica* 8:39-50.

Koch SH (1996). *Colletotrichum* spp. On dry bean and lupines in South Africa. Ph.D thesis. University of Pretoria, Pretoria.

Pastor-Corrales MA, Tu JC (1989). Anthracnose. Pages 77-144 IN: Schwartz, H.F and Pastor-Corrales, M.A., eds. Bean production problems in the Tropics. 2nd edition. CIAT, Cali, Colombia, 560 p.

Pastor-Corrales MA, Otoya MM, Molina A, Singh SP (1995). Resistance to *Colletotrichum lindemuthianum* isolates from Middle and Andead South America in different common bean race. Plant Dis. 79:63-67.

Pullman GS, DeVay JE, Garber RH, Weinhold AR (1979). Control of soil-borne pathogens by plastic tarping of soil. In: Schippers B, Gams W (eds) Soil-borne plant pathogens. Academic Press, London, pp. 439-446.

SAS (Statistical Analysis System) (2009). Software Version 9.2. Inc. Carry, North California.USA.

Schwartz RF, Pastor-Corrales MA, Singh SP (1983). New sources of resistance to anthracnose and angular leaf spot of bean (*Phaseolus vulgaris* L.), Euphytica, 31:741-754.

Sindhan GS, Bose SK (1981). Evaluation of fungicides against anthracnose of French bean caused by *Colletotrichum lindemuthianum*. Indian Phytopathol. 34:325.

Stapleton JJ, DeVay JE (1982). Effect of soil solarization on populations of selected soilborne microorganisms and growth of deciduous fruit tree seedlings. Phytopathol. 72:323-326.

Stapleton JJ (1997) Soil solarization: an alternative soil disinfestations strategy comes of age. UC Plant Protection Quarterly, 7:1-5.

Tesfaye B, Pretorius ZA (2005). Seed treatment and foliar application of fungicide for the control of bean anthracnose. Pest Manage. J. Eth. 9:57-62.

Wheeler JBEJ (1969). An Introduction to Plant Diseases. Wiley, London, P. 347.

23

Genetic diversity and association of physio-morphological traits for drought resistance in wheat (*Triticum aesitivum*)

Rudra Bhattarai[1], Surya Kant Ghimire[2], Dhruba Bahadur Thapa[3] and Madhav Prasad Pandey[4]

[1]Regional Agriculture Research Station, Nepal Agriculture Research Council, Tarahara, Nepal.
[2]Agricultural and Forestry University, Chitwan, Nepal.
[3] Agribatany Divisions, Nepal Agricultural Research Council, Kathmandu, Nepal.
[4]Institute of Agriculture and Animal Science, Trivuban University, Nepal.

Wheat (*Triticum aestivum* L.) occupied 17 percentage of the total cultivated land in the world whereas it has contributed major role to maintain the Nepalese agricultural gross domestic products (MOAC, 2012) but area of cultivation has been decreasing. Moreover, potential yield of popular varieties have reduced by the effect of drought. To combat from drought loss, tolerant variety is one of the major solutions to address that issue. From this point of view, an experiment was conducted in split plot design from November 2009 to May 2010. The result showed significant variations among both for genotypes as well as to both level of water management. The average leaf area were reduced by 8.80 cm^2, 0.2655 kg biomass yield, 6.64 g thousand kernel weight, 4.11 booting days, 3.23 heading days, 1.26 flag leaf senescence and 0.0706 kg yield in drought condition. Similarly, WK1123 has least drought susceptibility index for yield. However, relative water content and chlorophyll content were correlated significantly with yield. To conclude, genotypes WK1701, WK1444, WK1123, 3EBWYT513 and 3EBWYT512 considered as the most drought tolerant and might be used as a variety in drought prone area of Nepal.

Key words: Wheat, drought, stress chlorophyll content and relative water content.

INTRODUCTION

Wheat (*Triticum aestivum* L.) is the first important cereals crop in the world and staple food for 35% of the world's population and provides more calories and proteins in the world's diet than any other crop. In Nepal, total area of 765317 ha and 1846142 mt of production with productivity of 2412 kg/ha (MOAD, 2012/2013). Similarly, wheat has been grown both in irrigated and unirrigated conditions. However, 63.19% of wheat has been cultivated in completely irrigated condition with improved varieties but 34.44% of wheat grown in unirrigated Nepalese condition (MOAD, 2012/2013). Therefore, it will be wise to develop the drought tolerant varieties which would address the issue of yield loss through drought. The prevailing cropping system, rice after wheat would

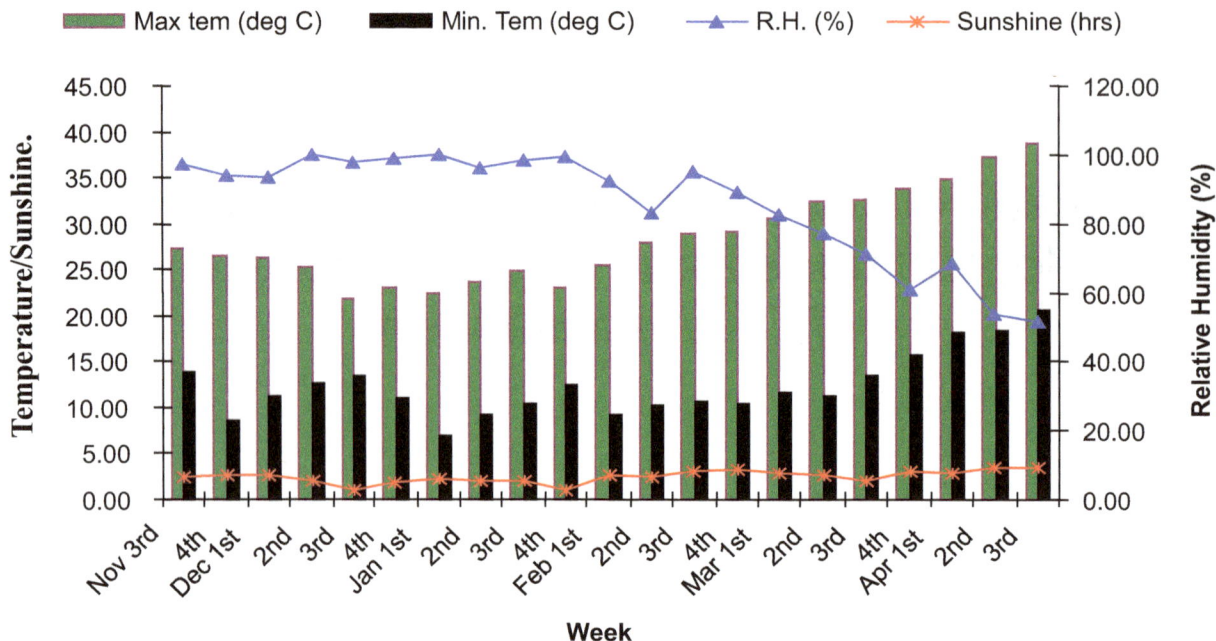

Figure 1. Agrometorological condition of the research site during 2009/2010.

enforced to the plantation of wheat in late sown condition that generally damaged the crop by combined effect drought and terminal heat stress. Drought stress has been recognized as one of the major abiotic factors limiting wheat production in Nepal (Sharma et al., 2008). Due to increasing summer temperature, uneven annual rainfall pattern and depleting water resource for irrigation, breeding wheat for drought tolerance will become an increasingly higher priority in this region (Joshi et al., 2007). Thus, wheat breeding for drought tolerance or higher water use efficiency is needed to supply food to the growing Nepalese population therefore, this study aimed to assess the genetic variability of drought adaptive traits and their association with drought tolerance which is vital for the development of drought tolerant wheat cultivars.

MATERIALS AND METHODS

The research site and experimental design

The field experiment was conducted at the research farm of the Institute of Agriculture and Animal Science (IAAS), Rampur, Chitwan, Nepal, from November 2009 to June 2010. The geography of this location is 27°64' N latitude and 84° 34' E longitude and at an altitude of 228 m above sea level, respectively. The research location is characteristics of subtropical climate and experiment was setup from November, 2009 to June, 2010 after that, field remained fallow until the next season, that is, November, 2010. Thirty genotypes were used in the research where most of the lines were Nepalese landraces and advanced lines derived from Agriculture Botany Division NARC, Khumaltar, Nepal and commercial varieties and check used was WK1204. The experiment

was laid out in a split plot design with optimum moisture and moisture stressed environments as main plot factors and 30 wheat genotypes as sub-plot factors. Each plot was 0.75 m². The rain-out shelter (stress) plot was erected over the plot at the start of tillering where trench had been made around the main plot, provided with a 1 m deep and 0.5 m wide ditch around the edge to prevent rain coming off to the shelter from seeping into the plot. Other intercultural operations and cultivation practices were completed according to the national recommendation for wheat cultivation.

Each set of experiment was replicated three times where composite soil sample had been taken from the soil after land preparation from the field at different depth (0 to 30 cm). The soil texture class was determined sandy loam where moisture content was found 75% of the field capacity at the time of seeding for both condition. For moisture stressed experiment, soil moisture content was maintained at 35% of the field capacity from heading to flowering. The maximum temperature recorded was 40°C in the month of April. The Minimum- maximum, standard deviation and Coefficient of variation for the quantitative traits were being measured. Differentiation between populations was usually quantified using the difference in mean expression (t-test) (Spagnoletti Zeuli and Qualset, 1987) (Figure 1). The following were the traits measured in this experiment:

Canopy temperature depression

The canopy temperature depression was calculated by using the following formula.

Canopy Temperature Depression (CTD) = Ambient temperature – Canopy Temperature

Relative water content

RWC was calculated from the equation of Schonfeld et al. (1988) as:

Table 1. Flag leaf area, plant height, panicle length, TKW, Booting days, heading days, Anthesis days, and yield as affected by drought condition in IAAS, Rampur, (2009/2010).

Condition	Leaf area	Plant height (cm)	Panicle length (cm)	TKW (gm)	Booting duration (days)	Heading duration (days)	Anthesis duration (days)	Yield (kg/ha)
Drought(mean)	54.4063[a]	113.075[a]	11.088[a]	34.825[a]	67.644[a]	75.100[a]	84.467[a]	2926[a]
SEM+or-for drought	2.58301	4.25584	0.3934	1.6209	2.5595	2.793	3.0992	0.01334
Irrigated	59.6154[b]	121.35[b]	11.765[b]	37.34[b]	70.044[b]	78.333[b]	88.233[b]	3365[b]
SEM+or- flor Irrigated	1.73924	1.4126	0.43526	1.581	1.6399	1.9585	2.159	0.0156
t-value	3.021**	3.137**	4.66**	3.133**	2.043*	2.812**	3.096**	3.380**

Means within the column with same letters are not significantly different. t- Value obtained from paired t-test are significantly different at 5% level of significance (*) and highly significantly different at 1% level of significance (**).

$$RWC = \frac{Fresh\ wt - Dry\ wt}{Turgid\ wt - Dry\ wt} \times 100$$

Flag leaf duration, days of booting, days of heading, Panicle length, seed weight per spike, leaf area, plant height, productive tiller per plot, thousand kernel weight (TKW) and drought susceptibility index (DSI).

Drought susceptibility indices for grain yield of each genotype were calculated as proposed by Fischer and Maurer (1992). DSI = [(1-Y/Yp)/D], Y = yield at normal sown condition, Yp = yield at drought sown condition, D = stress intensity = 1- X/Xp, X = mean Y of all genotypes, and Xp = mean Yp of all genotypes. Analysis of variance and calculation of means was done by using Cropstat 7.2. The main plot treatment comparison was done by paired t-test with SPSS. UPGMA clustering and PCA was done using Minitab-14.

RESULTS AND DISCUSSION

Days to booting and heading

The mean number of days to booting in irrigated normal sown condition was 70.044 where 67.6444 days for the stress conditions. Stress condition and normal condition were nonsignificant with dates to booting as seen by paired t-test (Table 1). In the irrigated conditions, WK1627 had

minimum days to booting days (61) whereas WK1719 was identified as late booted (82). The decrease in heading days in stressed plants as compared to non stress was reported by Reynolds et al. (1993).

Days to flag leaf senescence

The mean number of days to flag leaf senescence differed non significantly for the main plot treatment as shown by paired t-test (Table 1) but significantly different was seen through the Analysis of Variance. The mean numbers of days to flag leaf senescence were found 117.3 days and 114.67 days for Irrigated and drought conditions respectively. In the drought condition, WK1627 was senesced as earliest with 109 days whereas WK1123 senesced in later with 123 days. Later on in the growth cycle, water stress reduces the green leaf duration (GLD) from accelerated senescence by Reynold et al. (1993) (Table 2).

Thousand kernel weight

There was significant difference in the mean

thousand kernel weight for the Irrigated and drought condition as seen by paired t-test. The 37.34 and 34.825 g were mean thousand kernel weight (TKW) for irrigated and drought plot, respectively. The Godawari local had 20.847 g of thousand kernel weight which was found as minimum among the others and maximum (42.37 g) kernel weight was found to WK1700. However, genotypes WK1123 and WK1670 had least drought susceptibility index and considered as most useful for drought environment. Sharma et al. (2008) had identified reduction in kernel weight as a potential tool for indirect selection criteria for grain yield under drought stress.

Plant height

The drought and irrigated plots have had mean plant height of 113.07 and 121.35 cm respectively. The significant variations were shown among genotypes. In addition, shortest genotype was WK1204 (93.26 cm) where as Godawari local had been tallest among genotypes in drought condition. However, WK1912 was found as shortest (108.56 cm) and tallest was Godawari local with 134.4219 cm. Drought

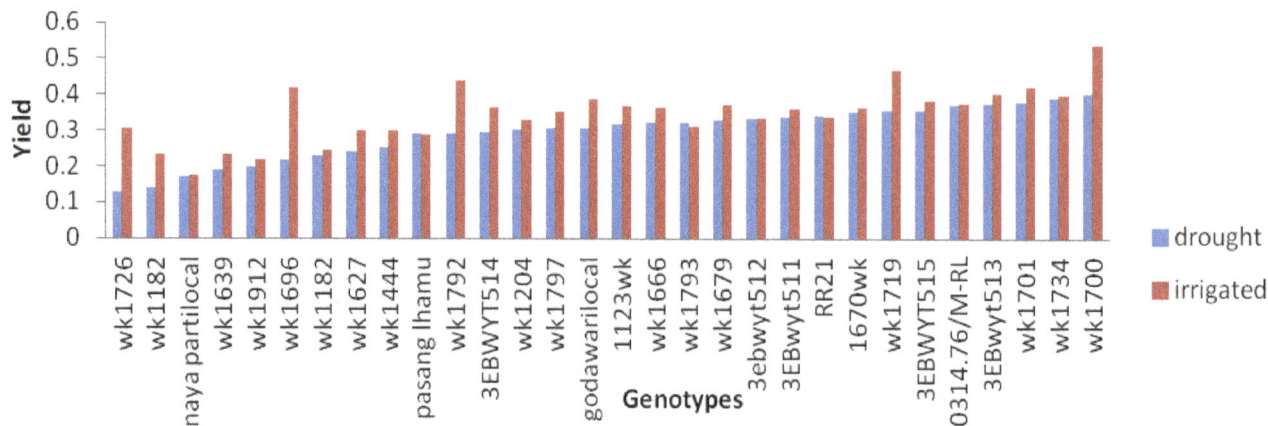

Figure 2. The effect of moisture level on the yield response of genotypes used in the experiment at IAAS, Rampur (2009/2010).

Table 2. Drought susceptibility index (DSI) for TKW, Yield/Plot, and Biomass/Plot of Five selected genotypes in IAAS, Rampur (2009/2010).

Genotypes	Biomass	Genotypes	TKW	Genotypes	Yield
WK1123	-7.017	WK1123	-0.4071	3EBwyt513	-10.198
F60314.76/M	-3.462	Wk1670	-0.1995	3ebwyt512	-8.195
Wk1912	-1.926	3EBwyt511	-0.0493	3EBwyt511	-5.324
Nayapartilocal	-0.4193	3EBwyt512	-0.003	WK1670	-0.0248
Wk1726	0.25625	3EBwyt513	0.0024	WK1123	0.254

significantly reduced internodes length and thus reduced the length of the main stem by Richards (1992). This decrease in plant height negatively correlated (-0.798) with high yield under stressed condition which guessed for the stem reserve mobilization might transferred towards the spikes.

Yield and yield component traits

There were significant differences between water management for mean per plot yield as seen by pair t-test (Table 1). In more clearly, mean per plot yield in drought and irrigated condition were 2563 and 3269 kg/ha. Moreover, highly significant yield differences among genotypes had been found for drought condition. In drought condition, WK 1701 and WK1700 had produced yield of 3737 and 3541 kg/ha respectively but, lowest yield found for Nayaparti local (1590.867 kg/ha). In contrast to drought condition, irrigated condition produced highest yield of 5504 kg/ha from WK1700 but least yield had been found from F60314.76/M-RL(1975 kg/ha) (Figure 2).

Drought susceptibility index

Drought susceptibility index was calculated for yield and yield related traits. This was important because we need to address higher yield potential against drought condition (Blum et al., 1989) (Table 3). Drought susceptibility indices are use either based on drought resistance or susceptibility of genotypes (Blum et al., 1989). Genotypes WK 1123, 3EBWYT513 and 3EBWYT512 were less influenced by the drought for biomass yield and thousand kernel weight.

Relative water content

Mean relative water content had valued for 1.370 in normal sown wheat while it was 1.244 in drought condition. Genotypes with highest RWC was WK1701 (2.19) but lowest was found to WK1123 (0.26) in drought therefore WK1701 might be more drought tolerance than rest of genotypes. Therefore genotypes which maintained higher RWC under stress conditions had maintained higher yield in spite of drought condition suggested for the existence of drought tolerance traits which has been supported by Araus et al. (1997) (Figure 3). The genotypic correlation between gain yield and RWC under water stress condition was positively significant (0.84[**]) suggested that under drought condition higher relative water content in leaf might contributed with the better yield (Blum et al., 1989).

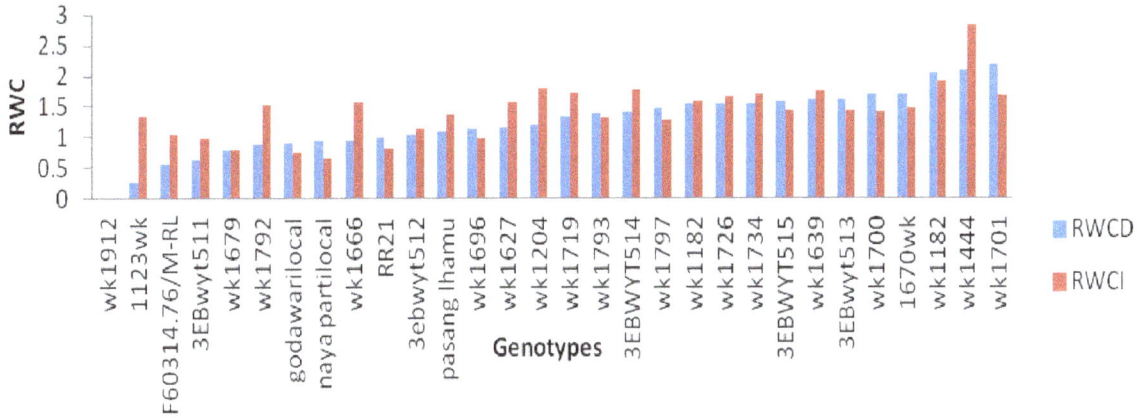

Figure 3. Wheat genotypes differing for RWC in both drought and irrigated condition (2009/2010).

Table 3. The correlation among different measured traits under drought conditions.

					Correlations				
	LA	Plht	DOFS	TKW	Y/Plot	PL	DOB	DOH	DOA
LA	1	0.372*	0.509**	0.583**	0.495**	0.693**	0.505**	0.452*	0.509**
Plht		1	0.903**	0.622**	-798**	-769**	0.836**	0.850**	0.857**
DOFS			1	0.693**	0.986**	0.886**	0.941**	0.949**	0.960**
TKW				1	0.720**	0.789**	0.537**	0.542**	0.602**
Y/Plot					1	0.884**	0.898**	0.915**	0.929**
PL						1	0.783**	0.793**	0.833**
DOB							1	0.992**	0.984**
DOH								1	0.988**
DOA									1

*. Correlation is significant at the 0.05 level (2-tailed), **. Correlation is significant at the 0.01 level (2-tailed).

Correlation studies

Yield had significant correlation with Thousand Kernel Weight and days to flag leaf senescence to each of drought and irrigated condition but negative correlation of plant height with grain yield was found under drought condition (Table 4) which was supported from Gill et al. (1989) and Patil and Jain (2002). On the other hand, spike length was positively correlated significantly with weight of grains per spike in both condition therefore selection based on panicle length would favor for the bold grain in drought condition. Similar findings supported by Shah et al. (1988). Briefly, highly significant positive association had been established for days to flag leaf senescence, thousand kernel weight, and panicle length, days of booting, days of heading and days to anthesis with higher yield per plot in drought condition. Ultimately, that would help to selection based on other traits rather than yield. The all the measured traits were significantly correlated with each other in irrigated condition (Table 5) suggested that selection based on other traits would

beneficial for the indirect selection of genotypes as a variety for future use.

Canopy temperature depression (CTD)

Average CTD in stress at three growth stages recorded highest for WK1444, WK1639 and WK1701 and had maintained lower canopy temperature. Relatively lower canopy temperatures under stress indicate the better plant water status and CTD were positively correlated with yield in stress condition (Blum et al., 1989). Similar kind of research result had been observed for correlation between CTD with yield advantage in stress condition.

Soil physical analysis and development meter (SPAD)

There were significant differences of SPAD values between drought and irrigated condition to illustrate,

Table 4. The correlation among different traits measured under irrigated conditions.

	LA	Plht	DOFS	TKW	Y/Plot	PL	DOB	DOH	DOA
					Correlations				
LA	1	0.530**	0.628**	0.587**	0.613**	0.740**	0.637**	0.450*	0.505**
Plht		1	0.32	0.639**	0.412*	-0.406*	0.431*	0.398*	0.495*
DOFS			1	0.506**	0.975**	0.749**	0.970**	0.851**	0.892**
TKW				1	0.531**	0.580**	0.446*	0.425*	-392*
Y/Plot					1	0.745**	-923**	0.763**	-813**
PL						1	0.768**	0.600**	0.663**
DOB							1	0.923**	0.946**
DOH								1	0.966**
DOA									1

*. Correlation is significant at the 0.05 level (2-tailed),**. Correlation is significant at the 0.01 level (2-tailed).

Table 5. Mean square from ANOVA of leaf area (LA), biomass (kg/Plot), plant height (cm), Flag leaf senescence (FLS), TKW, booting days, heading days, anthesis days as influenced by moisture stress of genotypes at IAAS Rampur (2009/2010).

Source	DF	LA	Biomass	Plant height (cm)	FLS	TKW
Water management	1	59.58	1.43	479.2	45.76	134.8
Repx WM	3	120.79	0.047	85.20	5.51	28.08
BlockxRepx WM	18	68.35	0.040	154.37	7.514	72.53
Genotypes	29	720.15**	0.50**	626.28**	85.24**	336.10**
Genotypesx WM	29	123.57*	0.017*	89.044*	6.46**	13.81**
Error	116	60.11	0.036	76.342	7.46	8.50
Total	196					

Source	DF	Booting days	SPAD at boot	Anthesis days	SPAD at Anthesis	Grain (ton/ha)
Water management	1	76.991	219.862	152.98	210.85	0.059
Repx WM	3	15.70	8.022	7.91	36.32	0.012
BlockxRepx WM	18	10.397	6.932	9.28	14.32	0.0070
Genotypes	29	289.18**	110.57**	340.90**	170.15**	0.0393**
Genotypesx WM	29	10.54*	9.96*	8.18**	11.38**	0.0088*
Error	116	7.24	10.24	6.066	11.43	0.0106
Total	196					

*, ** Significant at 0.05 and 0.01 probability levels, respectively. Water management (WM) refers to irrigated and drought conditions.

highest SPAD value had been found in normal irrigated condition with 24.82 and little less value was defined in drought condition worth 23.54. This decrease in value of SPAD was closely related with the decrease in value of chlorophyll and ultimately gave lower yield which has supported from Araus et al. (1997) (Table 6).

Multivariate analysis

UPGMA clustering

Four clusters were seen with a minimum of 50% similarity

level in UPGMA clustering where Cluster 1 consist of 4 genotypes WK1123, WK1182, F60314.76/M-RL and WK1444. The member of this cluster were characterized by their lowest drought susceptibility index for Biomass, TKW, Yield /Plot, Number of tillers per plot and Number of seed per panicle therefore these were selected and categorized as the most drought resistance. Cluster 2 was developed by the use of Biomass yield/plot, number of tillers per plot. Genotypes consisted WK1670, WK1639, WK1700, WK1701, 3EBWYT511, Pasang lhamu, WK1679, WK1726, 3EBWYT515, RR21, WK1792, WK1793, WK1794, WK1204 and WK1627. The members of this cluster also have mild drought tolerance

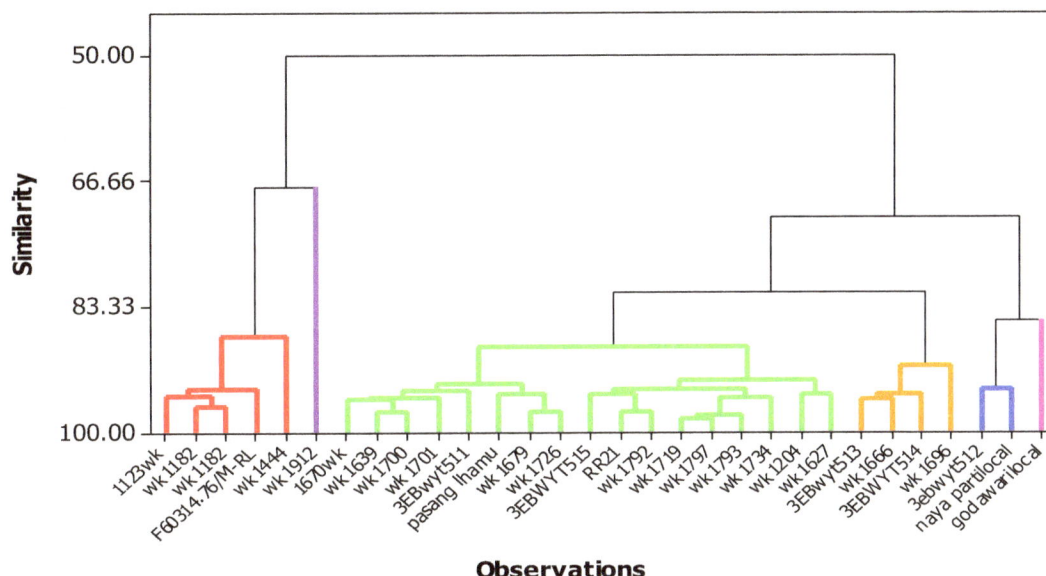

Figure 4. Different cluster groups in the dendogram while all the genotypes were grouped based on their average performance from irrigated environments.

Table 6. Eigen analysis and proportion of PCA to the different traits of the correlation matrix.

Eigenvalue	4.2937	1.8783	0.8686	0.6713	0.5034
Proportion	0.477	0.209	0.097	0.075	0.056
Cumulative	0.477	0.686	0.782	0.857	0.913
Variable	PC1	PC2	PC3	PC4	PC5

characters. Similarly, Cluster 3 consist of 3EBWYT513, WK1666, 3EBWYT514 and WK1696 which have categorized on the basis of similar type of plant height in addition to this, heights of these genotypes were highest among other and this group was also characterized on the basis of mild tolerance for drought adaptation on the basis of yield component traits. Cluster 4 was using the days to flag leaf senescence, days to booting, days to heading, days to anthesis, CTD at 13 February and CTD at 24th March. These clusters consist of 3EBWYT512, Nayaparti local and Godawari local (Figure 4).

Principle component analysis

The contribution of various parameters in the first three principle components is presented (Table 6). The first three principle component explained variation of 78.2% of the total variation. PC1 contributed 47.7% of the total variation and PC2 and PC3 contributed 20.9 and 9.7% of the total variation, respectively. Hence, the three components can be regarded the main contributing factors for the grouping of the genotypes across various groups. The PCA supported the results obtained by

cluster analysis. It clearly separated the four major groups recognized by the cluster analysis. This PCA analysis clearly showed that there were presences of diversity between and within group of the cluster for many quantitative traits (Figure 5).

Conclusion

The mean per plot yield in drought condition was 2563 kg/ha whereas it was 3269 kg/ha in irrigated condition. To conclude, all the measured traits were correlated medium to high correlation coefficient in both environments. Accordingly, possibility of indirect selection would increase rather than as direct yield component. In summary, least drought sensitive index for yield per plot were found to 3EBWYT513, 3EBWYT512, WK1444 and WK 1701 therefore these genotypes could be used either directly as a variety or used as a parent for hybridization.

Conflict of Interest

The authors have not declared any conflict of interest.

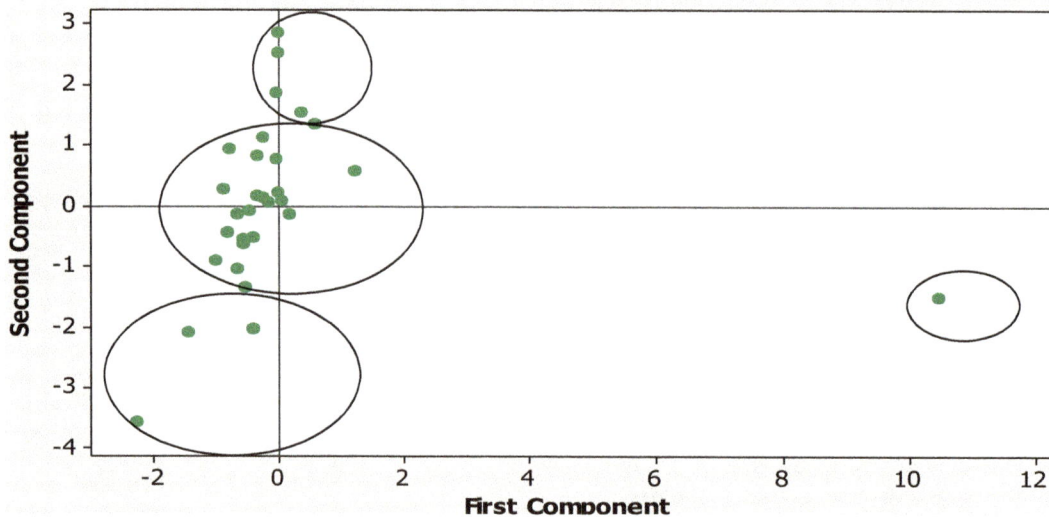

Figure 5. The score plot of first two components of 30 wheat genotypes in irrigated environment.

ACKNOWLEDGEMENTS

This research was possible due to the financial support from National Agricultural Research and Development fund, Government of Nepal. Similarly, seed materials were received from Agri-botany Division, Nepal Agricultural Research Council, and Government of Nepal. I hearty acknowledge to my major advisor Dr. Surya Kant Ghimire for his constant support and constant guidance throughout the whole research.

REFERENCES

Araus JL, Reynolds MP, Acevedo E (1997). Leaf posture, grain yield, growth, leaf structure, and carbon isotope discrimination in wheat. Crop Sci. 33:1273-1279.

Blum A, Shipiler L, Golan G, Mayer J (1989). Yield stability and canopy temperature of wheat genotypes under drought stress. Field Crops Res. 22:289-296.

Gill KS, Cox TS, Gill BS (1989). Variation in molecular markers among geographically diverse accessions of *Triticum tauschii*. Genome 34:354-361.

Joshi A, Mishra B, Chatrath R, Ortiz Ferrara G, Singh R (2007). 'Wheat improvement in India: present status, emerging challenges and future prospects. Euphytica 157(3):431-446.

MOAC (2012). Statistical Information on Nepalese Agriculture, Kathmandu: Ministry of griculture and Cooperatives/Agri-Business Promotion and Statistics Division.

Patil P, Jain MAJ (2002). Prospects for crop production under drought: research priorities and future directions. Ann. Appl. Biol. 147:211-226.

Reynolds MP, Balota M, Delgado MIB, Amani I, Fischer RA (1993). Physiological and Morphological traits associated with spring wheat yield under hot, irrigated conditions. Aus. J. Plant Physiol. 21:17-30.

Schonfeld MA, Johnson RC, Carwer BF, Mornhinweg DW (1988). Water relations in winter wheat as drought resistance indicators. Crop Sci. 28:526-531.

Shah M, Srivastava JP, Kumar A (1988). Effect of water on water potential components in wheat genotypes. Indian J. Plant Physiol. 33:312-317.

Sharma RC, Ortiz-Ferrara G, Bhatta MR (2008). Regional trial results show wheat yield declining in the eastern Gangetic plains of south Asia. Asian J. Plant Sci. 6:638-642.

Spagnoletti Zeuli PL, Qualset CO (1987). Geographical diversity for quantitative spike characters in a world collection of durum wheat. Crop Sci. 27:235-241.

24

Analysis of phenotypic responses influencing leaf growth rate and harvest parameters in cassava (*Manihot esculenta* Crantz) under hydrothermal stress

Ephraim Nuwamanya[1,2], Patrick R. Rubaihayo[2], Joseph F. Hawumba[2], Robert Kawuki[1] and Yona Baguma[1]

[1]National Crops Resources Research Institute (NaCRRI), P.O Box 7084, Kampala, Uganda.
[2]College of Agriculture and Environmental sciences, School of Agriculture, Makerere University Kampala, P. O. Box, 7062, Kampala-Uganda.

Plants under hydrothermal stress show a variety of responses aimed at reducing leaf growth and total leaf area, thereby reducing physiological processes such as transpiration and photosynthesis. The effects of hydrothermal stress on cassava plant growth parameters were thus investigated. Twenty cassava varieties representing a broad range of genetic diversity were used. Plants were grown in the field, physical measurements were made and percentage changes in growth parameters and biomass accumulation recorded bimonthly. Significant variation among varieties was found for the response to stress of leaf growth rate (losses of between 60 and 100% after stress) and duration (losses within 1 to 2 months) which equally resulted into reduction in leaf area. Variations were also observed in leaf retention (0 to 40%) and expansion rate (lamina width (25 to 33%) and length (14 to 58%) among different varieties. Differences were also observed for the time and rate of leaf loss during stress period. Based on observed differences, varieties were grouped under three distinct clusters including early recovering varieties, stay green varieties and the susceptible varieties. Alterations in leaf properties were highly correlated to harvest index where low harvest index (0.2 to 0.4) was observed for stay green and susceptible varieties compared to early recovering varieties (0.4 to 0.7). From the observation, different coping mechanisms, important in selection of drought stress tolerant genotypes were identified and pointed to specific genetic mechanisms for leaf retention/loss and biomass accumulation. The results suggest that rate of growth and duration differences are due to different physiological mechanisms, and can be combined to select for hydrothermal stress tolerant varieties.

Key words: Leaf expansion rate, leaf area, leaf development, recovery mechanisms.

INTRODUCTION

Plants respond to abiotic stresses, in part through the reduction of physiological processes (transpiration, photosynthesis and respiration) which can be can be achieved by morphological adjustments such as stomatal closure (Tardieu and Davies, 1993), leaf wilting and leaf rolling (O'Toole and Cruz, 1980). Understanding morphological trait adjustments during cassava growth is thus essential in elucidating how environmental stress affects plant growth and development components. In addition, common stress tolerance mechanisms involve reductions in leaf and shoots growth which are genetically determined, but are highly influenced by the environment (Kaplan, 2001). Therefore, their behavior can explain observed changes in plant structure in response to changes in environment. Notably, changes in leaf area under stress can be a consequence of reduced leaf number and/or reduced area of individual leaves (Aguirreza´bal, 2006; Kaplan, 2001) which in turn leads to reduction of leaf growth, a factor that is influenced by decreased cell division (Schuppler et al., 1998; Granier et al., 2000), cell wall hardening (Neumann, 1995), and decreased turgor (Hsiao et al., 1998).

The extent of leaf growth reduction in response to stress is an important factor in determining the adaptation of a certain plant or variety to an environmental scenario. As observed in some plants, prolonged drought leads to reduced leaf growth but plants are more likely to reach maturity with considerable yield. On the other hand, during short-term water deficits, a plant which maintains leaf growth is likely to have higher yields (Blum, 2005). This too is observed under thermal stress which results in total leaf morphology shift as observed by Guerin et al. (2012). Architectural changes during growths, which are related to such limitations as abiotic stresses or translocation of water and nutrients, can therefore decrease or, increase the effects of physiological changes on the functioning of entire plants (Valladares and Pearcy, 1999; Zotz et al., 2002). Such scenarios point to an internal genetic control mechanism that allows cassava to adapt to changes in environment. It also points to a well regulated gene expression network allowing plants to respond appropriately to hydrothermal stress.

Since the photosynthetic performance of vascular plants depends on plant and leaf size (Schmidt and Zotz 2001), changes in leaf related parameters such as leaf area, leaf size and leaf numbers are likely to influence net productivity of the plant and related plant processes such as cell expansion, physiology and development. Usually changes in plant size result into changes in leaf shape although changes in leaf sizes and leaf architecture may be mediated by the effects of the environment. In addition, changes in the leaf area index are not only affected by the age of the plant but by other environmental factors such as photosynthetic light flux and temperature (Zotz et al., 2002). This implies that occurrence of hydrothermal stress has a negative impact on leaf development and related processes such as nitrogen use efficiency and carbon partitioning which determine the photosynthetic potential of the leaf and hence its broad functioning (Katahata et al., 2007).

A plant can maintain its leaf area by maintaining its growth rate (Salekdeh et al., 2009; Blum, 2005) or by increasing the duration of leaf growth (Aguirreza´bal , 2006) both of which have the benefit of increasing the opportunity for recovery after offset of stress (Alves and Setter, 2004). However, changes in the moisture and temperature regimes over a certain period may result into failure of the plant to uphold its growth rate hence reducing the plant size. In a number of crops (Fasehun, 1979) low soil moisture results into dehydration or loss of cell water and hence cell death or inactivity (Osorio et al., 1998). Water deficits, important causes of dehydration, result into reduced branching, leaf production and leaf expansion which result into reduced biomass (Osorio et al., 1998) and reduced photosynthetic capabilities (Seyed et al., 2012). This could be attributed to dehydration that results into altered biomass allocation and changes in the translocation/evapo-transpiration processes. Thus, acclimatization to dehydration stress may trigger avoidance mechanisms which in turn protect the plant from dehydration stresses (Seyed et al., 2012). In particular, loss in turgor in the leaves results into surface area reduction hence reduced leaf size (Salekdeh et al., 2009), a stress tolerance/avoidance strategy. Hence, different plants will respond differently to such changes where the responses are mediated by the plants' genetic makeup. This genetic variability in leaf growth and development could be used to develop crop varieties adapted to specific stress scenarios. However, breeding for these traits is not a common approach for obtaining drought resistance in crop species because of a lack of well-characterized source of genetic variability. This study seeks to address this concern by providing relevant phenotypic information to explain the genetic variability observed.

In studies on the effect of temperature on leafy vegetative parts of the plants, it has been established that increase in temperature results into increased senescence of floral and leafy parts with more than 50% reductions observed in flowery parts compared to the leafy parts (Yin and Kropff, 1996). Temperatures slightly below the optimal growth temperatures result into increased leaf area, dry weight and thickness (Shin et al., 2001). For tropical plants, temperatures slightly above optimal growth temperatures may mean increased photosynthesis for the leaves although this may result into loss in dry weight (Boese and Huner, 1990) which later affects the plants ability to produce metabolites for storage or yield and accelerates aging. Thus, thermal stress negative effects on biomass and reductions observed are a result of modifications in the plants developmental organs which help the plant in offsetting the effects of the stress. This protects the plant from succumbing to cumulative stress as observed for most abiotic stresses (Salekdeh et al. 2009). In addition, when there is an interaction between water and heat stress (hydrothermal stress), plant growth and development may be severely compromised.

The changes in plant growth respond to water deficit and high temperatures as discussed above have not been studied in cassava. At this point and with the changes in climate anticipated, it is required that the analysis of cell- and leaf-level responses to water deficit and increased temperature in regard to plant growth traits be understood. Since radiation interception, as well as photosynthesis and transpiration, are largely affected by water availability and temperatures, the effects of these factors on primary morphological sites for physiological process need to be understood. Thus, the objective of this work was to analyze the response of leaf growth to water deficit and increased temperature in several cassava genotypes in order to identify and quantitatively describe sources of phenotypic variability for these drought tolerance determinant traits.

MATERIALS AND METHODS

Twenty varieties of cassava were selected based on their dry matter content, resistance to Cassava Mosaic Disease (CMD) and farmer preference and established in a Randomized Complete Block Design (RCBD) in Kasese, Western Uganda. The varieties included local farmer preferred varieties such as Nyaraboke, Kwatamumpale, Mpologoma, Bao, Bukalasa, Mercury, Magana, Rugogoma and Gwalanda. These varieties have good eating qualities, are very amenable to processing, show a specific level of tolerance to cassava mosaic disease and are said to be adapted to dry conditions. The elite varieties selected included NASE 1, NASE 2, NASE 3, NASE 12, TME 204, I/92/00067, MH96/0068 which combine tolerance to drought with CMD resistance, high dry matter content and good processing traits. In particular, variety MH96/0068 was selected in a study by Turyagyenda et al. (2013) as one of the stay green varieties with high levels of tolerance to water stress. The newly released varieties NASE 13, NASE 14, NASE 16 and NASE 19 are farmer preferred elite varieties recently released and have shown some level of resilience to drought stress. These varieties combine high dry matter contents and high yield with resistance to CMD, tolerance to CBSD and a recommendable level of resistance to abiotic stresses. However, they have not been screened thoroughly for their resistance to hydrothermal stress and no recommendations so far can be made in this effect to the farmers using these varieties.   The trial consisted of two experimental and two control blocks in 81M^2 plots, with up to 81 plants per plot. The plant responses to available conditions were monitored on a bimonthly basis where changes in leaf properties and plant growth parameters were recorded. Weather and location characteristics were also recorded during the trial period.

Soil properties, water properties and weather characteristics of the trial site

Properties including soil pH, organic matter, and minerals such as Nitrogen, Phosphorus, Calcium, Potassium, Boron, Zinc, Copper, Manganese, Iron, and Magnesium were investigated in terms of their quantities. Soil components such as silt, sand and clay were determined. Properties of the irrigation water to be used such as salinity and pH were also determined.

Weather patterns for the trial site were monitored during the trial period. This allowed the forecasting and determination of periods of critical soil moisture stress and increased air temperatures and factors that affect changes in these two properties that make hydrothermal stress. Important parameters affecting the relative humidity and soil water availability were considered. Daily rain fall patterns were recorded in addition to hourly temperature regimes throughout the stress period. The relative humidity was also recorded on an hourly basis. These were averaged on a monthly basis and used to make inferences on how the plant reacted to different environmental changes during its growth cycle.

Description of planting date and developmental cycles of cassava with rainfall pattern

In order to synchronize the data collection schedule with impending forecasted weather conditions, a specific date for planting the trial was chosen. The choice of the planting date was dependent on the developmental cycle of cassava plants and considerations for parameters to be studied at different times in the crop growth cycle. Planting was carried out in the first season representing a time of maximal rainfall during April to allow the plant to fully establish in the high moisture regimes and favorable temperatures at that time. Initial hydrothermal stress was experienced in the low rainfall period (June to August), allowing studies into the effect of stress and how it affects plant growth and development. This was followed by favorable moisture and temperature conditions between September and November allowing unaltered crop growth and root bulking coupled to maximum canopy development for the crop. The final stage of root bulking fell into a hydrothermal stress period between December and March affecting the plants canopy and its stored reserves.

Determination of leaf properties

During the plant growth cycle, the rate of growth and development was determined by direct measurements of leaf related properties and the changes thereof used to compare differences in plant growth and developments. Leaf properties were determined by direct measurements on the 5th most fully expanded leaf on each plant at a specific time. Leaf length, leaf width, petiole length, and plant height were measured using a measuring tape (Stanley 33-115, Power Lock, England) and the different measurements recorded in centimeters. For leaf lobe length and width, the centrally positioned leaf lobe on the 5th leaf was considered. This was meant to harmonize the different measurements taken. Numerical counting was used for leaf lobe numbers of the 5th fully expanded leaf and whole plant leaf numbers. Cumulative changes in leaf numbers and leaf characteristics were recorded over the 12 month growth period. These were compared across the three broad phenotypic manifestations observed during growth. In addition, changes in particular leaf properties with growth time were also recorded. Mean values for each property was determined from replicated observations on six plants per plot. Variations in these properties were then determined using the analysis of variance. Summary statistics (means, and standard deviations) were computed for leaf, soil and water properties. For each of the leaf properties, values for the fifteen replicate leaves in each replication were averaged to produce a mean for each of the varieties. Means were computed for cumulative changes in each leaf property to produce mean trends for each variety across the growing period. Subsequently, variables were averaged across all varieties and within each phenotypic class for trend evaluation. The significance ($P > 0.05$) of mean differences among variety groups was determined by analysis of variance (GenStat Discovery Edition, 2012).

Cumulative Leaf Percentage change (%CLP) in the different leaf properties was calculated using the formulae below:

$$\%CLP = \frac{100(V2 - V0)}{V0}$$

Where V0 = Variety group leaf property at preceding data taking period; and V2 = Variety leaf property two months after the preceding data taking period.

The percentage progressive change in total leaf numbers (%PCL) was calculated from two months after planting using the formulae:

$$\%PCL = \frac{100(Ln2 - Ln0)}{Ln0}$$

Where: Ln0 = Number of leaves in the preceding month of data collection, and Ln2 = Number of leaves two months after the preceding month of data collection.

Progressive change in plant height (PCH) was calculated from two months after planting using the formulae:

$$\%PCH = \frac{100(Ph2 - Ph0)}{Ph0}$$

Where: Ph0 = Plant height in the preceding month of data collection, and Ph2 = Plant height two months after the preceding month of data collection.

Harvest index and related parameters

To determine the overall effect of the hydrothermal stress on cassava yield and the differences among the varieties of cassava used in this study, the harvest index and related yield parameters were investigated. Harvest index was determined as the ratio of the weight of the root to the weight of the shoot for six plants in each plot in three replications. Total root number for each plant was recorded and cortex thickness determined for at least six commercial roots from each plot. In addition, the branching length and stem diameter were also determined by measurement.

Statistical analysis

Mean values for each of the properties were generated from the analysis of results using Microsoft excel software. The standard deviations and standard error were used to compare the variety characteristics across a given parameter. Analysis of variance and correlations among study parameters for all the test varieties were computed using the GenStat software (Genstat, discovery edition, 2012) and used to study the effect of above plant biomass on the yield components of the plant. Regression analyses and correlation analyses were employed to determine overall relationships among leaf properties and harvest parameters.

RESULTS

Soil, water and weather characteristics

Soil and water properties were determined for the trial site to understand the suitability of the site for a hydrothermal stress screening experiment. It was also important in understanding the water retention properties of the soil, mineral and nutrient availability and the suitability of the water for irrigating the control plots.

These properties are presented in Table 1. Soil pH ranged from 6.5 to 6.7 and was in the required range for growth and development of cassava as earlier suggested by Alves and Setter (2004). The water pH was also at neutrality ranging from 7.01 to 7.07. The average percentage of sand (53.47%) was twice as much as the percentage of clay (21.9%) and silt (18.63%) showing that the soil at the trial site could be easily drained and with very low capacity to hold water for a long time appropriate for drought screening. A high organic matter content (3.5 to 4.8%) and Nitrogen content 0.2 to 0.25% was observed which was appropriate for vegetative growth. Other elements such as phosphorus, Calcium, Magnesium, Potassium, Boron, Zinc, Copper, Manganese, and Iron fell in the recommended range for plant growth. The electrical conductivity of the irrigation water ranged from 104 to 111 µs/cm while the water contained low amounts of minerals such as calcium, magnesium, and potassium which ranged from trace to 0.17 ppm. The water also did not contain either carbonates or bicarbonates hence the near neutral pH observed.

Weather characteristics

The main determinant weather characteristics were recorded to show the relationships between weather characteristics and how they varied during the experimental period. The values are presented in Figure 2F. The trial was planted in April at a time when a considerable amount of rainfall and hence humidity was available to allow for crop establishment. By June (at two months after planting), the experiment was established. Between August and October, high rainfall hence increments in humidity were realized allowing the trial to fully establish and the plants to root and bulk. By the onset of stress in the month of December, the plants had already developed roots and had substantial reserves in the storage roots. Stress was characterized by reductions in rainfall and humidity and increments in temperature for a period of 12 weeks between the month of December and March (Figure 2F). Peak stress was characterized by increased temperature and reduced humidity experienced mid-January up to beginning of March for a period of 7 to 8 weeks. This was reversed with increased rainfall and moisture experienced in March and April (Figure 2F).

Phenotypic grouping of the varieties

Based on observed phenotypic characteristics, varieties were ranked according to their ability to retain leaves or recover leaves during and after the hydrothermal stress period (Plate 1). The groupings included varieties that maintained a moderately high Leaf Area Index (LAI) during hydrothermal stress or stay green varieties (SGV).

Table 1. Properties of Soil at the trial site and characteristics of irrigation water used.

Property (soil)	Units	Content	Range
pH	-	6.53	6.4-6.7
Sand	%	53.47	49.8-59.8
Clay	%	21.90	24.9-30.9
Silt	%	18.63	15.3-21.3
Organic matter	%	4.01	3.5-4.8
Nitrogen	%	0.22	0.2-0.25
Phosphorus	ppm	23.80	6.0-55
Calcium	Ppm	4090.26	2316.4-7573
Magnesium	ppm	882.3	346.3-1467.5
Potassium	ppm	383.98	194.8-675.1
Boron	ppm	17.50	12.87-26.8
Zinc	ppm	0.58	0.2-1.1
Copper	ppm	10.23	8.2-13.8
Manganese	ppm	156.34	82.0-235.7
Iron	ppm	457.57	380.7-1786
Water properties			
Water pH	-	7.05	7.02-7.07
Electrical conductivity	Us/cm	107	104-111
SAR	Meq/L	2.0	1.2-2.9
Calcium	ppm	Trace	Trace
Magnesium	ppm	0.11	0.08-0.15
Pottasium	ppm	0.09	0.04-0.17
Sodium	ppm	0.05	0.2-0.9
Carbonates	ppm	0.00	0
Bicarbonates	ppm	0.00	0

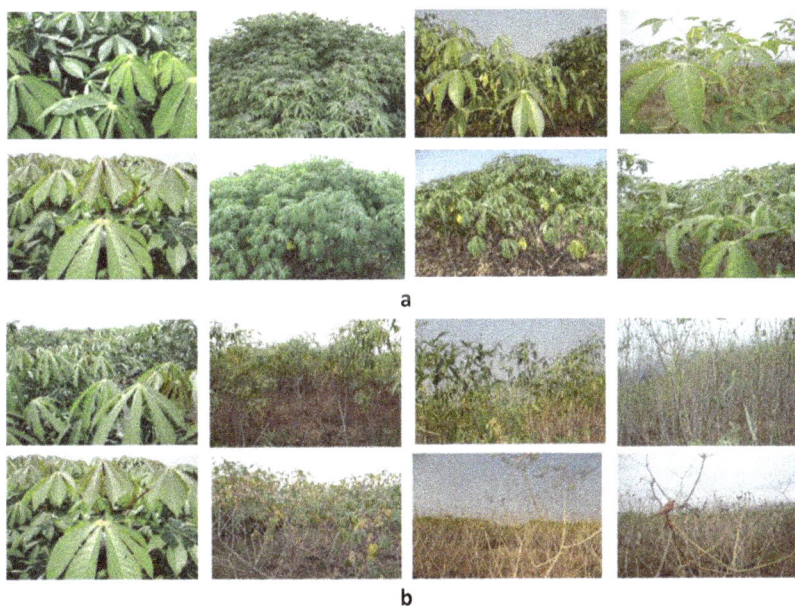

Plate 1. (a) Stay green varieties (NASE 3 [top] and 1/92/0067 [bottom]) as observed at 6 MAP, 8 MAP (on set of stress) and 9 to 10 MAP; (b) Some susceptible varieties (Mercury [top] and NASE 1 [bottom]) as observed 6 MAP, 8 MAP (on set of stress) and 9 to 10 MAP.

Plate 2. The early recovering trait observed in NASE 16, 19 and 12 two weeks before rains in response to increased relative humidity.

These included varieties such as NASE 2, NASE 3, MH96/0686, I/92/0067 and the local variety Magana. However, some varieties completely lost leaves as stress progressed and even the remaining leaves during stress were dechloropyllated and yellow signifying losses in chlorophyll and related pigments hence little or no capacity to photosynthesize. These varieties showed little or no capacity to recover easily after stress and they were labeled susceptible varieties (SV). They included varieties such as NASE 1, Rugogoma and Mercury. The other grouping included varieties that lost all their leaves immediately after onset of stress only to recover immediately with increase in relative humidity or early recovering varieties (ERV). They included NASE 16, NASE 19 and Bukalasa (Plate 2). Some varieties had both mechanisms but were not very pronounced in each case.

Cumulative changes in leaf properties over the growth period

For determination of the differences in the rate of growth and development among the cassava varieties used in this study, cumulative changes in leaf properties were determined and the results obtained are presented in Table 2. Significant differences were observed for petiole length among variety groups at 4 months after planting where the early recovering varieties had the highest cumulative petiole length growth observed. Stay green varieties had low cumulative petiole growth even at 4 months after planting. Negative cumulative increase in petiole length at 8 months after planting coincided with onset of stress. Highest reductions were observed for susceptible varieties (-22.4%) and the lowest for stay

green varieties (-13.4%) at the end of the critical stress period. At 9 months after planting, the early recovering varieties had lost almost all leaves only to recover them a month later (Table 2). At twelve months after planting, (one month post recovery), early recovering varieties had gained more new leaves hence positive cumulative percentage differences (231.4%) observed. However, the cumulative change was low in susceptible varieties (18.3%) while it was negative (-13.4%) for stay green varieties at twelve months after planting.

Significant variations (P<0.05) were observed in leaf lobe numbers at four months after planting with the early recovering varieties having the highest cumulative leaf lobe numbers (25.3%) compared to stay green varieties (4.2%) and susceptible varieties (14.7%). The pattern changed with onset of stress (eight months after planting) where severe reductions were observed for susceptible varieties (-48.6%) while low cumulative reductions were observed for stay green varieties (-14.3%) and early recovering varieties (-25.1). Differences occurred in the rate of reduction with stay green varieties having the highest cumulative reductions rate over the growth period. However, the early recovering varieties lost all the leaves as the stress period progressed making it difficult to follow the changes that occurred. Unlike petiole length, there were significant differences in the rate of reduction of leaf lobes between stay green varieties and early recovering varieties throughout the whole growth period. At twelve months after planting, stay green varieties had negative changes (-34.6%) while early recovering varieties had positive cumulative changes (234%) in leaf lobe numbers having regained a significant number of new leaves.

Changes in leaf lobe sizes were measured as leaf lobe width and length and decreased with growth time across

Table 2. Cumulative changes in leaf properties among different phenotypic groups for the 12 month growth period.

Time (months)	Leaf number	Petiole length	Leaf lobe No.	Leaf lobe width	Leaf lobe length
SGV4	238.9	6.22	4.15	29.33	4.32
ERV4	318.7	22.21	25.34	15.72	22.79
SV4	227.6	14.69	14.69	20	25.05
SGV6	115.6	-21.6	-17.4	-28.63	-24.98
ERV6	83.7	-35.61	-34.6	-27.85	-32.2
SV6	129	-35.26	-15.07	-32.91	-33.37
SGV8	131.4	-13.4	-14.3	-11.08	-26.47
ERV8	-59.8	-15.02	-25.09	-28.69	-38.69
SV8	66.2	-22.37	-48.61	-8.54	-20.04
SGV10	-59	-31.77	-5.24	-10.71	-14.25
ERV10	-67.9	-40.49	-49.63	-6.14	-21.15
SV10	-69.3	-23.73	26.88	-26.65	-17.99
SGV12	-58.3	-14.31	-1.89	-13.54	-13.35
ERV12	264	231.36	547.76	17.31	56.54
SV12	-47.3	18.3	61.69	21.08	1.54
CRC10 SGV	426.9	-60.58	-32.79	-21.09	-61.38
CRC10 ERV	274.72	-68.92	-83.99	-46.96	-69.25
CRC10 SV	353.52	-66.68	-22.11	-48.1	-46.35
CRC12 SGV	294.88	-59.88	-34.68	-34.63	-71.73
CRC12 ERV	430.96	129.96	463.78	-29.65	-12.71
CRC12 SV	244.96	-38.68	39.58	-27.02	-44.81

SGV = Stay green varieties, ERV = Early recovering varieties, SV = Susceptible varieties, 4, 6, 8, 10, 12 = months after planting (MAP), CRC10 = Total cumulative change 10 MAP (after peak stress), CRC12 = Total cumulative change 12MAP (at harvest).

the different varieties. At maximum vegetative growth between four and six months after planting, leaves were characterized into four broad categories including the short broad leaved (length between 14 and 17 cm and width between 6.5 and 7.5 cm), long broad leaved (length between 21 and 24 cm and width between 6.0 and 7.5 cm), short narrow leaved (length between 15 and 17 cm and width between 5.0 and 6.0 cm) and long narrow leaved (length between 21 and 23 cm and width between 4.5 and 6.0 cm). Most of the stay green varieties had short broad leaves and high cumulative increase in leaf lobe width and low cumulative increase in leaf lobe length (Figure 1). The early recovering varieties had intermediate characteristics with most of them having high cumulative leaf lobe length. On the other hand, the susceptible varieties had broad and long leaves. At onset of stress (eight months after planting), high cumulative reductions in leaf lobe length and width were observed for early recovering varieties (-38.7%) compared to stay green varieties (-26.5%) and susceptible varieties (-20.1%). Cumulative reductions were also observed to occur for both leaf length and leaf width at ten months after planting although major reductions were observed for leaf width.

Typically over 60% reductions for leaf lobe width compared to 35 to 45% reduction in leaf lobe length were observed (Figure 1). There were no significant (P>0.05)

differences for cumulative leaf lobe width between stay green and susceptible varieties much as considerable changes were observed for leaf lobe length between the two. After the plants had recovered from the stress and at twelve months after planting, positive cumulative differences for leaf lobe length (more than 55% gain in leaf lobe length) compared to leaf lobe width (17% gain in leaf lobe width) were recorded for early recovering varieties. However, negative cumulative differences were observed for stay green varieties (an average of -13.5% for both leaf lobe width and length) (Table 2). For leaf lobe length, positive cumulative differences were observed for early recovering varieties (56.5%) while the susceptible varieties had low positive changes (1.5%) implying that on recovery, susceptible varieties produced broad but shorter leaves, a general change in leaf morphology. Still negative differences were observed for stay green varieties in this instance (-13.5%) (Table 2).

Changes in leaf numbers, leaf lobe numbers and leaf sizes were related to plant biomass accumulation and development. Leaf lobe numbers for stay green varieties remained fairly constant over the growth period and were slightly affected after critical hydrothermal stress at ten months after planting. They ranged from 6.5 to 7.2 leaf lobes on average (Plate 3). Among the early recovering varieties, leaf lobe numbers peaked at four months after planting with some varieties having up to an average

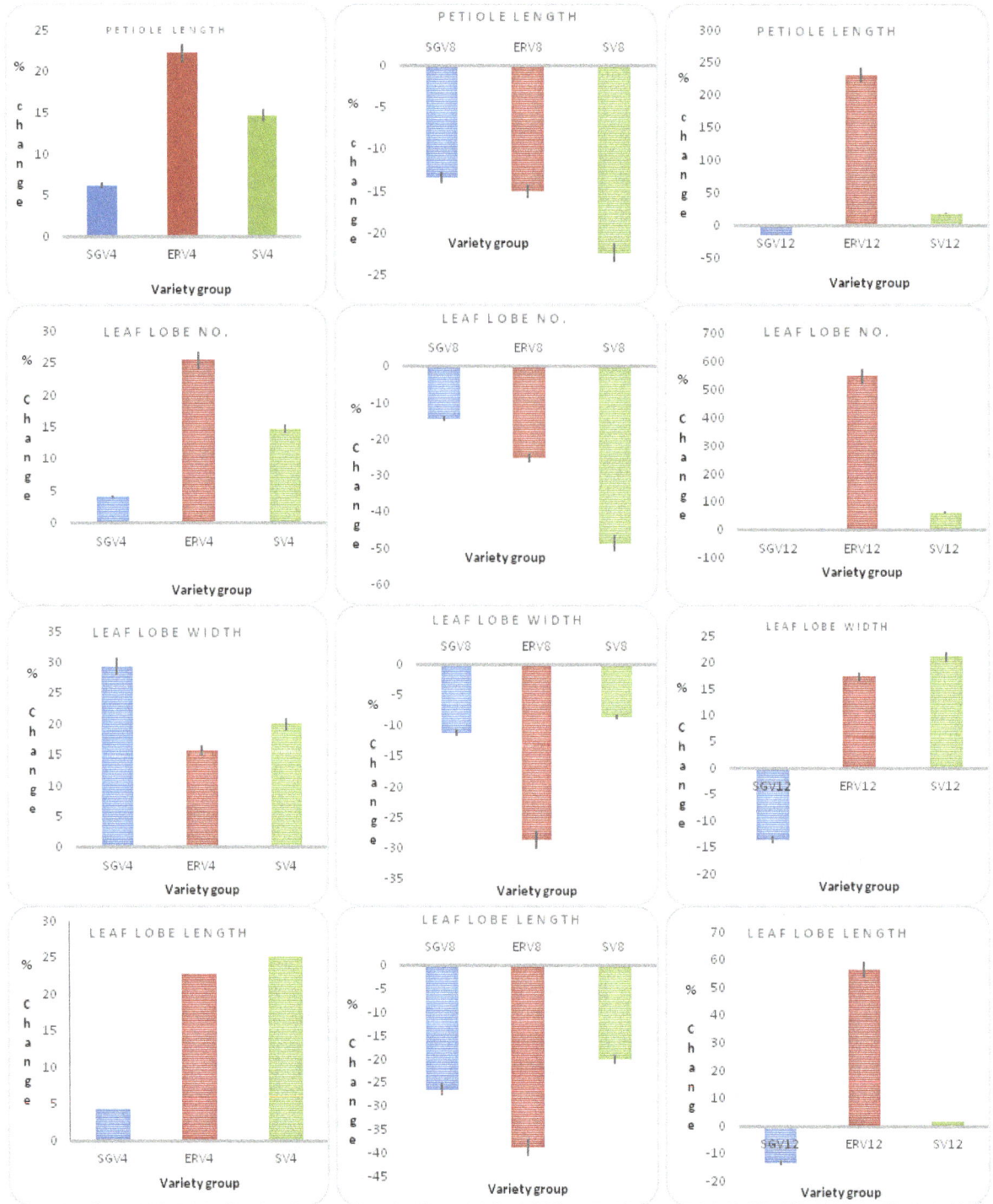

Figure 1. Percentage changes in different leaf properties at different times of crop growth and development compared to the preceding period.

of 9 leaf lobes. However, a drop was observed by six months after planting and with the onset of stress the plants shed off most of the leaves and leaf lobe numbers dropped significantly to no leaves or just one leaf lobe. On recovery, the average number of 7.4 leaf lobes was reinstated (Figure 2A) for these varieties. Susceptible varieties did not lose all the leaves during the stress period and had averages of about 3 dechlorophyllated leaf lobes with little or no capacity to carry out photosynthetic metabolism.

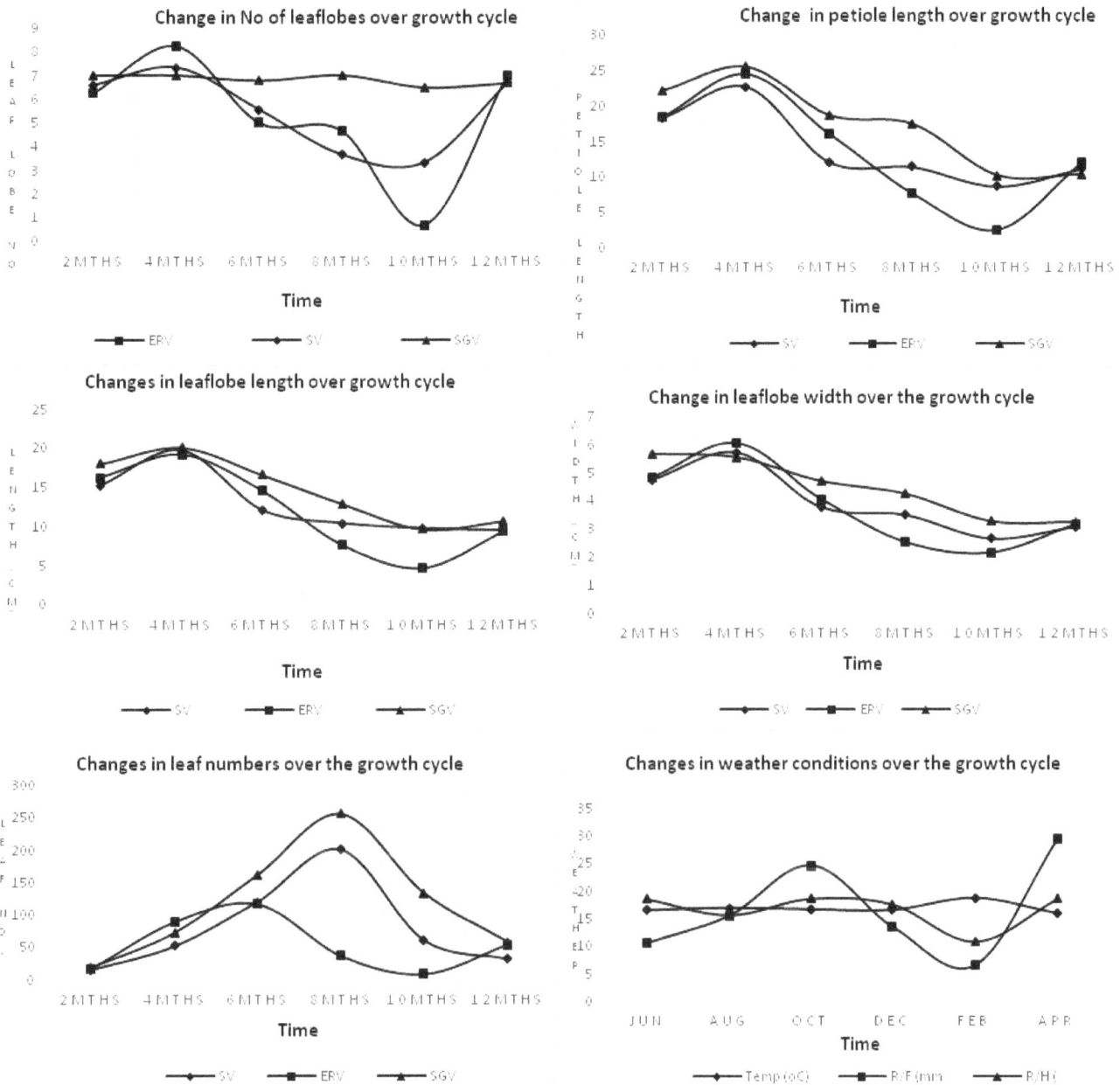

Figure 2. Comparisons for changes in leaf properties among the three broad variety groups tested from two to twelve months.

At two months after planting, petiole length ranged between 15.4 and 23.2 cm and the highest was recorded for the stay green varieties (Figure 2). The length of petioles increased significantly and by four months after planting, it ranged from 22.1 to 25.4 cm. Reductions in petiole growth were observed for all the varieties studied at six months after planting where petiole length ranged from 14.3 to 19.2 cm. In particular, the stay green varieties posted higher petiole lengths at all times compared to early recovering varieties and susceptible varieties. Reductions in moisture and increments

temperature between six to eight months after planting resulted into significant reductions in petiole length among the susceptible and stay green varieties. In stay green varieties, petiole length reduced from an average 17.2 cm on onset of hydrothermal stress to about 11.0 cm by the end of this stress period. No significant (P>0.05) increments in petiole length were observed after stress in these varieties with an average of 11.2 cm registered at 12 months after planting (Figure 2B). Among the susceptible varieties, petiole length dropped significantly from 13.7 cm at onset of hydrothermal stress

Plate 3. Photographic representation of changes in leaf numbers from 6 to 9 leaf lobes before the onset of stress to 3 to 4 leaf lobes on mild stress, to 1 leaf lobe or no leaves at maximum stress.

to about 2.3 cm by end of peak stress, a 70% reduction. The early recovering varieties had a recovery mechanism where the petiole increased in size as new leaves were produced when hydrothermal stress was relieved (Figure 2B). By two months after planting, leaf lobes lengths ranging between 15 and 18 cm were registered which increased up to 21.5 cm on average by four months after planting. However after that, a drop in size was observed ranging from 13 to 17 cm by six months after planting and 7 to 14 cm by eight months after planting just like other plant morphological properties. Onset of hydrothermal stress resulted into a more significant negative effects in leaf size growth with lengths ranging from 4 to 9 cm by the end of the stress period although increments were observed after recovery with up to about 11.5 cm registered. Differences occurred among varieties in terms of reduction in leaf size with more significant drops observed for stay green varieties (40% reduction)

compared to susceptible varieties (10% reduction) and early recovering varieties (24%) (Figure 2C).

The leaf lobe width followed the same pattern as length (Figure 2D).

Changes in total leaf numbers over the growth period

Accumulation of biomass in plants is as a result of leaf production which form the biggest part of the above ground biomass and are important in photosynthesis. However, plant leaf numbers vary due to genetic differences but most importantly due to changes in the environment. Thus, the variations in leaf numbers were determined and the results are presented in Figure 2E and Table 3. At two months after planting, most plants had average leaf numbers between 12 and 24 leaves which increased up to 40 to 94 leaves by 4 MAP.

Table 3. Progressive change in leaf numbers (LN) in comparison with percentage change (Δ %) over the growth cycle.

Accession	LN2	4Δ%	LN4	6Δ%	LN6	8Δ%	LN8	10Δ%	LN10	12Δ%	LN12	%CC10	%CC12	%TCC
Bao	16.42	236	55.17	164.2	145.8	35.487	197.5	-76.456	46.5	-39.1	28.33	359.2	320	-39
Nase 12	19.83	282.9	75.92	94.06	147.3	0.1154	147.5	-62.597	55.17	-60.7	21.67	314.4	254	-61
Nase 1	14.67	306.1	59.58	149.7	148.8	35.657	201.8	-87.698	24.83	-35.6	16	403.8	368	-36
72-TME-14	19.91	215.6	62.83	136.1	148.3	-51.743	71.58	-59.486	29	25.9	36.5	240.4	266	26
Nyaraboke	16.08	227.6	52.67	129.5	120.9	66.167	200.8	-69.292	61.67	-47.3	32.5	353.9	307	-47
Kwatamumpale	18.17	189.4	52.58	160.6	137	-7.1752	127.2	-43.383	72	-11.1	64	299.4	288	-11
Mpologoma	13.75	197.6	40.92	195.7	121	123.97	271	-86.041	37.83	-52.9	17.83	431.2	378	-53
Magana	21.58	326.7	92.08	143.3	224	9.4866	245.3	-68.33	77.67	-67.6	25.17	411.1	344	-68
Bukalasa	19.08	174.3	52.33	180.3	146.7	48.463	217.8	-90.664	20.33	32	26.83	312.3	344	32
Nase 2	16.08	284.5	61.83	73.23	107.1	288.46	416.1	-61.866	158.7	-85.3	23.33	584.3	499	-85
Rugogoma	19.08	245.9	66	69.03	111.6	156.14	285.8	-66.639	95.33	-85.5	13.83	404.4	319	-85
TME 204	18	227.8	59	170.2	159.4	14.806	183	-50	91.5	-79	19.17	362.7	284	-79
266Bam	17.17	421.8	89.58	31.23	117.6	-67.889	37.75	-76.159	9	348	40.33	309	657	348
Gwalanda	19.75	243.9	67.92	159.9	176.6	5.7774	186.8	-73.403	49.67	-30.9	34.33	336.2	305	-31
Nase 3	15.1	192.5	44.17	212.4	138	214.49	434	-73.541	114.8	-77.1	26.33	545.9	469	-77
MM96/4271	20.75	283.5	79.58	91.71	152.6	-8.613	139.4	-87.211	17.83	85.1	33	279.4	364	85
Mercury	17.33	225	56.33	125.4	127	31.207	166.6	-76.486	39.17	-33.2	26.17	305.1	272	-33
MM97/2961	20.91	214.1	65.67	128.6	150.1	-15.116	127.4	-81.69	23.33	77.9	41.5	245.8	324	78
O686	18.33	294.6	72.33	124.1	162.1	58.022	256.2	-47.625	134.2	-56.9	57.83	429.1	372	-57
I/92/0067	18.33	251	64.33	149.6	160.6	47.805	237.3	-67.417	77.33	-32.5	52.17	381	348	-33

Emphasis is on plant reaction 8MAP at the onset of peak stress up to 12MAP at harvest time. LN=Leaf numbers at different bimonthly periods. Δ%=change in leaf numbers as a percentage of the original number two months before.

Progressive increments were observed with time to an average of 107 to 177 leaves by six months after planting and 120 to 417 at eight months after planting representing a broad range of leaf numbers across different varieties. After this, leaf numbers fell significantly to as low as only 9 leaves per plant in early recovering varieties and 158 leaves in some of the stay green varieties at ten months after planting during the critical stress period. However, after critical hydrothermal stress, continued growth resulted into more leaves depending on the mechanism of tolerance to stress displayed.

The number of leaves increased significantly reaching the peak at 6 MAP for early recovering varieties (125 leaves on average) and at 8 MAP for susceptible varieties (200 leaves on average) and stay green varieties (250 leaves on average) per plant. In early recovering varieties, there were losses in the number of leaves at onset of critical hydrothermal stress at eight months after planting. Such losses continued throughout the stress period reaching a minimum of an average of two leaves per plant by peak stress at 10 MAP representing an average loss of 80% in the total leaf numbers (Table 3). However on recovery, early recovering plants regained most of the leaves per plant at the end of the growth period representing an average gain of over 95% and an overall net gain of 15% of the leaves compared to the period before critical hydrothermal stress period. For the stay green varieties, drops in leaf numbers started later in the critical hydrothermal stress period at about 9 MAP. Significant (p<0.05) losses in leaves were observed from about 250 leaves registered at 8 MAP to a loss of about 48%

of all the leaves by 10 MAP. With no recovery mechanism presented in these varieties, continued leaf loss was observed up to about 78 leaves per plant by the end of the growth cycle representing an average net loss of about 98% of the total leaves compared to the period before critical hydrothermal stress. The same pattern was observed for susceptible varieties although more losses (70%) were registered by 10 MAP and with up to 34 leaves per plant at the end of the growth period representing about 100% loss in leaves compared to the leaf number before the stress period (Figure 2E). While the early recovering variety plants shed off their leaves, the stay green and susceptible varieties maintained their leaves, significantly reducing the leaf size and slowly reducing leaf numbers through normal leaf senescence. However, on recovery, the ERV easily and spontaneously gained photosynthesizing leaves and by harvest time had the same leaf numbers as stay green varieties. Variations in leaf numbers among the different variety groups were significant (p<0.05) between six and ten months after planting but specifically at eight months after planting. This coincides with the period of peak stress and the variations were more pronounced between the stay green varieties and the early recovering varieties. While stay green varieties like Magana, NASE 3, NASE 2, I/92/0067 and MH96/0686 maintained high leaf numbers even during the stress period with the highest average leaf number being 416 leaves at eight months after planting, varieties like NASE 16 NASE 19, and Bukalasa lost most of the leaves with more than 76.2% reductions immediately after onset of critical hydrothermal stress (Table 3). Percentage total cumulative changes (%TCC) in leaf number at ten months after planting was high among the stay green varieties (350 to 580%) and low for the early recovering varieties (200 to 310%) showing that the stay green varieties maintained leaves over a long period of time during stress compared to other varieties. Percentage total cumulative change at twelve months after planting was almost equal for both stay green and early recovering varieties but low for the susceptible varieties (Table 3). Total cumulative percentage change for the entire growth period was high and positive for the early recovering varieties (260 to 348%) and low and negative for susceptible varieties and stay green varieties (-80%). Based on this, the stay green varieties maintained some of the leaves although a negative cumulative percentage was observed after critical hydrothermal stress. Some of the test varieties recovered increasing the leaf numbers by eleven and twelve months after planting. Such an increment defines recovery of the plant after stress.

Thus, depending on the number of leaves on the plant, selections for drought tolerant varieties can be made easily where high leaf numbers before stress depicts a characteristic stay green variety and low leaf numbers lost immediately at onset of stress depict an early recovering variety. Among the varieties that showed a

recovery mechanism, the time of recovery varied hence differences in the ability to recover during stress. The value of the total cumulative change (%TCC) at harvest can also be an important selection indicator for tolerance to stress especially if plant yield components are taken into consideration. For selection of the stay green phenotype, varieties which showed a high positive percentage between 100 and 290% eight months after planting can be important

Changes in plant height as a determinant of growth rate

Above ground biomass increments are related to the height of the plant and hence there was need to assess the variations in plant height to supplement the selection methods of hydrothermal stress tolerant varieties. Results for changes in plant height as a factor of plant growth rate are presented in Table 4. High growth rates were observed for all varieties between two and four months after planting where some varieties achieved more than 350% plant height changes compared to the height after germination. In particular high progressive changes in the growth rate were observed in NASE 16(397%), NASE 2(345%) and Mpologoma (323.7%). At six months after planting, reduced growth rates were observed with limited growth increments ranging between 30 and 80% except for the stay green variety NASE 3 with up to 135.2% growth increments. By eight months after planting, exceedingly low growth rates were observed with some varieties exhibiting negative growth rates such as NASE 16(-1.49%) due to loss of leaves and shoots at the onset of critical hydrothermal stress. At ten months after planting, a few varieties posted positive growth changes especially the stay green varieties such as I/92/0067 and Magana and some early recovering varieties such as Bukalasa. Most of the other varieties had negative growth increments due to the effect of stress on the growth and development of shoots during the stress period. Severe growth retardation was observed in varieties such as Rugogoma, Mercury, NASE 1 and Mpologoma, susceptible varieties which had most of their shoot drying up during the critical stress period and hence reductions in growth. On a cumulative basis and up to ten months after planting, high growth rates were observed for mainly stay green varieties such as Magana, MH96/0686, NASE 3, NASE 2 and the early recovering variety NASE 16 while at twelve months after planting, high cumulative growth was observed for NASE 2, NASE 16 and MH96/0686. Over all high growth rates for the whole growth period were observed for Magana, NASE 16, and MH96/0686 while low growth rates were observed for Nyaraboke, NASE 1 and NASE 3. Varieties with overall high growth rates and yield after recovery such as NASE 16 and Magana can be selected and deployed as moisture and temperature stress tolerant varieties.

Table 4. Progressive change in plant height (PH) in comparison with percentage change (Δ%) over the growth cycle on a bimonthly basis.

Accession	PH2	PH4	4Δ%	PH6	6Δ%	PH8	8Δ%	PH10	10Δ%	PH12	12Δ%	CR10 mths	CR12mths	FCAR
Bao	37.42	133.8	257.64	203.6	52.11	228.33	12.17	234	2.48	225	-3.8	324	321	-3.85
Nase 12	30.73	91.25	196.94	162.9	78.51	180.5	10.81	179.17	-0.7	155	-13	286	272	-13.4
Nase 1	32.36	104.4	222.68	179.7	72.06	217.42	21.01	209.33	-3.7	167	-20	312	292	-20.1
72-TME-14	34.32	99.75	190.64	186.6	87.03	241.08	29.22	245.33	1.76	198	-19	309	289	-19.2
Nyaraboke	32.32	121.3	275.15	187.7	54.77	210.67	12.26	219.16	4.03	148	-33	346	314	-32.7
Kwatamumpale	41.41	127.8	208.5	225.4	76.47	245.17	8.75	252.5	2.99	234	-7.2	297	290	-7.2
Mpologoma	29.99	127.1	323.74	215.9	69.88	266.17	23.29	282.67	6.2	232	-18	423	405	-17.9
Magana	38.05	136.7	259.18	224.1	63.97	256.25	14.34	273.67	6.8	279	1.83	344	346	1.83
Bukalasa	46.56	153	228.61	219	43.14	252.67	15.37	270.67	7.12	264	-2.6	294	292	-2.59
Nase 2	24.08	107.2	345.06	169.4	58.1	218.5	28.95	231.67	6.03	164	-29	438	409	-29.4
Rugogoma	40.06	145	261.95	214.1	47.66	252.42	17.89	251.67	-0.3	224	-11	327	316	-11.1
TME 204	45.75	157.3	243.89	230.6	46.55	263.92	14.47	274	3.82	261	-4.7	309	304	-4.68
266Bam	32.23	160.3	397.47	234.4	46.22	231.17	-1.39	243	5.12	252	3.57	447	451	3.57
Gwalanda	52.73	147.8	180.21	227	53.63	267.42	17.81	274	2.46	236	-14	254	240	-13.9
Nase 3	23.83	70.58	196.18	166	135.2	211.5	27.41	188	-11	133	-29	348	318	-29.3
MM96/4271	37.02	128	245.76	215	67.96	237.67	10.54	226.67	-4.6	182	-20	320	300	-19.6
Mercury	41.48	159.1	283.63	183.3	15.21	214.25	16.87	217.67	1.6	186	-15	317	303	-14.7
MM97/2961	42.67	150.6	252.89	208.8	38.65	250.17	19.82	246.67	1.4	200	-19	313	294	-19.1
O686	33.83	130.6	285.98	230.8	76.73	276.25	19.7	266.67	3.47	285	6.75	386	393	6.75
I/92/0067	38.23	134.6	252.02	203.6	51.25	254.83	25.19	283.67	11.3	255	-10	340	330	-10.1

FCAR = Final change after recovery, PH = Plant height, CR10 = Cumulative percentage change at peak stress (10 months), CR12 = Cumulative percentage change after recovery (12months). Bolded = Varieties showing recovery mechanisms/stay green mechanisms. Bolded/italicized = susceptible variety, Nyaraboke.

Harvest index related parameters

The ratio of the below ground biomass (cassava yield components) to the above ground biomass was estimated and compared to root related properties such as Dry matter yield, cortex thickness and root number. Variations among genotypes were observed in morphological parameters such as root cortex thickness, and the branching level and number of branches. The cortex thickness was positively correlated to root number (r = 0.84), root weight (r = 0.62) and stem weight (r = 0.51). The early recovering varieties had a smaller cortex thickness (0.2 to 0.3 cm) compared to stay green varieties which had cortex thickness of 0.35 to 0.55 cm and the susceptible varieties which had cortex thickness of between 0.3 and 0.4 cm. A bigger cortex was thus related to high levels of resilience to stress as earlier observed by (Okogbenin et al., 2013). Plant height in the different genotypes was positively correlated to the stem diameter (r = 0.33) and root weight (r = 0.28). The harvest index varied significantly within the different genotypes and was negatively correlated to plant height (r = -0.19) and stem weight (-0.42). High harvest index was recorded for early recovering varieties while low harvest index was recorded for the stay green varieties. The number of roots varied significantly in different cassava varieties although no significant differences were observed for root weight (p > 0.05). Generally stem related properties were not significantly (p>0.05) different among different cassava varieties although shoot weight was positively correlated to root number (r = 0.53) and root weight (r = 0.69). Likewise stem

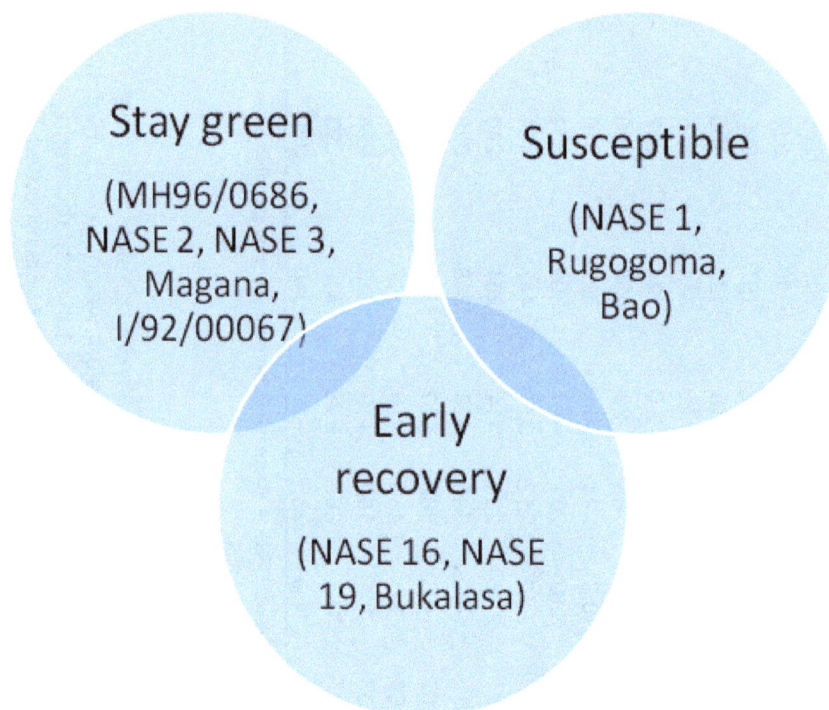

Figure 3. Varieties selected under the stay green and early recovering in relation to the susceptible varieties.

diameter was positively correlated to root weight (r = 0.57) and root number (r = 0.68). Since no differences were observed for stem related parameters, the yield components that varied significantly between different varieties are the main determinants of stress resilience in plants.

Selection of hydrothermal stress tolerant and avoidant varieties

This study was mainly conducted to understand the phenotypic mechanisms for tolerance to hydrothermal stress in cassava varieties and from the results above, a number of plant properties were identified that were key in the selection of tolerant varieties. These selections and their mechanisms are presented in Figure 3. Selections for the stay green mechanism were mainly based on the ability of the plant to accumulate biomass during the vegetative stage of growth from two to six months after planting. Plants which accumulated a lot of biomass at this time maintained substantively higher biomass contents during stress, had higher rate of leaf retention and higher growth rates. In addition, they had lower harvest indices and they were identified as stay green varieties. However, cassava plants displaying the stay green mechanism did not necessary maintain high leaf lobe numbers but instead reduced the number of leaf lobes at a lower rate compared to other varieties as the

critical hydrothermal stress period progressed (Figure 2). On the other hand, selection for the early recovering mechanism were based on the loss of leaves immediately at onset of critical hydrothermal stress, the ability to produce new shoots immediately on removal of hydrothermal stress, leaf re-growth and maintenance immediately after hydrothermal stress and higher harvest indices (Figure 2). Like in the stay green mechanism, the early recovering mechanism was mainly based on leaf lobe number maintenance after hydrothermal stress but not maintenance of individual leaves. Plants which displayed no mechanism and easily succumbed to stress were described as susceptible plants.

DISCUSSION

Changes in biomass allocation and accumulation during the cassava plant growth cycle are expected. These changes describe the inherent plant characteristics which are mediated by both the genetic component of the plant and its interaction with the environment. Such changes determine the yield potential of the plant in period of normal growth where as in stress conditions; the environment plays a big role in influencing the rate of growth and ultimately affecting the yield potential of the cassava plant as earlier observed by Osorio et al. (1993). In this study, the effects of stress resulted into significant changes in plant morphology resulting into altered yield

Table 5. Variations observed for different parameters at harvest.

Source of variation	d.f.	s.s.	m.s.	v.r.	F pr.	CV	LSD
Cortex Thickness 19	19	0.347	0.021	6.52	<.001	21.3	0.121
Branch length	19	1034	5174.	4.74	<.001	1.30	77.16
Harvest Index	19	1.658	0.083	3.14	0.001	25.9	0.379
Plant Height 19	19	55909.	2795.	2.57	0.006	11.7	77.07
Root Number 19	19	210.09	10.504	2.24	0.015	23.8	5.057
Root Weight	19	57.55	2.878	0.69	0.812	41.3	4.77
Stem diameter	19	12.52	0.6260	1.16	0.334	19.4	1.714
Stem weight	19	411.	20.55	1.01	0.468	52.9	10.51

compared to the control conditions. The stress conditions also resulted into reduced growth rates and the plants ability to attain a particular growth height or leaf area index as observed (Tables 2 and 3). The results also showed that low moisture and elevated temperature stress had significant effects on dry matter accumulation, root numbers and size, plant height and also leaf lobe length and leaf width. Such significant differences in shoot based biomass affected the plants ability to effectively increase its root based biomass resulting into reductions in yield and yield components.

Reductions in leaf size and leaf expansion were observed as the main factors that cause loss in biomass during stress. Shoot and leaf biomass in most plants accumulates in a curve-linear fashion increasing to maximum by the time the plant gets into the reproductive stage (Ranawake et al., 2011). This was observed in cassava between two to six months after planting as the plants increased their biomass content linearly though variety based differences were observed showing that some of these traits are partly under genetic control (Figure 1 and Table 3). Differences observed in petiole length and leaf size (drought related traits) among variety groups at the onset of hydrothermal stress show that tolerance mechanisms are related to plant genetic makeup. Such variations result into differences in accumulation of biomass in different varieties describing a stress tolerance criterion for a particular variety. Negative changes in leaf size properties and the petiole length at eight months after planting shows the negative influence of hydrothermal stress on cassava plants. This had an impact on the harvest index as observed later on in the growth cycle of the plant and even the harvest parameters as shown in Table 5. High rates of biomass accumulation observed earlier in the plants growth cycle may be a preparatory measure for carbon sequestration preparing for periods of low carbon uptake. Hydrothermal stress increases the imbalance between carbon supply and demand resulting into reduced biomass and loss of reproductive organs (Blum, 2005). Cassava plants which have mechanisms of increasing above ground biomass for purposes of increased photosynthetic and growth metabolism may be more prepared to counteract effects

of stress compared to the others which slowly accumulate biomass as has been observed in other plants by Ranawake et al. (2011). Specific increments in biomass earlier before stress was observed in the early recovering and stay green varieties which by six months after planting had significant biomass in the above ground parts hence were prepared for onset of stress later in the growth cycle

Correlation between the above ground biomass especially the stem characteristics with root biomass points to the influence of the above ground biomass on the sink (root) organs. It also points to the importance of the root cortex as one of the sink organs in stress tolerance as observed by Cohen et al. (2005) and Okogbenin et al. (2013). This also explains the correlation between harvest index and biomass parameters such as leaf size and plant height (Table 5). Higher values for cortex thickness in stay green varieties also point to the use of the storage reserves in times of stress. Thus, breeding for stress tolerance requires a full understanding of the relationships between biomass and root mass for selection of plants that effectively accumulate photosynthetic biomass and offset the effects of stress. The differences observed before onset of stress among the different variety groups but especially for the stay green and early recovering varieties point to the fact that stress tolerance or avoidance mechanism are a function of plants genetic framework as was earlier confirmed by Thornley (1972).

The positive cumulative increments in leaf sizes observed post stress in early recovering varieties were due to production of new vigorous leaves compared to the stay green varieties which maintained old non vigorous leaves. This implies that avoidance mechanisms are desired to allow the plant to attain a certain level of metabolic activity after stress which helps the plant to grow faster and revive the photosynthetic processes necessary for increasing plant biomass (Peroni et al., 2012). Observed changes in leaf morphology after stress for leaf size had an implication on the form and type of metabolism carried out by these leaves as suggested by Aguirreza'bal (2006). Leaves produced due to heterotrophic metabolism by using the plants own resources have less biomass and hence low photosynthetic capacity as observed by Hedayati et al. (2013) in other species of Euphorbiaceae. However, such leaves were able to regain normal metabolism and normal morphology although in the stay green varieties the morphological state of the leaves did not change a lot, an indicator of compromised photosynthetic metabolism. The efforts put forward by the plants to regain normal leaf morphology after recovery was observed for early recovering varieties suggesting that leaf and plant morphology need to be reinstated for optimal functioning of the cassava plants metabolic apparatus. This may also be related to particular genetic mechanisms meant to help the plant attain the former metabolic state as observed by Blum (2005).

The maintenance of a fairly constant leaf area index by stay green varieties during growth period and even in the stress period describes a tolerance mechanism by cassava plants as suggested by Asadi et al. (2012) and Turyagyenda et al. (2013). Such a mechanism may allow plants to minimally photosynthesize and maintain basic growth and physiological activities throughout the growth period. However, it comes with basic changes in rate, type and direction of translocated material in such plants. This explains the variations in root parameters among different varieties (Table 5) which also point to the importance of stored material in the root in allowing plants to cope with stress.

The differences observed at harvest in the different cassava varieties for root properties are an indicator of differences in yield in addition to the use of stored material as reserve metabolites to help the plant counteract stress as suggested by Zotz et al. (2002). Plant growth rate was probably dependent on increased rates of metabolism that increase the available plant resources meant for growth and development and hence helping the plant to achieve relevant metabolites that usher it into the reproductive phase as was observed by Blum (2005). This is due to increase in the rates of both respiratory and photosynthetic metabolism (Thornly, 1972). Interference in any of these forms of metabolism by either biotic or abiotic stress results into reduced growth hence reduction in plant growth properties as the critical hydrothermal stress conditions are maintained. The significant differences between the varieties tested for plant height, shoot dry weight and root based properties showed that some varieties were more sensitive while others were more tolerant. This has been reported in cereals such as wheat (Pauk et al., 2010) and *Jatropha curcas* (Hedayati et al., 2013) where plant height also decreased with increasing levels of water stress conditions.

Understanding how biomass reductions are photosynthetically driven is very important in determining their effect on the partitioning of photo assimilates to the significant parts of the plant at different growth times (Seyed et al., 2012). Since biomass accumulation during stress is mainly heterotrophic and is dependent on carbon resources in the root for growth support (Holzapfel et al., 2010), there is a significant effect of stress on harvest index as was observed in different cassava varieties (reducing in susceptible and stay green varieties due to use of stored resources to maintain basic metabolic processes) in effect reducing crop yield. Thus, selection of tolerant plants should account for the impact of hydrothermal stress on storage organs of cassava as a food resource. Loss in root weight and hence yield reduces the food and industrial value of cassava.

Apart from the leaves, the branching ability (determinant of shoot biomass) of the plant also has roles to play as far as stress tolerance/resistance is concerned. An increase in root to shoot ratio (or root to total dry matter ratio), as proposed by Steinberg et al. (1990) and attributable mainly to a reduction in shoot growth, was observed when water was limiting. This was a main mechanism for the stress avoidant plants of the early recovering varieties which maintained a low LAI. The size-dependent changes in biomass partitioning from direct changes in the carbon allocation process observed for these cassava plants reveal that water deficits probably act on biomass allometry by slowing down growth and adjusting plant size to the reduced amounts of carbon assimilated. This may be as a result of stomatal closure and decreased leaf area (Osorio et al., 1993). Employment of such mechanisms by cassava plants during abiotic stresses in different ways (hence the different phenotypes observed during stress) shows that cassava varieties differ in their ability to utilize the critically reduced amount of water available to the plant during drought incidence. If biomass production in the different varieties and the trend of water use efficiency are taken into account, it follows that when water is available to the plant, transpiration is unimpeded by stomata and lesser changes in biomass allocation and plant morphology may be expected as suggested by Bergantine et al. (2004).

Based on the results and observations from the field, it was realized that biomass based selection is important in the selection of hydrothermal stress tolerant plants. However, in earlier studies, (Okogbenin et al., 2013; Turyagyenda et al., 2013), selections were based on biomass properties such as high leaf retention during stress, high biomass retention during stress, high growth rates and high harvest indices. Such a selection mechanism based on leaf retention is not exhaustive since in this case even plants with mixed reactions will be selected even if the leaf morphology is severely compromised. In addition, selections only based on growth rate during stress will not easily identify drought avoidant plants that display a recovery mechanism after stress since these plants will be in a state of dormancy during this time. Thus, the author propose that selections for hydrothermal stress tolerant and avoidant plants be taken into account by considering phenotypic parameters such as leaf lobe retention during stress, cumulative leaf lobe retention during the growth cycle of the plant, and cumulative rate of growth during the plant growth cycle coupled to high harvest indices (Table 5). This will allow for selection of hydrothermal stress tolerant and avoidant plants falling under two broad stress tolerance mechanisms identified as stay green and early recovering mechanisms. It will also allow for possible understanding of plants that have mixed mechanisms and fall under the intersection between stress tolerant and avoidant plants in addition to identifying the stress susceptible plants as shown in Figure 3.

Conclusions

Results from this work showed that hydrothermal stress reduced biomass due to changes in the plants' morphological properties. The reductions were mainly in leaf based biomass and leaf related properties compared to the whole plant. Since leaves are the most sensitive organs to drought stress and respond quickly to low water availability, such a study is imperative in understanding the responses of cassava plants to water deficit and increased temperature, factors that mainly contribute to drought. Modification of plant morphological properties in response to stress does not only have an effect in controlling water loss but also provides an energy saving mechanism including saving stored resources in the plant. In cases where this is not possible, the plant either utilizes its resources to tolerate the stress or completely succumbs to the effects of stress and is hence susceptible. Identifying this in cassava is possible by studies on cassava plant phenotypic properties. And thus in this study, hydrothermal stress tolerant and avoidant plants were selected based on leaf lobe retention, biomass accumulation based selection during the vegetative stage of growth, leaf and leaf lobe recovery immediately after elimination of hydrothermal stress, selection based on number of fibrous roots/root cortex and selection based on harvest index after stress. However, relating these phenotypic properties to their genetic controls may provide a much more reliable basis for selection of tolerant varieties of cassava.

Conflict of Interests

This work was carried out by NARO scientists in collaboration with partners and funders. None of the parties have raised any concerns regarding this work and there is no conflict of interest in regard to the material contained in this study.

ACKNOWLEDGEMENTS

This work was supported by the Bio-Innovate project, a subsidiary of SIDA grant under National Agriculture Research Organization (NARO).

REFERENCES

Alves AC, Setter TL (2004). Abscisic acid accumulation and osmotic adjustment in cassava under water deficit. Environ. Exp. Bot. 51(3):259-271.

Asadi S, Lebaschy MH, Khourgami A, Mirrad AH (2012). Effect of Drought Stress on the Morphology of Three Salvia sclarea Populations. Ann Biol. Res. 3(9):4503-4507.

Bergantine R, Yamauchi A, Pardales J,Bolatete D (2004). Screening cassava genotypes for resistance to water deficit during crop establishment. Philippine J. Crop Sci. 29(1):29-39.

Blum A (2005). Drought resistance, water-use efficiency, and yield potential—are they compatible, dissonant, or mutually exclusive. Aus. J. Agric. Res. 56:1159-1168.

Boese RS, Huner NPA (1990). Effect of Growth Temperature and Temperature Shifts on Spinach Leaf Morphology and Photosynthesis. Plant Physiol. 94:1830-1836.

Cohen S, Raven E, Li Y, Grava A, Goldschmidt EE (2005). Physiological responses of leaves, tree growth and fruit yield of grapefruit trees under reflective shade screens. Scientia Hort. 107:27-35.

Fasehun FE (1979). Effect of soil matric potential on leaf water potential, diffusive resistance, growth and development of Gmelina arborea L. seedlings. Biologia Plantarum 21(2):100-104.

Granier C (2000). PHENOPSIS, an automated platform for reproducible phenotyping of plant responses to soil water deficit in A. thaliana....ion with low sensitivity to soil water deficit. New Phytologist, 169(3):623-635.

Guerin GR, Wen H, Lowe A (2012). Leaf morphology shift linked to climate change. Biol. Lett. 23 8(5):882-886.

Holzapfel BP, Smith JP, Field SK, Hardie WJ (2010). Dynamics of carbohydrate reserves in cultivated grapevines. Hortic. Rev. 37:143-211.

Hsiao TC, O'Toole JC, Yambao EB, Turner NC (1998). Influence of Osmotic Adjustment on Leaf Rolling and Tissue Death in Rice (Oryza sativa L.). Plant Physiol. 75(2):338-341.

Hedayati A, Alizadeh O, Sharafzadehs AM, Azarpanah A (2013). Physiological and morphological responses biodiesel plant (Jatropha curcas L.) to water stress condition. Int. J. Agric. Crop Sci. (IJACS) 5(7):695-703

Katahata S, Naramoto M, Kakubari Y, Mukai Y (2007). Photosynthetic capacity and nitrogen partitioning in foliage of the evergreen shrub Daphniphyllum humile along a natural light gradient. Tree Physiol. 27(2):199-208.

Kaplan DR (2001). The science of plant morphology: Definition, history, and role in modern biology. Am. J. Bot. 88(10):1711-1741.

Osório J, Pereira JS (1993). Genotypic differences in water use efficiency and ^{13}C discrimination in Eucalyptus globulus Tree Physiol. 14(7-8-9):871-882.

O'toole JC, Cruz RT (1980). Response of leaf water potential, stomatal resistance, and leaf rolling to water stress. Plant Physiol. 65(3):428-432.

Okogbenin E, Setter TL, Ferguson M, Mutegi R, Ceballos H, Olasanmi B, Fregene M (2013). Phenotypic approaches to drought in cassava: review. Front Physiol. 4:93.

Pauk J, Cseuz L, Lantos C, Mihály R, Vass I, Dudits D (2010). Drought stress and the response of wheat: nursery and complex stress diagnostic experiments. Association of PflanzenzÄchter and Seed Merchants Austria, pp. 15–18

Peroni I, Pagliarani C, Lovisolo C, Chitarra W, Roman F, Schubert A (2012). Recovery from water stress affects grape leaf petiole transcriptome. Planta 235(6):1383-96.

Ranawake AL, Amarasingha UGS, Rodrigo WDRJ, Rodrigo UTD, Dahanayaka N (2011). Effect of water stress on growth and yield of mung bean (Vigna radiate L). Trop. Agric. Res. Ext. 14(4):76-79

Salekdeh GH, Reynolds M, Bennett J, Boyer J (2009). Conceptual framework for drought phenotyping during molecular breeding.

Trends Plant Sci. 14:488-496.

Schmidt G, Zotz G (2001). Ecophysiological consequences of differences in plant size: in situ carbon gain and water relations of the epiphytic bromeliad, *Vriesea sanguinolenta*. Plant Cell Environ. 24:101-112.

Schuppler U, He PH, John PCL, Munns R (1998). Effect of water stress on cell division and cell-division-cycle 2-like cell-cycle kinase activity in wheat leaves. Plant Physiol. 117(2):667-678.

Seyed YS, Lisar RM, Hossain MM, Rahman IMM (2012). Water Stress in Plants: Causes, Effects and Responses. Water Stress, ISBN: 978-953-307-963-9

Shin K, Lieth H, Kim S (2001). Effects of temperature on leaf area and flower size in rose. Proc. III IS Rose Research Acta Hort. 547:185-191.

Steinberg SL, Miller JC, McFarland MJ, (1990). Dry matter partitioning and vegetative growth of young peach trees under water stress. Aust. J. Plant Physiol. 17:23-26.

Turyagyenda L, Kizito EB, Ferguson M, Baguma Y, Agaba M, Harvey J, Osiru D (2013). Physiological and molecular characterization of drought responses and identification of candidate tolerance genes in cassava. *AoB PLANTS*: plt007 doi: 10.1093/aobpla/plt007

Tardieu F, Davies WJ (1993). Integration of hydraulic and chemical signalling in the control of stomatal conductance and water status of droughted plants. Plant Cell Environ.16:341-349.

Thornley JHM (1972). A balanced quantitative model for root: shoot ratios in vegetative plants. Annals Bot. 36:431-441.

Neumann PM (1995). The role of cell wall adjustment in plant resistance to water deficits. Crop Sci. 35:1258-1266.

Valladares F, Pearcy RW (1999). The geometry of light interception by shoots of *Heteromeles arbutifolia*: morphological and physiological consequences for individual leaves. Oecologia 121:171-182.

Yin X, Kropff MJ (1996). The effect of temperature on leaf appearance in rice. Ann. Bot. 77:215-221.

Zotz G, Reichling P, Valladares FA (2002). Simulation study on the importance of size-related changes in leaf morphology and physiology for carbon gain in an epiphytic bromeliad. Ann. Bot. 90(4):437-443.

Permissions

The contributors of this book come from diverse backgrounds, making this book a truly international effort. This book will bring forth new frontiers with its revolutionizing research information and detailed analysis of the nascent developments around the world.

We would like to thank all the contributing authors for lending their expertise to make the book truly unique. They have played a crucial role in the development of this book. Without their invaluable contributions this book wouldn't have been possible. They have made vital efforts to compile up to date information on the varied aspects of this subject to make this book a valuable addition to the collection of many professionals and students.

This book was conceptualized with the vision of imparting up-to-date information and advanced data in this field. To ensure the same, a matchless editorial board was set up. Every individual on the board went through rigorous rounds of assessment to prove their worth. After which they invested a large part of their time researching and compiling the most relevant data for our readers.

The editorial board has been involved in producing this book since its inception. They have spent rigorous hours researching and exploring the diverse topics which have resulted in the successful publishing of this book. They have passed on their knowledge of decades through this book. To expedite this challenging task, the publisher supported the team at every step. A small team of assistant editors was also appointed to further simplify the editing procedure and attain best results for the readers.

Apart from the editorial board, the designing team has also invested a significant amount of their time in understanding the subject and creating the most relevant covers. They scrutinized every image to scout for the most suitable representation of the subject and create an appropriate cover for the book.

The publishing team has been an ardent support to the editorial, designing and production team. Their endless efforts to recruit the best for this project, has resulted in the accomplishment of this book. They are a veteran in the field of academics and their pool of knowledge is as vast as their experience in printing. Their expertise and guidance has proved useful at every step. Their uncompromising quality standards have made this book an exceptional effort. Their encouragement from time to time has been an inspiration for everyone.

The publisher and the editorial board hope that this book will prove to be a valuable piece of knowledge for researchers, students, practitioners and scholars across the globe.

List of Contributors

Rouhollah Rouzbeh
Islamic Azad University of Takestan branch, 10 km Quzwin-Zanjan highway, Azad University of Takestan, Faculty of Agriculture, Iran

Jahanfar Daneshian
Islamic Azad University of Takestan branch, 10 km Quzwin-Zanjan highway, Azad University of Takestan, Faculty of Agriculture, Iran
Seed and Plant Improvement Institute (SPII), P. O. Box: 4119, Mardabad Road, Karaj 31585, Iran

Hossein Aliabadi Farahani
Islamic Azad University of Shahr-e-Qods Branch, Iran

İlkay Öztürk Cali
Department of Biology, Faculty of Art and Science, Amasya University, Amasya-Turkey

Faten Omezzine
Department of Biology, Higher Institute of Biotechnology of Monastir, University of Monastir, 5000, Tunisia Unit of search for Mycotoxins, Phycotoxins and associated Pathologies 03/UR/08-14

Rabiaa Haouala
Department of Biology, Higher Institute of Biotechnology of Monastir, University of Monastir, 5000, Tunisia Unit of search for Mycotoxins, Phycotoxins and associated Pathologies 03/UR/08-14

Asma El Ayeb
Department of Biology, Higher Institute of Biotechnology of Monastir, University of Monastir, 5000, Tunisia Unit of search for Mycotoxins, Phycotoxins and associated Pathologies 03/UR/08-14

Neziha Boughanmi
Department of Biology, Faculty of Sciences of Bizerte, Tunisia

Hossein Aliabadi Farahani
Islamic Azad University of Shahr-e-Qods branch, Iran

Sayed Alireza Valadabadi
Islamic Azad University of Shahr-e-Qods branch, Iran

Mohammad Ali Khalvati
Department of Plant-Microbe Interactions, German Research Centre for Environment and Health, Munich (HelmholtzZentrum München), Ingolstaedter 1, Neuherberg Germany

Woldeyesus Sinebo
Ethiopian Institute of Agricultural Research, P. O. Box 2003, Addis Ababa, Ethiopia

Berhane Lakew
Ethiopian Institute of Agricultural Research, P. O. Box 2003, Addis Ababa, Ethiopia

Abraham Feyissa
Ethiopian Institute of Agricultural Research, P. O. Box 2003, Addis Ababa, Ethiopia

T. Neeraj
Department of Biotechnology, C. C. S. University, Meerut -250004, India

Sachin
Department of Biotechnology, C. C. S. University, Meerut -250004, India

S. S. Gaurav
Department of Biotechnology, C. C. S. University, Meerut -250004, India

S. C. Chatterjee
Department of Biotechnology, C. C. S. University, Meerut -250004, India

S. Poongothai
Department of Biotechnology, Bharath University, Chennai-600 073, India

R. Ilavarasan
Captain Srinivasa Murthi Drug Research Institute for Ayurvedha and Siddha, Chennai-600 106, India

C. M. Karrunakaran
Department of Biotechnology, Bharath University, Selaiyur, Chennai-600 073, India

M. I. B. Efombagn
Institute of Agricultural Research for Development (IRAD), P. O. Box 2067 or 2123, Yaoundé, Cameroon

S. Nyassé
Institute of Agricultural Research for Development (IRAD), P. O. Box 2067 or 2123, Yaoundé, Cameroon

D. Bieysse
CIRAD-CP, TA 80/02, 34398 Montpellier Cedex 5, France

O. Sounigo
Institute of Agricultural Research for Development (IRAD), P. O. Box 2067 or 2123, Yaoundé, Cameroon
CIRAD-CP, TA 80/02, 34398 Montpellier Cedex 5, France

E. K. Eifediyi
Department of Crop Science, Ambrose Alli University, P. M. B. 14, Ekpoma, Edo State, Nigeria

S. U. Remison
Department of Crop Science, Ambrose Alli University, P. M. B. 14, Ekpoma, Edo State, Nigeria

S. Das
Department of Plant Breeding and Genetics, College of Agriculture, Orissa University of Agriculture and Technology, Bhubaneswar, Orissa, India

R. C. Misra
Department of Plant Breeding and Genetics, College of Agriculture, Orissa University of Agriculture and Technology, Bhubaneswar, Orissa, India

S. K. Sinha
Department of Plant Breeding and Genetics, College of Agriculture, Orissa University of Agriculture and Technology, Bhubaneswar, Orissa, India

M. C. Pattanaik
Department of Plant Breeding and Genetics, College of Agriculture, Orissa University of Agriculture and Technology, Bhubaneswar, Orissa, India

Nehla Labidi
The Laboratry of Plant Adaptation to Abiotic Stress (LAPSA), Biotechnology Center at the Technopark of Borj-Cedria (CBBC), BP 901, Hammam-Lif 2050, Tunisia

Henda Mahmoudi
The Laboratry of Plant Adaptation to Abiotic Stress (LAPSA), Biotechnology Center at the Technopark of Borj-Cedria (CBBC), BP 901, Hammam-Lif 2050, Tunisia

Messedi Dorsaf
The Laboratry of Plant Adaptation to Abiotic Stress (LAPSA), Biotechnology Center at the Technopark of Borj-Cedria (CBBC), BP 901, Hammam-Lif 2050, Tunisia

Ines Slama
The Laboratry of Plant Adaptation to Abiotic Stress (LAPSA), Biotechnology Center at the Technopark of Borj-Cedria (CBBC), BP 901, Hammam-Lif 2050, Tunisia

Chedly Abdelly
The Laboratry of Plant Adaptation to Abiotic Stress (LAPSA), Biotechnology Center at the Technopark of Borj-Cedria (CBBC), BP 901, Hammam-Lif 2050, Tunisia

Nafiseh Katooli
Islamic Azad University, Mashhad Branch, Young Researchers Club, Iran

Raheleh Maghsodlo
Department of Plant Protection, University of Agricultural Science and Natural Resources, Gorgan, Iran

Seyed Esmaeil Razavi
Department of Plant Protection, University of Agricultural Science and Natural Resources, Gorgan, Iran

Abdul Rehman
Rice Program, Crop Sciences Institute, National Agricultural Research Centre Islamabad, Pakistan

Wilf Powell
IACR-Rothamsted, Harpenden, Herts., AL5 2JQ, UK

E. D. Oruonye
Department of Geography, Taraba State University, P. M. B. 1167, Jalingo, Taraba State, Nigeria

E. Okrikata
Department of Biology Education, School of Science Education, Federal College of Education, Technical, P. M. B. 1013, Potiskum, Yobe State, Nigeria

F. Fernández
SYMBORG S.L. Campus de Espinardo No 7. Edificio CEEIM. Murcia. España. CP 30100. Instituto Nacional de Ciencias Agrícolas. (INCA), Habana, Cuba

J. M. Dell'Amico
Instituto Nacional de Ciencias Agrícolas. (INCA), Habana, Cuba

M. V. Angoa
Centro Interdisciplinario de Investigación para el desarrollo Integral Regional. (CIIDIR) IPN, Justo Sierra No. 28, CP 59519, Jiquilpán Michoacán, México

I. E. de la Providencia
Université catholique de Louvain, Unité de microbiologie, 3 Place Croix du Sud, 1348 Louvain la Neuve, Belgium

T. Ghotbi
Department of Plant Virus Research, Iranian Research Institute for Plant Protection (IRIPP), Agricultural Research and Education Organization, Tehran, Iran

N. Shahraeen
Department of Plant Virus Research, Iranian Research Institute for Plant Protection (IRIPP), Agricultural Research and Education Organization, Tehran, Iran

Sukristiyonubowo
Indonesian Soil Research Institute, Jalan ir. Juanda 98 Bogor 16123, Indonesia

Ibrahim Adamy Sipahutar
Indonesian Soil Research Institute, Jalan ir. Juanda 98 Bogor 16123, Indonesia

Tagus Vadari
Indonesian Soil Research Institute, Jalan ir. Juanda 98 Bogor 16123, Indonesia

Agus Sofyan
Direktorat Jendral Prasarana dan Sarana Pertanian, Jalan M.T. Harjono, Jakarta Selatan

Ping Zhang
College of Horticulture, Nanjing Agricultural University, Nanjing, 210095, China

Zhi-Ying Yu
College of Horticulture, Nanjing Agricultural University, Nanjing, 210095, China

Zong-Ming Cheng
College of Horticulture, Nanjing Agricultural University, Nanjing, 210095, China

Zhen Zhang
College of Horticulture, Nanjing Agricultural University, Nanjing, 210095, China

Jian-Min Tao
College of Horticulture, Nanjing Agricultural University, Nanjing, 210095, China

E. Anim-Kwapong
Cocoa Research Institute of Ghana, P. O. BOX 8, New Tafo-Akim, Ghana

G. J. Anim-Kwapong
Cocoa Research Institute of Ghana, P. O. BOX 8, New Tafo-Akim, Ghana

B. Adomako
Cocoa Research Institute of Ghana, P. O. BOX 8, New Tafo-Akim, Ghana

Mohamed Abd El-Himed Sayed Ahmad El-Bramawy
Agronomy Department, Faculty of Agriculture, Suez Canal Univ., 41522 Ismailia, Egypt

R. J. Rukarwa
School of Agricultural Sciences, Makerere University, P.O. Box 7062, Kampala, Uganda

S. B. Mukasa
School of Agricultural Sciences, Makerere University, P.O. Box 7062, Kampala, Uganda

A. Sefasi
International Potato Center, Applied Biotechnology Laboratory, P. O. Box 1558 Lima, Peru

G. Ssemakula
National Crop Resources Research Institute, P. O. Box 7084, Uganda

R. O. M. Mwanga
International Potato Center – Kampala, P. O. Box 7065, Kampala, Uganda

M. Ghislain
International Potato Center, Nairobi, P. O. Box 25171, Nairobi 00603, Kenya

Mohammed Amin
Department of Plant Sciences and Horticulture, College of Agriculture and Veterinary Sciences, Ambo University, P.O. Box No 19 Ambo, Ethiopia

Ayalew Amare
School of Plant Sciences, College of Agriculture, Haramaya University, P. O. Box No: 138, Dire Dawa, Ethiopia

Thangavel Selvaraj
Department of Plant Sciences and Horticulture, College of Agriculture and Veterinary Sciences, Ambo University, P.O. Box No 19 Ambo, Ethiopia

Rudra Bhattarai
Regional Agriculture Research Station, Nepal Agriculture Research Council, Tarahara, Nepal

Surya Kant Ghimire
Agricultural and Forestry University, Chitwan, Nepal

Dhruba Bahadur Thapa
Agribatany Divisions, Nepal Agricultural Research Council, Kathmandu, Nepal

Madhav Prasad Pandey
Institute of Agriculture and Animal Science, Trivuban University, Nepal

Ephraim Nuwamanya
National Crops Resources Research Institute (NaCRRI), P.O Box 7084, Kampala, Uganda
College of Agriculture and Environmental sciences, School of Agriculture, Makerere University Kampala, P. O. Box, 7062, Kampala-Uganda

Patrick R. Rubaihayo
College of Agriculture and Environmental sciences, School of Agriculture, Makerere University Kampala, P. O. Box, 7062, Kampala-Uganda

Joseph F. Hawumba
College of Agriculture and Environmental sciences, School of Agriculture, Makerere University Kampala, P. O. Box, 7062, Kampala-Uganda

Robert Kawuki
National Crops Resources Research Institute (NaCRRI), P.O Box 7084, Kampala, Uganda

Yona Baguma
National Crops Resources Research Institute (NaCRRI), P.O Box 7084, Kampala, Uganda